Enhancing Learning Experiences With Digital Tools:

AI, ChatGPT, and Virtual and Augmented Reality

Mageswaran Sanmugam
Universiti Sains Malaysia, Malaysia

Bosede Iyiade Edwards
Universiti Sains Malaysia, Malaysia

Nurul Maziah Mohd Barkhaya
Universiti Sains Malaysia, Malaysia

Zuheir Khlaif
An-Najah National University, Palestine

IGI Global
Scientific Publishing
Publishing Tomorrow's Research Today

Published in the United States of America by
IGI Global Scientific Publishing
701 East Chocolate Avenue
Hershey, PA, 17033, USA
Tel: 717-533-8845
Fax: 717-533-8661
Website: https://www.igi-global.com E-mail: cust@igi-global.com

Library of Congress Cataloging-in-Publication Data

Names: Sanmugam, Mageswaran, 1982- editor. | Edwards, Bosede, 1971- editor. | Barkhaya, Nurul Maziah Mohd, 1988- editor. | Khlaif, Zuheir, 1971- editor.
Title: Enhancing learning experiences with digital tools : AI, ChatGPT, and virtual and augmented reality / edited by Mageswaran Sanmugam, Bosede Iyiade Edwards, Nurul Maziah Mohd Barkhaya, Zuheir Khlaif.
Description: Hershey, PA : IGI Global Scientific Publishing, [2025] | Includes bibliographical references and index. | Summary: "The purpose of this book is to investigate the ways in which these cutting-edge digital tools can be utilised to effectively enhance learning in a series of different educational settings"-- Provided by publisher.
Identifiers: LCCN 2024054546 (print) | LCCN 2024054547 (ebook) | ISBN 9798369398111 (hardcover) | ISBN 9798369398128 (paperback) | ISBN 9798369398135 (ebook)
Subjects: LCSH: Artificial intelligence--Educational applications. | Virtual reality in education. | Education--Effect of technological innovations on.
Classification: LCC LB1028.43 .E487 2025 (print) | LCC LB1028.43 (ebook) | DDC 371.33/4--dc23/eng/20241101
LC record available at https://lccn.loc.gov/2024054546
LC ebook record available at https://lccn.loc.gov/2024054547

British Cataloguing in Publication Data
A Cataloguing in Publication record for this book is available from the British Library.

Table of Contents

Detailed Table of Contents

In recent years, the application of Artificial Intelligence (AI) and Augmented Reality (AR) in music education has shown great potential to promote personalised, interactive and immersive teaching methods. However, current research on AI and AR in music education lacks a systematic analysis of their independent roles and cross-fertilisation. This study aims to comprehensively sort out the current research status, hotspots and development trends of AI and AR in music education through the methods of systematic literature review and bibliometric analyses. The literature will be classified into three categories; AI and music education, AR and music education, and AI and AR in music education. The results will reveal the main application scenarios of AI and AR in music education and challenges of their combination in education. The review will provide academic references to explore the future direction of AI and AR in music education, and provide practical insights for innovation in digital music teaching and learning.

This chapter explores the advantages and disadvantages of university students using ChatGPT, an AI-powered chatbot. By synthesizing existing research, it aims to provide a comprehensive understanding of the topic. Using thematic analysis, the study identifies key themes: the benefits of ChatGPT, its drawbacks, and its potential integration into university education. The findings suggest that ChatGPT enhances learning experiences, offers personalized support, and promotes self-directed learning. However, concerns such as overreliance on AI, ethical issues, and privacy challenges are also highlighted. The chapter emphasizes the need for further research to address these issues and guide effective integration into academic settings.

The advent of AI in education provides new resources with the potential to revolutionize how we teach and learn. This chapter thoroughly summarizes artificial intelligence (AI) technologies, their possible uses in education, and the challenges that arise from using them. Researchers examine chatbots and similar algorithms that can mimic human interaction and create writing that reads naturally given linguistic input. While innovative chatbots like ChatGPT have many benefits, they also present significant ethical and practical concerns when used in the classroom. The chapter aims to promote AI's responsible and ethical use in education by providing insightful information on how AI may be effectively introduced into the educational context to benefit instructors and students.

Teaching children about animals is a fundamental aspect of early childhood education and is universally valued across different cultures and societies. Understanding animals fosters a sense of curiosity and wonder in young minds, encouraging them to explore the natural world and enhance their comprehension of the interdependence of life on our planet. Picture books, flashcards, stories, songs, and animal toys are some of the traditional ways that kids are introduced to animals. These resources make learning interesting and enjoyable for kids. However, if AI can be deployed for the enhancement of the current method, it can benefit the overall educational system. To make that a reality, "Scanimals" aims to introduce a novel and engaging educational experience for children through a mobile application focused on terrestrial animals. The main objective of this project is to establish an interactive learning and image recognition-based digital platform to help young learners gain a greater grasp of the animal kingdom (terrestrial animal) in this digital age of technology.

The potential of artificial intelligence (AI) in fostering inclusive education, especially in classroom learning, is well-documented by current educational trends. This study aims to explore students' acceptance of AI in higher education as a preliminary investigation. It identifies the challenges faced from the students' perspectives. Data will be collected through interviews with three students enrolled in instructional technology course, focusing on their experiences and acceptance levels. Thematic analysis will be used to uncover recurring themes and patterns in the interview transcripts. Expected findings include a diverse range of student attitudes, with some highlighting the enhanced learning opportunities and personalization provided by AI. Overall, this research addresses fostering the inclusivity of AI among students to create an effective learning environment.

This chapter examines the transformation of education with Virtual and Augmented Reality (VR and AR), highlighting their remarkable capability to create immersive learning experiences that significantly enhance student engagement and knowledge retention. AR and VR technologies, with their interactive and dynamic interfaces, overcome traditional educational challenges, thereby enhancing the inclusivity and personalisation of learning. The chapter analyses the current state of the education system, presents use cases of AR and VR in education, and discusses the steps to incorporate AR and VR into educational settings for immersive learning experiences. As AR and VR progress, their integration into education offers substantial prospects to connect theoretical knowledge with practical implementation, establishing these technologies as crucial catalysts for the future of learning.

Lagging executive functioning skills are common amongst individuals diagnosed with ADHD (Adler et al., 2017; Ustun et al., 2017) and research indicates, these

This study investigates the role of ChatGPT in enhancing the creative writing skills of English language students from the perspectives of instructors and postgraduate students at Palestinian universities. Using a mixed-methods research design, the researchers employed both quantitative and qualitative descriptive analytical approaches, including questionnaires and interviews for data collection. The study population consisted of 89 instructors and 169 postgraduate students. Quantitative data were analyzed using SPSS V.23, while qualitative data were analyzed thematically. The findings indicate that both instructors and doctoral students have positive perspectives on ChatGPT's role in improving creative writing. The study also found a high level of implementation of ChatGPT as an AI tool in higher education for creative writing enhancement. No statistically significant differences were observed among participants' perspectives.

This paper explores the potential applications of ChatGPT, an advanced language generation model developed by OpcnAI, in sustainable research within the field of Library and Information Science (LIS). ChatGPT has attracted attention due to its ability to generate text that resembles human language and engage in interactive conversations, making it a valuable tool for LIS researchers. Previous studies have demonstrated its effectiveness in assisting librarians with reference queries and enhancing user experiences. However, some ethical concerns and challenges must be carefully addressed to ensure the responsible and sustainable utilization of ChatGPT. These include addressing biases, ensuring information accuracy, promoting ethical use, protecting user privacy, and improving explainability. This paper discusses the benefits of integrating ChatGPT into various aspects of LIS research, such as information retrieval, knowledge organization, user interaction, collaboration, and decision-making. It emphasizes the importance of considering ethical considerations and proposes future directions for integration, such as domain-specific training, knowledge base integration, collaboration, ethical guidelines, evaluation methodologies, user-centred design, and interdisciplinary collaboration.

By embracing these directions, ChatGPT can be effectively integrated into sustainable LIS research, driving advancements in the field and empowering researchers to address the complex challenges of the evolving information landscape.

Preface

Enhancing Learning Experiences With Digital Tools: AI, ChatGPT, and Virtual and Augmented Reality presents an exploration of the transformative power of digital tools in modern education. As technology increasingly permeates our everyday lives, the traditional educational landscape is experiencing a dynamic shift. This book seeks to illuminate how artificial intelligence, virtual and augmented reality, and other digital resources are being leveraged to reimagine learning environments, elevate student engagement, and foster innovative teaching methods.

The rapid development of artificial intelligence (AI) and immersive technologies has introduced unprecedented possibilities for education, making this an exciting yet challenging time for educators, researchers, and students alike. This book brings together a collection of research on these technologies, reflecting both the promise and the complexities they bring to the educational sphere. The studies span various applications and contexts, illustrating a global and interdisciplinary approach to understanding these innovations. In exploring AI and immersive tools, we gain a deeper insight into how these technologies can empower learners, support educators, and potentially create more inclusive and accessible educational experiences.

A key theme throughout the book is the examination of AI's role, particularly ChatGPT, in educational settings. ChatGPT, as an AI language model, offers students and instructors unique opportunities, such as aiding in creative writing, providing personalized tutoring, and even serving as a resource for independent research. However, as the chapters delve into its application, it becomes evident that ChatGPT presents a double-edged sword. While it facilitates enhanced learning experiences and enables greater autonomy in education, concerns surrounding ethical usage, dependency, and potential biases in AI responses emerge. This balance of advantages and limitations encourages readers to consider ChatGPT as a powerful tool that requires mindful integration and ongoing evaluation.

Chapter 1

In this chapter, we explore the transformative potential of Artificial Intelligence (AI) and Augmented Reality (AR) in music education. Despite their growing application, research lacks a systematic analysis of their independent roles and synergy in the field. By conducting a thorough literature review and bibliometric analysis, we categorize the findings into three distinct areas: AI in music education, AR in music education, and the integration of both technologies. This chapter reveals the main application scenarios and challenges of combining AI and AR, providing valuable insights for future innovations in digital music teaching and learning.

Chapter 2

This chapter delves into the use of ChatGPT, an AI-powered chatbot, among university students. It synthesizes existing research to provide a comprehensive understanding of the benefits and drawbacks of this technology. The study identifies key themes such as the enhancement of learning experiences, personalized support, and self-directed learning through ChatGPT. However, it also highlights concerns like overreliance on AI, ethical issues, and privacy challenges. The chapter calls for further research to address these challenges and guide the responsible integration of AI tools like ChatGPT in higher education.

Chapter 3

The chapter provides an in-depth analysis of the role of Artificial Intelligence (AI) in education, examining both its advantages and the challenges it presents. The authors focus on innovative AI technologies, such as chatbots, which can simulate human interaction and create natural written content. While AI tools like ChatGPT offer significant educational benefits, they also raise ethical and practical concerns, particularly in classroom settings. The chapter emphasizes the need for responsible and ethical AI use in education, offering guidance on how to introduce these technologies effectively to enhance both teaching and learning.

Chapter 4

Focusing on early childhood education, this chapter explores the importance of teaching children about animals as a way to foster curiosity and understanding of the natural world. While traditional methods like picture books and songs have proven effective, the chapter proposes an innovative approach using AI and image recognition technology. Through the development of a mobile application, "Scanimals," children can engage in an interactive, digital learning experience about terrestrial animals. This chapter highlights how AI can enhance traditional educational tools, making learning more engaging and informative for young children.

Chapter 5

This chapter examines students' acceptance of AI in higher education, focusing on the challenges they face when integrating this technology into their learning environments. Through interviews with students enrolled in an instructional technology course, the study identifies a range of attitudes toward AI, from enthusiasm about personalized learning to concerns about its implementation. The chapter explores the need for fostering inclusivity and creating an environment that supports the effective use of AI to enhance educational opportunities and experiences.

Chapter 6

The integration of Virtual and Augmented Reality (VR and AR) into education has the potential to significantly enhance student engagement and knowledge retention. This chapter examines how these immersive technologies create interactive and dynamic learning experiences, offering substantial opportunities for inclusive and personalized education. By analyzing current trends and presenting real-world use cases, the chapter explores how VR and AR can transform the education system, bridging theoretical knowledge with practical application to prepare students for future challenges.

Chapter 7

In this chapter, the authors focus on the potential of assistive technology and AI tools to enhance executive functioning skills in individuals with ADHD. Given that emotional regulation, organizational skills, attention, and memory are often challenging for those with ADHD, the chapter highlights various AI-powered tools, such as Summit, Headspace, and Goblin Tools, that can support the development of these skills. By incorporating these tools, individuals with ADHD can improve their independence, confidence, and overall self-efficacy, which ultimately leads to better outcomes in daily life and academic settings.

Chapter 8

This chapter investigates the use of immersive Virtual Reality (VR) to enhance English speaking instruction for Chinese university students. With the growing popularity of VR technology in education, this study explores how immersive VR can offer new opportunities to improve the quality of language learning. The chapter emphasizes the practical value of integrating VR into English speaking instruction, providing insights into how this technology can create a more engaging and effective learning environment for students.

Chapter 9

Artificial Intelligence (AI) is reshaping education, and this chapter explores its role in enhancing English as a Foreign Language (EFL) teaching in Palestine. By analyzing the perspectives of 43 teachers on AI and AI-based drama techniques in language learning, the study reveals a strong optimism regarding AI's potential to improve the EFL classroom experience. Through statistical analysis, the chapter sheds light on both the benefits and challenges of AI integration, offering a glimpse into how these technologies could transform language education across various cultural contexts.

Chapter 10

This chapter delves into the complexities of Large Language Models (LLMs) like ChatGPT, focusing on the inherent challenges of using these tools due to their opaque nature. While LLMs can generate content quickly, their lack of transparency raises concerns about the accuracy of their outputs. The chapter discusses the limitations of LLMs, urging users to critically assess the content they generate. It advocates for human validation to ensure the safe and effective use of AI-generated content in educational and professional settings.

Chapter 11

Immersive technologies, including Virtual Reality (VR), Augmented Reality (AR), and Extended Reality (XR), are transforming education by providing highly interactive and context-rich learning environments. This chapter synthesizes research on the potential of VR and AR to foster essential 21st-century skills such as critical thinking, self-directed learning, and teamwork. Despite the promising benefits of these technologies, the chapter highlights the challenges related to cost, accessibility, and pedagogical integration. It advocates for a balanced approach to the deployment of these technologies in education to maximize their impact on student learning outcomes.

Chapter 12

This chapter explores the role of ChatGPT in enhancing the creative writing skills of English language students in Palestine. By employing a mixed-methods approach, including both quantitative and qualitative research, the study captures the perspectives of instructors and postgraduate students on the use of ChatGPT in creative writing instruction. The findings show a positive reception toward ChatGPT as a tool for improving writing skills, and the chapter emphasizes the growing implementation of AI tools in higher education to enhance the creative writing process.

Chapter 13

In this chapter, the authors examine the applications of ChatGPT in sustainable research within the field of Library and Information Science (LIS). ChatGPT's ability to generate human-like text and engage in conversations makes it a valuable tool for LIS researchers, especially in assisting with reference queries and enhancing user experiences. However, the chapter addresses critical ethical concerns related to the use of AI in LIS, including biases, accuracy, privacy, and explainability. It calls for a responsible approach to integrating ChatGPT in LIS research, ensuring its sustainable and ethical utilization in the field.

In addition to AI, the book explores the application of augmented and virtual reality in diverse educational fields. For instance, a systematic review of augmented reality in music education highlights how this technology can make complex music concepts more accessible and engaging for students, promoting active learning and creativity. Similarly, immersive virtual reality environments offer new dimensions for language learning, as demonstrated in the study on English-speaking instruction for Chinese college students. Virtual reality's ability to create lifelike scenarios and simulations is revolutionizing language instruction by enabling learners to practice in realistic yet controlled settings. This capability highlights the immersive nature of VR, where students can actively engage, interact, and practice skills that extend beyond the classroom.

Furthermore, digital tools designed to support specific cognitive needs are also explored. One study examines digital tools for individuals with Attention-Deficit Hyperactivity Disorder (ADHD), showing how technology can support executive functioning and improve learning outcomes. Digital aids that cater to individual needs reflect the promise of personalized education, providing students with tailored support to reach their full potential.

In considering the ethical and practical implications of these technologies, the book also addresses the broader challenges of AI and immersive tools in education. A preliminary study on the acceptance and challenges of AI integration in classrooms reveals the necessity of building trust among educators, students, and parents. As AI and digital tools evolve, educators are faced with the task of understanding and adapting these tools in ways that are pedagogically sound and ethically responsible. Similarly, the study on discuss output risks from imperfect language models (LLMs) discusses the need for careful handling of AI outputs, especially in high-stakes educational environments. It serves as a reminder that while technology offers extraordinary benefits, its risks must be managed through well-informed practices and guidelines.

Another notable theme is the potential of these tools to promote sustainability in education and research. A study on ChatGPT's potential for sustainable library and information science research discusses how AI can assist researchers by streamlining processes, managing large data sets, and even supporting in-depth analyses. AI can thus become a tool for sustainable knowledge management, provided that it is used thoughtfully and responsibly.

This book aims to offer both a comprehensive understanding and a critical perspective on the ways digital tools are reshaping education. By collecting diverse research and experiences, we hope to provide a balanced view that acknowledges the tremendous potential of technology while addressing the practical, ethical, and pedagogical considerations that come with it. Each chapter invites readers to engage with the complex realities of AI, virtual reality, and digital aids in education, encouraging further dialogue on best practices, responsible use, and innovative applications.

Enhancing Learning Experiences with Digital Tools: Artificial Intelligence, ChatGPT, and Virtual and Augmented Reality is a resource that is envisioned into being of great value to educators, students, researchers, and policymakers. This book serves as both a roadmap and an invitation to imagine a future in which digital tools empower all learners, support meaningful learning experiences, and inspire a new generation of educators to harness the best that technology has to offer. As we navigate the intersection of technology and education, this book serves as both a roadmap and an invitation.

At a time when the pace of digital transformation is rising at a rapid pace, may this work serve as a guide to the appropriate and meaningful integration of technology in educational settings. We have high hopes that readers will be motivated to investigate, customise, and make contributions to the fascinating new frontier of digital learning experiences.

CHIEF EDITOR,

Mageswaran Sanmugam

Centre for Instructional Technology and Multimedia, Universiti Sains Malaysia, Malaysia

Acknowledgment

We extend our heartfelt gratitude to the dedicated reviewers whose insightful feedback and expertise have been invaluable in bringing *Enhancing Learning Experiences With Digital Tools: AI, ChatGPT, and Virtual and Augmented Reality* to life. Your thoughtful comments and constructive guidance have greatly enriched the quality and depth of this work. The list of reviewers are:

1) Wang Ying
2) Nusaila Johari
3) Jinya Liu
4) Krishna Pandey
5) Darren Yie
6) Arvin Cheong Kee Sin
7) Sobihatun Abdul Salam
8) Zhang Jingru
9) Anusha Achuthan
10) Jagneet Kour
11) Raino Bhatia
12) J. Joshua Thomas
13) Damola Olugbade
14) Yasmeen Abu Mukh
15) Ilham Chniet
16) Seema Yadav
17) Terri Jackson
18) Purushothman Munusamy
19) Sahil Lal
20) Mehdi Kaddouri
21) Girija Ramdas
22) Li Jiang
23) Wei Lun Wong

Thank you for your time, expertise, and unwavering support in shaping this book. Your contributions have been essential in making this project a success, and we are deeply appreciative of your commitment to advancing knowledge in the field of educational technology.

With sincere appreciation,

The Editors

Chapter 1

A Systematic Review of Augmented Reality in Music Education

Zhang Xuejing
https://orcid.org/0009-0003-9455-699X
Universiti Sains Malaysia, Malaysia

Wan Ahmad Jaafar Wan Yahaya
https://orcid.org/0000-0002-8605-0062
Universiti Sains Malaysia, Malaysia

ABSTRACT

In recent years, the application of Artificial Intelligence (AI) and Augmented Reality (AR) in music education has shown great potential to promote personalised, interactive and immersive teaching methods. However, current research on AI and AR in music education lacks a systematic analysis of their independent roles and cross-fertilisation. This study aims to comprehensively sort out the current research status, hotspots and development trends of AI and AR in music education through the methods of systematic literature review and bibliometric analyses. The literature will be classified into three categories; AI and music education, AR and music education, and AI and AR in music education. The results will reveal the main application scenarios of AI and AR in music education and challenges of their combination in education. The review will provide academic references to explore the future direction of AI and AR in music education, and provide practical insights for innovation in digital music teaching and learning.

DOI: 10.4018/979-8-3693-9811-1.ch001

INTRODUCTION

Augmented Reality (AR) is a technology that overlays computer-generated virtual information, such as images, sounds, and haptic feedback, onto the real world (Azuma, 1997). In education, AR is recognized as a transformative technology capable of revolutionizing traditional teaching and learning through its interactive and immersive features, making the process more vivid, intuitive, and engaging (Rovithis et al., 2019). The educational applications of AR extend beyond enhancing students' interest and participation; they also significantly improve comprehension and memorization. By integrating virtual experiments, 3D models, and dynamic simulations into the learning process, AR enables students to better understand abstract concepts and complex content (Cook, 2019; Lu et al., 2022; Zünd et al., 2015).

The introduction of AR technology has opened up new possibilities for music education through virtual instrument simulations, interactive teaching and other means, which allow students to learn in a dynamic and immersive environment, thus improving their musical understanding and performance skills (Keebler et al., 2014; Lu et al., 2022; Mei & Yang, 2021). In the field of music education, traditional teaching methods often rely on teachers' verbal explanations and written materials, and students' learning experiences may be limited by limited resources and insufficient interactivity.

AR technology in music education not only enhances students' motivation, but also improves their learning outcomes and creativity (Zünd et al., 2015). For example, AR applications can visualise abstract music concepts and help students better understand complex music theories through visualisation and interaction. Traditional Malaysian musical instruments were developed through AR to increase learning interest through interaction and engagement (Tan & Lim, 2018). However, most of the existing research focuses on small-scale trials and short-term applications, and there are relatively few systematic studies on the long-term impact of AR in music education and its application to music learning in primary and secondary schools schools. In addition, the implementation of AR technology faces challenges such as technology cost, device compatibility and teacher training, which need to be further explored in future studies.

Through a systematic literature review, this study aims to analyse the current status of AR technology in music education, explore how it can enhance music learning and teaching experiences, and identify the strengths and challenges in its application. This paper will focus on answering the following research questions:

i. What is the current state of research on the use of augmented reality in music education?

ii. How does augmented reality technology enhance the music learning and teaching experience?
iii. What are the advantages and challenges of applying augmented reality in music education?
iv. What are the gaps in current research and what are the potential future research directions?

METHODOLOGY

This study adopted a systematic literature review approach and selected two major academic databases, Scopus and Web of Science (WOS), for the literature search. The researchers used Boolean operators to search for 'augmented reality' or 'AR' and 'music' as keywords in the title, abstract, and keywords. As the number of documents retrieved using 'music education' as a keyword alone was low, 'music' was selected to obtain a broader collection of documents. In the abstract screening process, irrelevant articles were excluded. Based on these search criteria, a total of 573 documents were retrieved from Scopus and 66 from WOS.

To ensure that the screening process was transparent and reproducible, the researchers used the PRISMA criteria for literature screening (Page et al., 2021). The screening process was documented through a PRISMA flowchart. The researchers felt that selecting literature from the last ten years would reflect the latest research findings and technological advances and avoid outdated information, and limited the timeframe to the last ten years, i.e., 2014-2024. Screening criteria included excluding review articles and literature that was not open access, and limiting the language of publication to English. After the initial screening, 97 documents were obtained from the Scopus database and 33 from the WOS database. All literature was imported into Zotero literature manager, where duplicates were first removed and titles, abstracts and full texts were carefully read to exclude literature not related to the topic of this study, resulting in the identification of 61 relevant literature. The detailed process of literature screening is shown in Figure 1.

Figure 1. PRISMA Literature Screening Flowchart

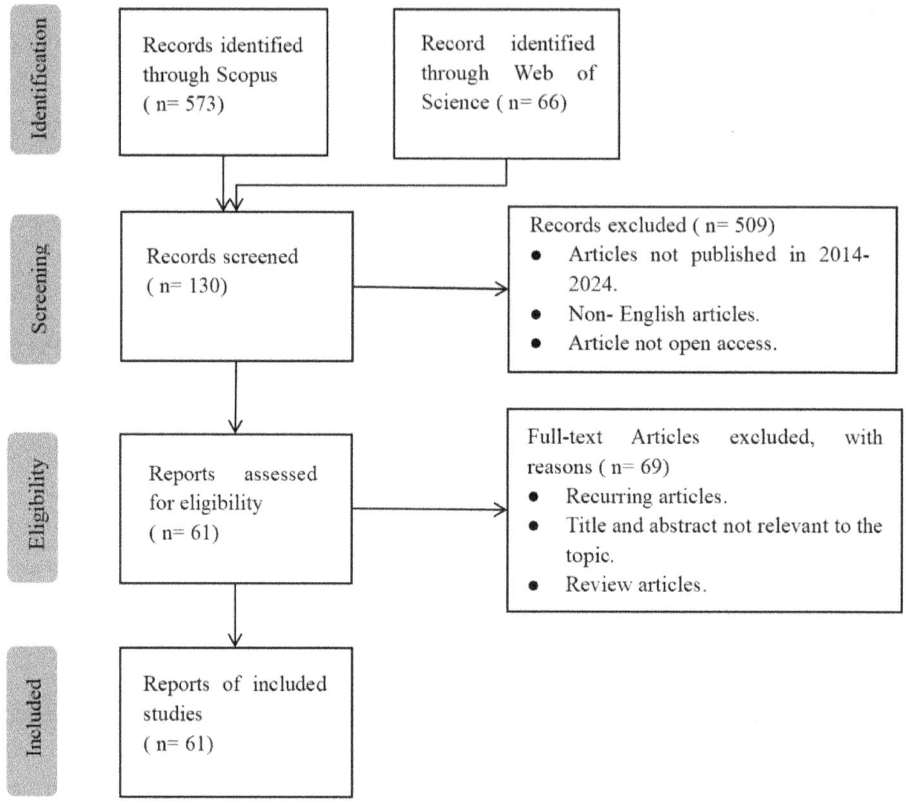

The researcher conducted econometric analysis of the 61 screened literature and keyword co-occurrence analysis using VOSviewer software (Eck & Waltman, 2010). After importing the literature information, 'co-occurrence' was selected as the type of analysis, and the unit of analysis was set as 'all keywords'. The minimum number of occurrences of keywords was set as '2' to better understand the research hotspots and connections in the field. Due to the small number of documents matching the theme and the fact that the research field is still in the development stage, the researchers hope that by lowering the threshold of keyword occurrence, the researchers can explore the connection between keywords in the field, rather than focusing on the high frequency of a single keyword. And remove irrelevant or less relevant keywords. The results of the keyword co-occurrence analysis were presented in three different visualisations showing the bibliometric mapping: network visualisation (Figure 2), overlay visualisation (Figure 3) and density visualisation (Figure 4).

Figure 2. Keyword Co-occurrence Analysis, Network Visualisation

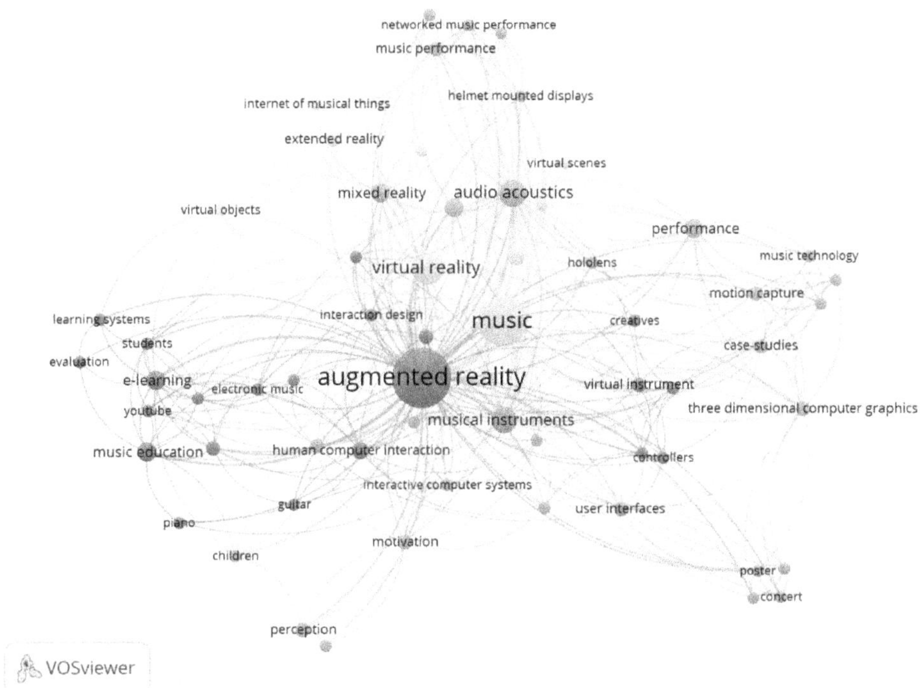

Figure 3. Keywords Co-occurrence Analysis, Overlay Visualisation

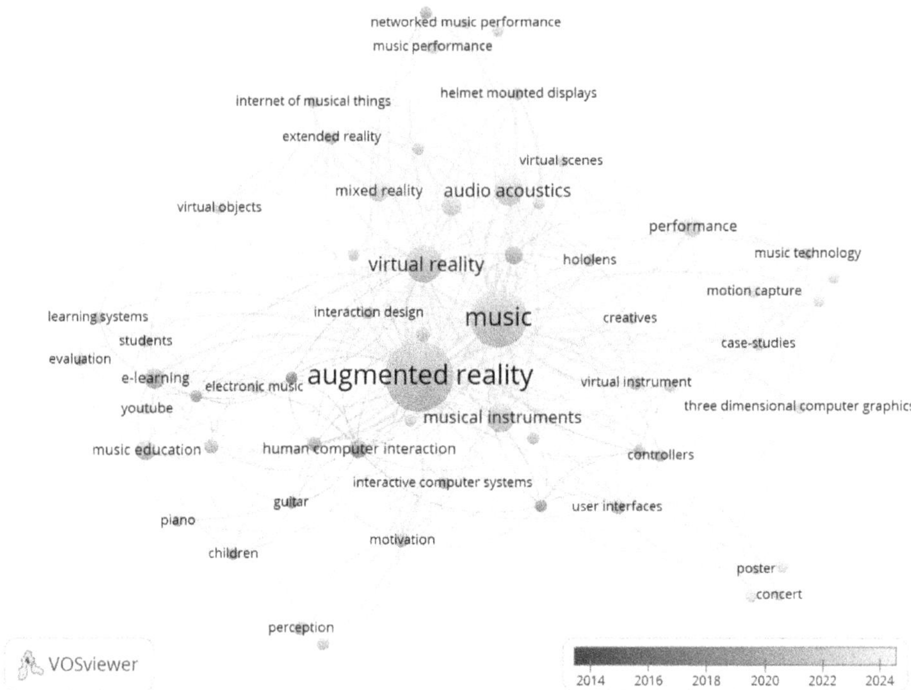

Figure 4. Keywords Co-occurrence Analysis, Density Visualisation

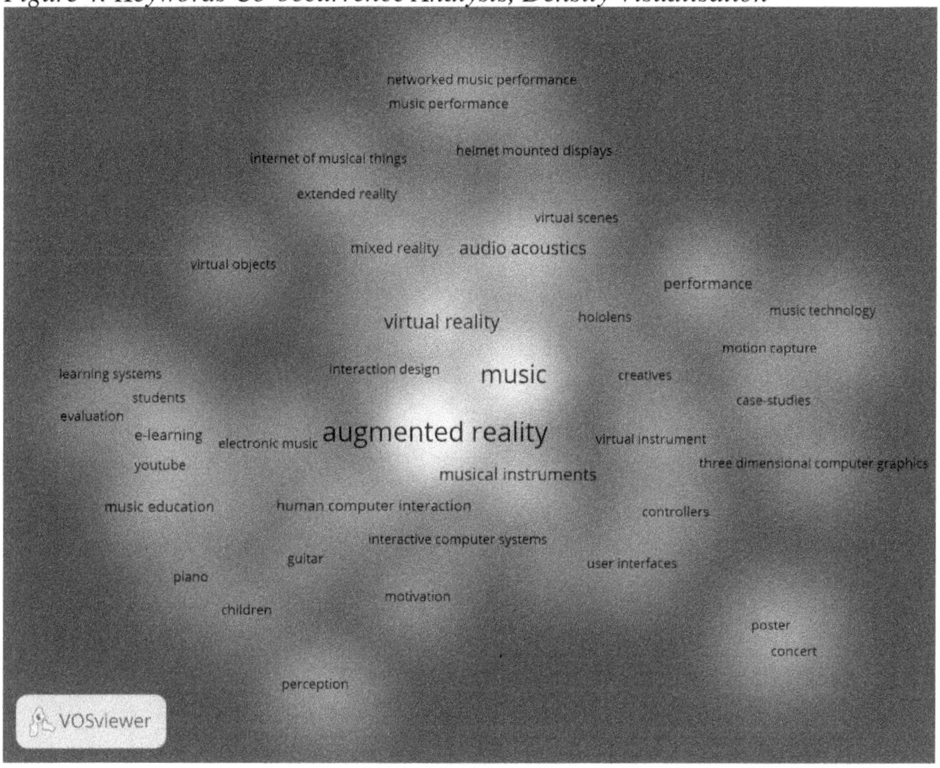

In addition, the researchers used Excel software to create bar charts for the 'Country of Authorship' and line charts for the 'Year of Journal Publication,' providing an intuitive visualisation of the geographic and temporal distribution of research on AR and music education. The bar charts highlight the contributions of researchers from various countries and regions, showcasing the global participation in this field. Meanwhile, the line charts illustrate research trends over time by displaying the number of journal publications per year. These visualisations enable readers to clearly understand how AR and music education research has expanded and evolved across different regions and periods. This information is crucial for comprehending the global distribution and historical progression of this research area, serving as a valuable reference for future studies. Detailed graphical representations are shown in Figures 5 and 6.

Figure 5. Country of Authorship Bar Chart

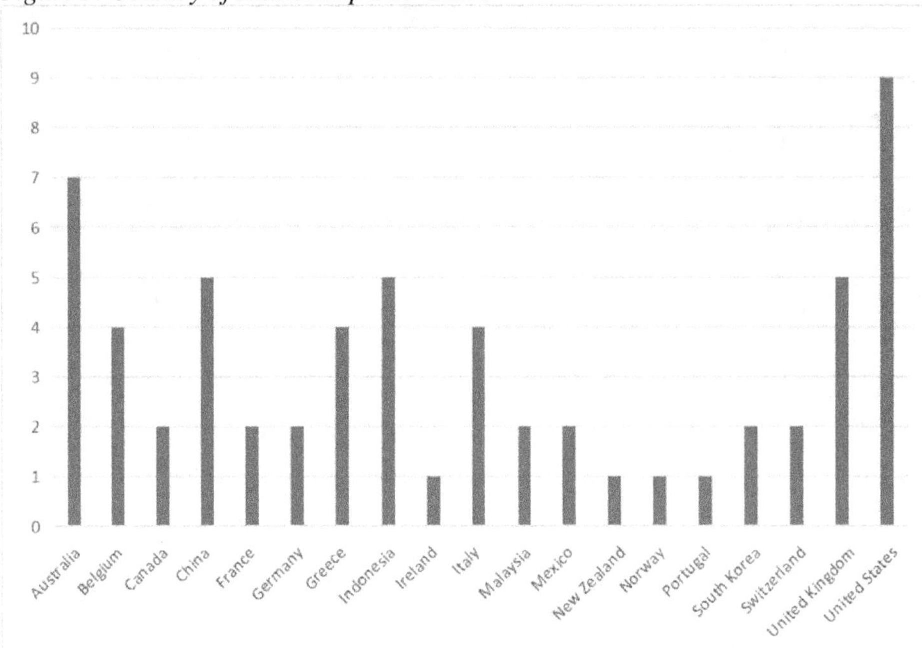

Figure 6. Year of Journal Publication Line Chart

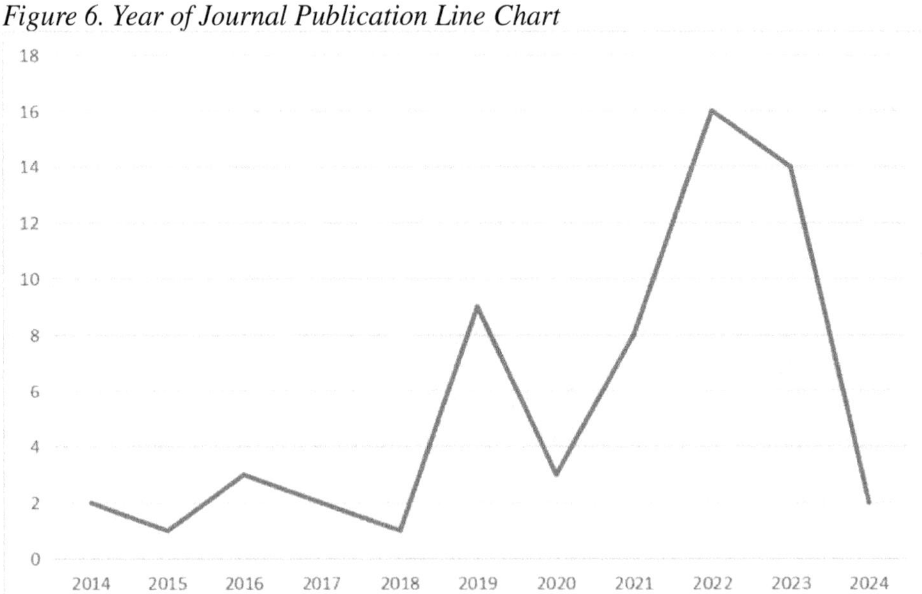

RESULTS

Keyword Co-occurrence Analysis

In this study, the researcher conducted a detailed co-occurrence analysis of keywords in literature related to AR and music education using VOSviewer software. The results are presented in three formats: network visualization (Figure 2), overlay visualization (Figure 3), and density visualization (Figure 4).

Network Visualisation

The network visualisation graph (Figure 2) clearly demonstrates the complex relationships and the cluster structure of interactions between the keywords within the research area. Keywords are marked as coloured circles in the graph, and the size of the nodes in the graph intuitively reflects the frequency of occurrence of each keyword in the literature, with a positive correlation between the size of the circle and the number of occurrences of the keyword. Meanwhile, the thickness of the lines connecting these nodes indicates the co-occurrence intensity between the keywords; the thicker the line, the more often the two keywords co-occur in the same literature, suggesting a close association between them. Thus, the size of the letters and circles depends on the frequency and importance of the keywords: the more occurrences, the larger the size of the circles and letters, a form of visualisation that allows the researcher to identify the core keywords of the research area at a glance.

In the generated maps, prominent nodes such as 'augmented reality', 'virtual reality', 'music', 'music education', 'musical instruments' and 'audio acoustics' occupy the centre of the graph, showing that these topics have significant research status and influence in the current literature. This indicates that the research around these keywords is the main research focus in the field of AR and music education, reflecting the current research trends and hotspots in academia. This network diagram not only reveals the relationship between the keywords, but also helps researchers to better understand the knowledge structure and future research directions in the field.

Around these core themes, several dense clusters of keywords were formed. The results are divided into eight clusters: cluster 1 is depicted in red, cluster 2 is depicted in yellow, cluster 3 is depicted in green, cluster 4 is shown in blue, cluster 5 is shown in purple, cluster 6 is shown in orange, cluster 7 is shown in pink, and cluster 8 is shown in brown. Each cluster shows the relationship between one term and another.

The red and yellow clusters have more nodes with larger relevant connecting lines, and these two colours include retrieved keywords, so the correlation is stronger radiating a larger range. The red cluster focuses on the technological application of ‘augmented reality’ to ‘music education’, includes ‘e-learning’, ‘ electronic music’, ‘learning systems’, and also includes specific items used, such as ‘piano’ and ‘guitar’ (Martin-Gutierrez et al., 2020; Keebler et al., 2014; Klanjscek et al., 2021). It points to innovations in music education methods and the use of digital learning tools. The yellow cluster focuses on ‘music’ and ‘virtual reality’, ‘extended reality’ and ‘internet of musical things’ (Weinel, 2020; Zhang, 2022), highlighting the convergence of musical performance enhancement and technological innovation in educational practice, which are becoming core research directions in the field of music education.

The green cluster includes ‘audio acoustics’, ‘immersive’, ‘music performance’, ‘virtual scenarios’, and ‘music internet of things’(Lu et al., 2022; Klanjscek et al., 2021), reflecting the importance of these technologies and concepts in creating immersive music experiences and the trend towards their application in music performance and education. The blue cluster focuses on ‘musical instruments’, as well as ‘user interfaces’ and ‘smart print documents’, ‘concerts’ and ‘posters’ (Campo et al., 2023; Hopkins et al., 2022; Keebler et al., 2014; Patias et al., 2023; Zünd et al., 2015), reflecting current research that is exploring how technology can revolutionise traditional musical instruments or create new virtual ones. The formation of these clusters not only identifies key research directions within the research community, but also demonstrates trends in technology-driven innovation in education. The purple cluster focuses on ‘virtual instruments’, ‘musical expression’, ‘creatives’, ‘controllers’ and ‘feedback’ (Campo et al., 2023; Klanjscek et al., 2021; Zünd et al., 2015), suggesting that research is exploring ways to enhance the variety and interactivity of musical creation and expression through the use of virtual instruments and controllers.

The orange cluster focuses on ‘music technology’, ‘performance’, ‘three-dimensional computer graphics’, ‘motion capture’ and ‘imitation learning’ (Campo et al., 2023; Klanjscek et al., 2021). This cluster reflects the application of technology in music performance, which makes musical performances more vivid and interactive. This cluster reflects the use of technology in music performance, making it more vivid and interactive. The pink cluster focuses on ‘games’, ‘motivation’, ‘computer music’, ‘interactive computer systems’ and ‘visual communication’ (Preka & Rangoussi, 2019; Zünd et al., 2015), suggesting that research is exploring how to motivate learners through gamification elements, computer music and interactive computer systems. The brown cluster has smaller dots and fewer lines, and focuses mainly on ‘interaction design’, ‘sonic interactions’ and ‘user experience’ (Klanjscek et al., 2021; Rovithis et al., 2019), suggesting that these research themes may be in an exploratory phase. The number of studies is relatively small, but they focus on how to enhance the user experience through interaction design and sound interaction.

Overlay Visualisation

The researcher systematically analysed the literature on AR in the field of music education using the Overlay Visualization graph generated by VOSviewer (Figure 3). The graph demonstrates the temporal evolution of research in this field and the associations between topics through variations in colour, node size and connecting line thickness.

The gradation of colours from blue to yellow in Figure 3 represents the focus of research over time, with blue corresponding to earlier research time periods (e.g., 2014) and yellow representing the latest research trends (e.g., 2024). This colour gradient allows us to visually observe how topics within the research field have evolved over time. For example, in the lower right part of the graph, keywords such as 'poster', 'concert', 'three-dimensional computer graphics', 'motion capture', and 'imitation learning' are highlighted in yellow, indicating that these research themes have become the focus of research in the last few years. This indicates that these research topics have become the focus of research in recent years. This trend shows that academics have begun to pay more attention to the application and potential of these emerging technologies in music education in recent years. In contrast, keywords such as 'mobile devices', 'human computer interaction' and 'visual communication' are predominantly in blue or purple, indicating that research in these areas has been more active in earlier periods.

Keyword links related to 'augmented reality', 'virtual reality', 'music' and 'musical instruments' are shown in different colour levels, indicating that research in these areas is more active at earlier times. different colour levels, suggesting that technologies or methods for enhancing musical performance continue to be used across multiple time periods. This suggests that the research focus may have shifted from earlier widespread applications to specific technological applications and innovations. Over time, researchers (Keebler et al., 2014; Klanjscek et al., 2021; Martin-Gutierrez et al., 2020) have moved from early exploration of the basic applications of these techniques to progressively deeper into how they can be effectively applied in music and instrumental education. This also suggests that the applications of these technologies have become more prominent over time and have developed closer associations in academic research.

Density Visualisation

The Density Visualization chart is a visualisation tool in VOSviewer that is used to show the density of individual topics or keywords within a research area. This chart helps researchers to identify hotspots and possible research gaps in a research area.

The yellow areas in the Density Visualization chart (Figure 5) indicate hotspot areas with high keyword density, while the green and blue areas indicate low keyword density. This means that within the yellow areas, these keywords appear frequently in the relevant literature, indicating that these topics are the core areas of current research. In addition to the keywords 'augmented reality', 'music' and 'music education'. The keywords 'virtual reality', 'musical instruments' and 'audio acoustics' are in the yellow area of the graph. are in the yellow area of the graph, indicating that these topics are the focus of research in the field of music education and AR. The very large nodes and the yellow colour of these nodes also indicate that these keywords appear more frequently in the literature and are strongly associated with multiple other keywords.

In contrast, keywords such as 'learning systems' and 'motivation', 'perception', 'concert', 'music technology' and 'music performance' are located at the edge of the graph and are coloured blue-green. These keywords may represent emerging areas or directions that have received less research attention. In addition, although 'e-learning' and 'performance' are also important research areas, their nodes are larger, but their colours are closer to green or blue, which implies that although they are important, they may be relatively minor or more mature in this area of research. This means that although they are important, they may be relatively minor or more mature in this area of research.

Analysis by Country of Authorship

This bar chart (Figure 5) shows the number of articles published by country in a given research area. As can be seen from the graph, the USA and Australia lead the way in terms of research output in this area, demonstrating the strong activity and efficient output of research organisations in these two countries. The UK, Canada, France and Germany are in the middle of the pack in terms of the number of publications, reflecting the solid foundation and stable performance of these countries in the global research arena.

In contrast, countries such as China, Italy and Mexico have relatively low numbers of research publications, possibly indicating that these countries are less active than others in this area of research. These data show an uneven distribution of research activity across the globe and also hint at differences in research resources and financial support between different countries. Overall, this information provides a preliminary understanding of the global distribution of research power and the level of activity of countries in specific research areas.

Analysis of the Number of Publication Years

This line graph (Figure 6) shows the trend in the number of publications in a given research area from 2014 to 2024. The data in the graph shows that the number of publications in the literature starts low in 2014 and then gradually rises, experiencing periods of fluctuation in 2018 and 2019. In particular, the number of literatures shows large fluctuations between 2018 and 2019, possibly reflecting the volatility of research activities.There is a significant drop in the number of literatures in 2020, but it reaches its highest point in 2022 for the entire time period.

It is further observed that after the peak in 2022, the number of literatures declined sharply in 2023 and 2024, showing a significant decrease in research activities. This rapid decline may be related to research funding, research interest, or external economic and social factors. Overall, these data points provide direct evidence of the changing dynamics of the research field, revealing fluctuations in research activity across years and providing a basis for further analyses of trends in the field.

DISCUSSION

Researchers discuss key trends in research on AR and music education, the unevenness of global research distribution, and fluctuations in research dynamics. By combining the results of keyword co-occurrence analyses, national author distribution analyses, and year of publication analyses, we will reveal the core hotspots in the current research field, the differences in global research strengths, and their potential impact on future research directions.

Research Hotspots and Trends

This study analyses the application of AR technology in music education, revealing its important role in changing the way teaching content is presented and optimising the interactivity of the learning environment. Keyword co-occurrence analyses show that AR technology is particularly significant in enhancing interactivity and engagement in music learning (Rovithis et al., 2019). The application of this technology not only increases students' immersion in the learning content, but also provides a richer and more realistic learning experience through emerging technologies such as motion capture and 3D computer graphics (Lu et al., 2022), suggesting that the field of music education is rapidly adopting these new technologies to enhance teaching and learning.

In further exploring the use of AR and virtual reality (VR) in music education, it is important to consider the impact of these technologies in specific educational practices in addition to identifying the core keywords. Research has shown that AR and VR not only improve traditional teaching methods, but also provide students with a more immersive learning experience that can stimulate their interest in learning. For example, AR technology allows students to see and manipulate musical instruments up close in a virtual environment through 3D modelling and interactive features, an experience that far exceeds that of the traditional classroom (Jeon et al., 2016; Martin-Gutierrez et al., 2020; Tan & Lim, 2018). The spread of such technologies may lead to a fundamental change in music education, especially in areas with limited resources. AR and VR can fill the gaps in educational resources, enabling more students to have access to high-quality music education.

Furthermore, although these technologies have shown clear advantages in music education so far, future research needs to focus on how to overcome technical and ethical challenges in their application. For example, how to ensure the safety of students in virtual environments and how to balance the relationship between technology and traditional educational methods remain pressing issues. These challenges suggest that although AR and VR technologies are expected to be important tools for music education in the future, their application still requires further research and optimisation.

Imbalance in the Global Distribution of Research

In analysing the global distribution of research, it is worth noting that the reason why countries such as the United States and Australia are leading in this area reflects their advantages in terms of financial investment, research infrastructure and technological innovation. It may also be closely related to their technological development and education policies strengths. Not only do these countries have adequate research funding, but they also have policies that support the widespread use of emerging technologies in education. For example, some educational policies in the United States actively promote the integration of technology in the classroom, which provides strong support for the application of AR and virtual reality technologies in music education.

On the other hand, the poor performance of research in this field in developing countries reveals an imbalance in global educational resources. Some countries with more scarce research resources may face technological innovation and application diffusion challenges. The relatively low level of research activity in these countries not only limits the modernisation of local education but may also lead to biased and uneven global research results. Therefore, the international academic community must pay more attention to how such gaps can be bridged through transnational collaboration. This phenomenon suggests that, along with the promotion of global educational technologies, it is crucial to foster international cooperation to bridge the education gap. For example, educational equity and technology sharing on a global scale can be achieved through international collaborative projects, research grants, or knowledge-sharing platforms that help these countries in the field of AR and music education.

Fluctuations in Research Dynamics and Future Prospects

Statistical analyses of publication years indicate that the number of publications in the research area experienced fluctuations between 2018 and 2019, possibly due to changes in research interest triggered by certain key events or major discoveries. The significant decline in the number of publications in 2020 may be associated with the COVID-19 pandemic, indicating the broader impact of external events on research activity. The increase in the number of publications in 2022, on the other hand suggests that technological innovations or research breakthroughs are driving research output. However, the sharp decline in publications in 2023 and 2024 suggests that the research field may be experiencing new challenges, such as reduced funding or shifting research interests.

Fluctuations in research dynamics not only reflect the impact of the external environment but may also reveal changes in academic enthusiasm for certain technologies. For example, the decline in literature in 2020 may be related to research disruptions due to the global epidemic, a phenomenon that suggests that research continuity may be affected by uncontrollable factors in the face of global events. Many academic conferences were cancelled or postponed during the epidemic, and research funding may have been reallocated to more pressing areas, all of which may have led to a short-term downturn in specific research areas.

However, this volatility may also provide new opportunities for future research. In the wake of the epidemic, the use of AR and virtual reality technologies in distance education has become more promising as the demand for online education increases (Jeon et al., 2016; Martin-Gutierrez et al., 2020; Young et al., 2020). These technologies not only overcome the time and space constraints of traditional education, but also provide students with a personalised and interactive learning experience. Therefore, future research could explore how these technologies can be more effectively integrated into online education platforms to develop more flexible and adaptable education models.

In addition, new research directions are emerging as global research dynamics change. For example, the combination of AR, VR, MR and AI may provide smarter solutions for music education (Campo et al., 2023;Klanjscek et al., 2021; Lu et al., 2022; Martin-Gutierrez et al., 2020; Schlagowski et al., 2023). The introduction of these technologies will not only enhance the learning experience, but also improve performance by providing students with tailored learning paths through big data analytics and personalised feedback. This trend suggests that future research needs to focus more on technology integration and innovation to drive further development in music education.

Future research should also delve into the specific applications of AR technologies in music education, especially how to effectively integrate these technologies to improve teaching quality and learning outcomes. Attention also needs to be paid to how to reduce the cost of AR technologies and improve teacher training to facilitate the widespread application of these technologies. These explorations will help to optimise the use of AR in education and maximise its potential, thereby informing efforts to address gaps in current research and guide future research directions.

CONCLUSION

Research Status of Using AR Technology in Music Education

The co-occurrence analysis of the keywords in this study shows that AR has become a research direction that has received a lot of attention in the field of music education. Keywords such as 'musical instruments', 'virtual reality' and 'audio acoustics' take centre stage in the network visualisation graph. This indicates that these topics have attracted a great deal of academic research at the intersection of AR and music education.

In addition, the overlaid visualisations reveal a gradual evolution of research in this area over time, from 2014 to 2024, with a gradual shift in research themes from basic applications to the application of specific technologies and innovations. This trend suggests that the use of AR in music education and instrumental learning is rapidly evolving, nevertheless, there is still room for expansion of existing research in terms of the integration of technology with educational practice. By analysing the literature and keyword co-occurrence data, the application of AR not only enriches teaching and learning content, but also improves learner interactivity and engagement, especially with the support of emerging technologies such as motion capture and 3D computer graphics, music education is able to provide learners with a more immersive and personalised learning experience. This is further evidence that the researcher has a positive attitude towards the potential of AR in music education and that the field has a promising future.

Ways in which AR enhances the Music Learning and Teaching Experience

The results of this study show that AR technologies significantly enhance the experience of learning and teaching music by creating immersive and interactive learning environments. Clusters of keywords such as 'audio acoustics', 'immersive experience' and 'musical performance' suggest that these technologies play a key role in enhancing the musical experience and performance. Meanwhile, the analysis of the density visualisation maps further validates this conclusion, showing that hot keywords such as 'augmented reality', 'virtual reality' and 'musical instruments' have become core research themes in the field of music education. This shows that hot keywords such as 'augmented reality', 'virtual reality' and 'musical instrument' have become core research themes in the field of music education. This suggests that students are able to engage more deeply in music learning in an AR environment, which improves learning motivation.

Educational applications of AR technology are not limited to increasing students' interest and engagement in learning, but also significantly enhance students' comprehension and memorisation performance. By providing virtual instruments and simulated playing environments, AR technology helps students learn in dynamic and immersive environments, enabling them to better grasp abstract concepts and complex content, which in turn enhances musical understanding and performance skills. In addition, AR apps make complex music theories more accessible through visualisation and interaction, which not only increases students' motivation but also enhances their learning outcomes and creativity. Thus, AR technology not only plays an important role in enriching the content of teaching and learning, but also significantly improves the overall effectiveness of teaching and learning by increasing student engagement, comprehension and creativity.

Advantages and Challenges of Applying AR in Music Education

The use of AR in music education has significant advantages but also faces some challenges. The advantages lie in the fact that AR technology can be seamlessly integrated into all aspects of music education, such as e-learning and learning systems, providing strong support for the innovation of educational methods and the application of digital tools. This technology not only provides students with innovative teaching methods, but also enriches the learning experience, making teaching more interactive and livelier. The studies shown in the red cluster demonstrate that the use of AR is driving innovation in music education, particularly in the provision of richer learning content and interactive learning environments. However, challenges remain, and areas such as 'interaction design' and 'user experience' are still in the exploratory phase, as reflected in the less researched brownfield clusters. This illustrates that there are still many unresolved issues in optimising the user experience and designing effective interaction patterns.

In addition, the rapid iteration of technology requires teachers to constantly learn and adapt to new tools, which puts constant pressure on educators. The keywords in the blue and purple regions show that early research has focused more on basic applications, and how to extend these basic applications to a wider range of educational scenarios remains a challenge that needs to be explored in depth. Meanwhile, the high cost and maintenance requirements of AR technologies may limit their application in resource-limited educational settings. These challenges need to be overcome through rational resource allocation and continuous technical support to ensure that the technology can be effectively applied in a wider range of educational settings to maximise its potential.

Gaps in Current Research and Potential Future Research Directions

Despite important advances in research on the use of AR technologies in music education, the findings also reveal a number of research gaps. Firstly, the temporal evolution of the overlaid visualisation maps shows a gradual shift in research focus from early widespread applications to deeper applications of specific technologies, yet research on the long-term impact and practical effects of these technologies in education remains limited. Secondly, the research themes in the brown cluster show that relatively little research has been conducted on interaction design and user experience, which may be a research direction that needs to be strengthened in the future.

In addition, the analysis of the density visualisation maps shows that certain emerging areas such as the 'internet of music things' and 'virtual musical instruments' have not yet been fully researched and that these areas have great potential for future research. Therefore, future research could focus more on how to achieve deeper interaction and personalised learning through AR and explore the application of these emerging technologies in music education. In addition, enhancing teacher training to improve the effectiveness of technology applications, as well as exploring how AR technologies can contribute to inclusive education, are also directions that deserve attention.

REFERENCES

Azzuma, R. (1997). A Survey of Augmented Reality Presence. *Presence (Cambridge, Mass.)*, 6(4), 355–385. DOI: 10.1162/pres.1997.6.4.355

Campo, A., Michałko, A., Van Kerrebroeck, B., Stajic, B., Pokric, M., & Leman, M. (2023). The assessment of presence and performance in an AR environment for motor imitation learning: A case-study on violinists. *Computers in Human Behavior*, 146, 107810. DOI: 10.1016/j.chb.2023.107810 PMID: 37663430

Cook, M. J. (2019). Augmented reality: Examining its value in a music technology classroom. Practice and potential. *Waikato Journal of Education*, 24(2), 23–38. DOI: 10.15663/wje.v24i2.687

Dai, D. D. (2021). Artificial intelligence technology assisted music teaching design. *Scientific Programming*, 2021(1), 9141339. DOI: 10.1155/2021/9141339

Hopkins, T., Weng, S. C. C., Vanukuru, R., Wenzel, E. A., Banic, A., Gross, M. D., & Do, E. Y. L. (2022). Ar drum circle: Real-time collaborative drumming in ar. *Frontiers in Virtual Reality*, 3, 847284. DOI: 10.3389/frvir.2022.847284

Jeon, J., Hong, M., Yi, M., Chun, J., Kim, J. S., & Choi, Y. J. (2016). Interactive authoring tool for mobile augmented reality content. *Journal of Information Processing Systems*, 12(4), 612–630.

Keebler, J. R., Wiltshire, T. J., Smith, D. C., Fiore, S. M., & Bedwell, J. S. (2014). Shifting the paradigm of music instruction: Implications of embodiment stemming from an augmented reality guitar learning system. *Frontiers in Psychology*, 5, 471. DOI: 10.3389/fpsyg.2014.00471 PMID: 24999334

Klanjscek, N., David, L., & Frank, M. (2021). Evaluation of an e-learning tool for augmented acoustics in music education. *Music & Science*, 4, 20592043211037511. DOI: 10.1177/20592043211037511

LeCun, Y., Bengio, Y., & Hinton, G. (2015). Deep learning. nature, 521(7553), 436-444.

Lu, Y., Wang, X., Gong, J., & Liang, Y. (2022). ChordAR: An educational AR game design for children's music theory learning. *Wireless Communications and Mobile Computing*, 2022(1), 5268586. DOI: 10.1155/2022/5268586

Martin-Gutierrez, J., Del Rio Guerra, M. S., Lopez-Chao, V., Soto Gastelum, R. H., & Valenzuela Bojórquez, J. F. (2020). Augmented reality to facilitate learning of the acoustic guitar. *Applied Sciences (Basel, Switzerland)*, 10(7), 2425. DOI: 10.3390/app10072425

Mei, B., & Yang, S. (2021). Chinese pre-service music teachers' perceptions of augmented reality-assisted musical instrument learning. *Frontiers in Psychology*, 12, 609028. DOI: 10.3389/fpsyg.2021.609028 PMID: 33613388

Page, M. J., McKenzie, J. E., Bossuyt, P. M., Boutron, I., Hoffmann, T. C., Mulrow, C. D., Shamseer, L., Tetzlaff, J. M., Akl, E. A., Brennan, S. E., Chou, R., Glanville, J., Grimshaw, J. M., Hróbjartsson, A., Lalu, M. M., Li, T., Loder, E. W., Mayo-Wilson, E., McDonald, S., & Moher, D. (2021). The PRISMA 2020 statement: An updated guideline for reporting systematic reviews. *PLoS Medicine*, 18(3), e1003583. DOI: 10.1371/journal.pmed.1003583 PMID: 33780438

Pati, K. A., Gururani, S., & Lerch, A. (2018). Assessment of student music performances using deep neural networks. *Applied Sciences (Basel, Switzerland)*, 8(4), 507. DOI: 10.3390/app8040507

Patias, P., Georgoula, O., Kaimaris, D., Georgiadis, C., Roustanis, T., Klimantakis, K., Fragkoulidou, V., Chalkidou, S., Papadopoulos, G.-J., Giannopoulos, D., Poulopoulos, K., & Vassilakopoulou, A. (2023). Ar and Chamber Music: Musical Kaleidoscope. *The International Archives of the Photogrammetry, Remote Sensing and Spatial Information Sciences*, 48, 1165–1171. DOI: 10.5194/isprs-archives-XLVIII-M-2-2023-1165-2023

Preka, G., & Rangoussi, M. (2019). Augmented Reality and QR Codes for Teaching Music to Preschoolers and Kindergarteners: Educational Intervention and Evaluation. In CSEDU (1) (pp. 113-123).

Rovithis, E., Moustakas, N., Floros, A., & Vogklis, K. (2019). Audio legends: Investigating sonic interaction in an augmented reality audio game. *Multimodal Technologies and Interaction*, 3(4), 73. DOI: 10.3390/mti3040073

Schlagowski, R., Nazarenko, D., Can, Y., Gupta, K., Mertes, S., Billinghurst, M., & André, E. (2023, April). Wish you were here: Mental and physiological effects of remote music collaboration in mixed reality. In *Proceedings of the 2023 CHI conference on human factors in computing systems* (pp. 1-16). DOI: 10.1145/3544548.3581162

Tan, K. L., & Lim, C. K. (2018, September). Development of traditional musical instruments using augmented reality (AR) through mobile learning. In AIP Conference Proceedings (Vol. 2016, No. 1). AIP Publishing.

van Eck, N. J., & Waltman, L. (2010). Software Survey: VOSviewer, a Computer Program for Bibliometric Mapping. *Scientometrics*, 84(2), 523–538. DOI: 10.1007/s11192-009-0146-3 PMID: 20585380

Weinel, J. (2020). Cyberdreams: Visualizing music in extended reality. Technology, Design and the Arts-Opportunities and Challenges, 209-227.

Zhang, J. (2022). Practical research on the assistance of music art teaching based on virtual reality technology. *Wireless Communications and Mobile Computing*, 2022(1), 8479040. DOI: 10.1155/2022/8479040

Zünd, F., Ryffel, M., Magnenat, S., Marra, A., Nitti, M., Kapadia, M., . . . Sumner, R. W. (2015). Augmented creativity: bridging the real and virtual worlds to enhance creative play. In SIGGRAPH Asia 2015 Mobile Graphics and Interactive Applications (pp. 1-7). DOI: 10.1145/2818427.2818460

Chapter 2
Advantages and Disadvantages of Access to ChatGPT Among University Students:
A Systematic Literature Review

Tahani R. K. Bsharat
Universiti Sains Islam Malaysia, Malaysia

Ismail Sheikh Ahmad
https://orcid.org/0009-0001-9735-5339
International Islamic University Malaysia, Malaysia

Zuheir Khlaif
https://orcid.org/0000-0002-7354-7512
An Najah National University, Palestine

ABSTRACT

This chapter explores the advantages and disadvantages of university students using ChatGPT, an AI-powered chatbot. By synthesizing existing research, it aims to provide a comprehensive understanding of the topic. Using thematic analysis, the study identifies key themes: the benefits of ChatGPT, its drawbacks, and its potential integration into university education. The findings suggest that ChatGPT enhances learning experiences, offers personalized support, and promotes self-directed learning. However, concerns such as overreliance on AI, ethical issues, and privacy challenges are also highlighted. The chapter emphasizes the need for further research to address these issues and guide effective integration into academic settings.

DOI: 10.4018/979-8-3693-9811-1.ch002

1. INTRODUCTION

The advancement of natural language processing (NLP) and artificial intelligence (AI) technologies has led to the emergence of conversational agents that can simulate human-like conversations. One such prominent conversational agent is the Chat Generative Pre-Trained Transformer (ChatGPT), developed by OpenAI, which employs a deep learning model to generate human-like text responses (Roumeliotis & Tselikas, 2023). In recent years, ChatGPT has gained significant attention and has been widely explored in various domains, including customer service, healthcare, and entertainment. However, its potential applications in higher educational settings have received relatively less attention (Ray, 2023). This systematic literature review aims to examine the use of ChatGPT in higher educational settings and explore its benefits, challenges, and implications for teaching and learning.

This systematic literature review explores the advantages and disadvantages of university students accessing ChatGPT, an AI-powered chatbot, aiming to identify its impact. Using NVivo software, the review synthesizes existing research, revealing key themes such as enhanced learning experiences, improved access to information, and potential drawbacks like ethical concerns. Furthermore, the study highlights the need for further research to address concerns and establish best practices for integrating ChatGPT into university settings. Moreover, the significance of this study lies in its comprehensive examination of ChatGPT's impact on university students, offering insights into the benefits and challenges involved. This study establishes the importance of responsible implementation by addressing concerns such as over-reliance on artificial intelligence and privacy concerns. The recommendations so far emphasize the need for further research, balanced integration, and collaboration among stakeholders, as well as guidance for effective use by universities. The goal of this study is to integrate ChatGPT effectively, balancing AI with human interaction, inform students about its benefits and drawbacks, collaborate with AI researchers, prioritize privacy, foster stakeholder collaboration, and gather feedback from stakeholders. A collaborative approach is required to address ethical considerations when integrating AI into education. AI, like ChatGPT, must be used responsibly to preserve human skills while reaping its benefits. The study also emphasizes the need to use AI cautiously in educational assessments, recognizing its limitations and potential dangers. To integrate AI responsibly, collaboration is required to harness technology while addressing ethical concerns.

2. LITERATURE REVIEW

This review of the selected articles has illuminated the multifaceted impact of large language models like ChatGPT across various educational domains and industries. The published studies collectively demonstrate the potential benefits and limitations of integrating AI technologies into education and training contexts. From healthcare to statistics, chemistry, and beyond, ChatGPT's role in reshaping pedagogical practices and industry standards is undeniable. However, as these studies reveal, careful consideration of domain-specific requirements, ethical concerns, and user perceptions is essential for harnessing the full potential of AI in education and beyond.

Therefore, this study explores the various applications and implications of ChatGPT in different educational domains, including medical education, statistics and data science education, chemistry education, construction safety education, nuclear medicine and radiography education, parasitology, Chinese academic settings, English as a Foreign Language (EFL) education, and event planning within an event management class. Each section provides insights from relevant studies and highlights the strengths and limitations of ChatGPT in these educational contexts. The details of the reviewed articles are presented in Table 1.

Table 1. Details of Authors, Research Context, and Source Journals

Study	Context	Source(s)
(Chukwuere,2024).	The study looks at ChatGPT, an AI tool created by OpenAI, and its application in higher education. It evaluates research from January 2023 to July 2023 to determine the possible advantages and disadvantages of ChatGPT. Among the benefits are enhanced access to learning resources, timely feedback, individualized instruction, and more student interaction. However, there are drawbacks, including the risk of relying too much on AI, social interaction deficits, difficulty understanding emotions, and technological limitations. The study recommends integrating it with conventional teaching techniques and considering ethical considerations.	Computer Science Journal
(Garcia Castro, et al., 2024)	The integration of ChatGPT into basic education is examined in this study, with particular attention paid to the benefits, drawbacks, and effects on academic assignments. The quick uptake of ChatGPT in educational environments begs concerns regarding its drawbacks and efficacy. ChatGPT is positioned as a revolutionary tool in education due to its sophisticated features. This study aims to close the gap between ChatGPT's alleged advantages and its real-world applications in the classroom. To understand how ChatGPT influences educational practices, a varied group of educators and students.	Contemporary Educational Technology,
(Ali, Barhom, Tamimi, & Duggal, 2023)	The performance evaluation of ChatGPT in the context of contemporary healthcare education.	European Journal of Dental Education
(Chan & Hu, 2023)	Exploration of university students' perceptions of Generative AI (GenAI) technologies, such as ChatGPT, in higher education. The research indicated that students are generally familiar with GenAI technologies, but concerns about their use were not necessarily related to knowledge about the technology.	International Journal of Educational Technology in Higher Education
(Currie, Singh, Nelson, & Nabasenja, 2023)	Performance evaluation of ChatGPT among nuclear medicine and radiography students. The study found that ChatGPT consistently performed below average in written tasks but outperformed students in foundational subjects where shallow responses sufficed.	Radiography Journal
	This highlights the model's potential as a resource for certain types of assessments.	
(Ellis & Slade, 2023)	Utility investigation of ChatGPT in statistics and data science education. The study recognized its role in developing course materials and providing guidelines for responsible interaction, but it also pointed out limitations in ChatGPT's common-sense knowledge and contextual understanding, which may hinder its effectiveness in certain contexts.	Journal of Statistics and Data Science Education
(Emenike & Emenike, 2023)	ChatGPT's potential assessment in chemistry education. The findings highlighted limitations in its ability to answer algorithmic questions accurately. Chemistry's reliance on visual concepts like Lewis structures and protein representations makes it challenging for text-only systems like ChatGPT to provide comprehensive support in this domain.	Journal of Chemical Education

continued on following page

Table 1. Continued

Study	Context	Source(s)
(Hung & Chen, 2023)	Discussion on the implications of Chinese students using ChatGPT for academic activities. The study highlighted concerns about the negative impact on critical thinking skills and the risk of plagiarism. It underscored the need for a well-designed regulatory system to mitigate these issues.	Social Sciences-Basel Journal
(Iskender, 2023)	Discussion on the multifaceted impacts of ChatGPT on education and the tourism and hospitality industry. While ChatGPT proved valuable in various educational aspects, there were concerns about over-reliance on AI hindering critical thinking. In the tourism and hospitality sector, ChatGPT showed potential but should not replace human creativity and personalized service.	European Journal of Tourism Research
(Kasneci, et al., 2023)	Discussion on the potential of large language models like ChatGPT in medical education and clinical decision-making. The study also identified the importance of thoughtful integration, considering factors such as age and background. It suggested that ChatGPT can enhance medical education by providing resources and assistance tailored to individual needs.	Learning and Individual Differences
(Keiper, 2023)	Examination of ChatGPT applications in event planning of an event management class. The study revealed its effectiveness in content creation tasks, such as generating bios, promotional materials, and event descriptions. ChatGPT's ability to provide templates and reduce faculty time commitments can enhance students' acquisition of employability skills in event management.	Journal of Hospitality Leisure Sport & Tourism Education
(Uddin, Albert, Ovid, & Alsharef, 2023)	Examination of ChatGPT's integration into safety education for construction students. The study observed a substantial improvement in hazard recognition among students after using ChatGPT. It suggested	Sustainability
	that AI can significantly enhance safety education, a critical aspect of preparing future construction professionals.	
(You, Chen, You, Zhang, & Cao, 2023)	Utilization of evolutionary game analysis to explore the role of AI in education. The study discovered strategies that could improve educational effectiveness in an AI- driven era. Schools were recommended to adopt positive supervision to standardize AI usage, while students were encouraged to engage with AI technology actively. The research demonstrated the potential of AI to reshape the education landscape.	Sustainability

continued on following page

Table 1. Continued

Study	Context	Source(s)
(Young & Shishido, 2023)	Investigation of ChatGPT's ability to generate reference dialogues for an EFL chatbot system. The study found that ChatGPT's dialogues were suitable for elementary and intermediate-level students, offering comprehensible content and vocabulary, suggesting its potential in EFL education.	International Journal of Advanced Computer Science and Applications
(Šlapeta, 2023)	Transformative potential analysis of large language models like ChatGPT in parasitology. While acknowledging limitations in providing authentic experiences, the integration of AI in this field can enhance research, education, and communication.	Trends in Parasitology

By conducting a systematic literature review, we aim to provide a comprehensive overview of the existing research on the use of ChatGPT in higher education. This methodology is driven by its merits in the form of systematic, transparent, and replicable review, and it is also inspired by prior review articles (Montenegro-Rueda, Fernández-Cerero, Fernández-Batanero, &

López-Meneses, 2023). Therefore, this review will help in identifying the current state of knowledge and research gaps, as well as in providing insights into the potential future directions of using ChatGPT. Through the analysis of the selected articles, the current systematic literature review provides a comprehensive understanding of the current state of research on the advantages and disadvantages of using ChatGPT in higher educational settings. This review explores the potential benefits of employing ChatGPT in teaching and learning. Additionally, it examines the challenges and limitations associated with its implementation.

3. METHODOLOGY

The systematic review methodology applied in this study was developed based on recent education publications (Shahrol, Sulaiman, Samingan, Mohamed, 2020; Azman, Maat, 2021; Mohamed Nor & Sihes, 2021). The methodology was driven by its merits, as it allowed for systematic, transparent, and replicable reviews. Additionally, the study drew inspiration from prior review articles (Pahlevan Sharif, Mura, Wijesinghe 2019; Stracke ; Trisolini, 2021 &Garcia Castro, et al., 2024)). A systematic search strategy to ensure the rigor and reliability of the review was employed, and managed to identify 24 relevant studies from the Scopus academic database. Keywords such as "ChatGPT", "advantages of ChatGPT", "disadvantages of ChatGPT", "university students", and other related terms were used to retrieve these relevant articles while limiting the search to Open-Access category. The inclusion criteria to screen the published studies were made to gather the articles published within the 2023 timeframe, written in English, and focused on the use of ChatGPT in higher educational settings. Subsequently, out of the 24 articles, 12 were excluded following the exclusion criteria, i.e., the articles that do not specifically discuss the use of ChatGPT in higher education or those that are not directly related to teaching and learning. The remaining 12 selected-for-review articles went through a systematic screening process based on their titles, abstracts, and full texts. These articles were subjected to data extraction, where relevant information such as research objectives, methodologies, findings, advantages and disadvantages of ChatGPT, educators' and students' perception of the use of ChatGPT, use of ChatGPT, how best to use ChatGPT, ways of identifying a text written by ChatGPT, and recommendations were extracted and synthesized. These 12 papers were then selected for further analysis, as listed in Table 1.

4. RESULTS

Figure 1. Comparation Among Researchers that Discussed the Advantages and Disadvantages of ChatGPT

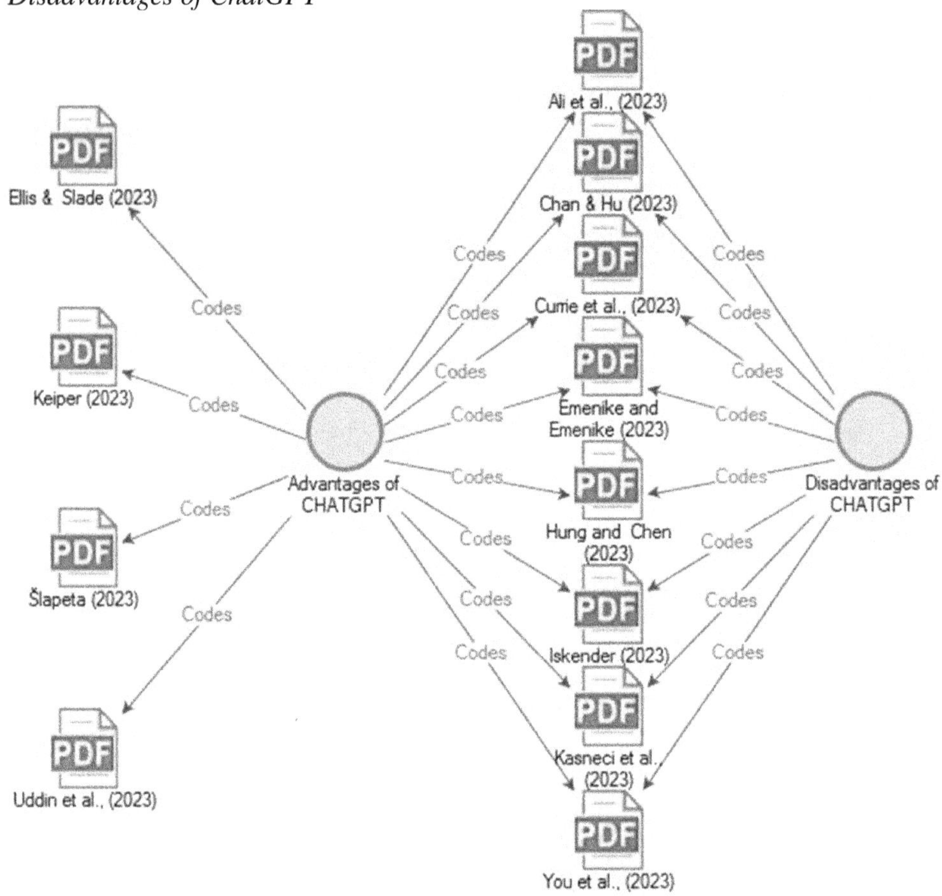

a. Discussion on the Advantages of ChatGPT

Figure 1 illustrates a comparison among researchers who have explored the advantages of ChatGPT utilization for students.

Interactive Learning with ChatGPT

One of the distinctive strengths of ChatGPT, as highlighted by Ali et al. (2023), lies in its conversational and interactive nature. Unlike traditional web search engines, ChatGPT offers a platform where learners can engage directly, receiving immediate responses to their queries.

This facet empowers students not only to consume information but also to dissect it critically and question its authenticity and sources. AI technology transcends physical barriers, providing equal access to education for students with diverse needs (You et al., 2023). In essence, ChatGPT functions as a virtual tutor, available around the clock and free of cost, which is likely to witness increased utilization in higher education, including medical and science education.Furthermore, the benefits are enhanced access to learning resources, timely feedback, individualized instruction, and student interaction (Chukwuere,2024).

Personalized Learning Support

ChatGPT's utility extends to personalized learning, empowering students with tailored content to meet their individual needs (Ali et al., 2023; Iskender, 2023; You et al., 2023). Through adaptation and personalization, it enhances comprehension and retention, benefiting both students and educators. Moreover, the system provides valuable explanations and examples, which are pivotal in elucidating complex concepts (Currie et al., 2023). ChatGPT's summarization capabilities further streamline the learning process, allowing students to condense and simplify intricate information for better understanding and retention (Emenike & Emenike, 2023; You et al., 2023). Additionally, students can harness these systems to access a wealth of practice materials generated by GPT (the type of machine learning model used by ChatGPT), aiding in exam preparation and reinforcing their grasp of course content (Iskender, 2023; You et al., 2023). By leveraging AI algorithms, Generative AI or GenAI (the type of AI that can create a wide variety of data, in which ChatGPT is included) can dynamically adjust instruction in real-time to address each student's unique strengths and weaknesses (Chan & Hu, 2023).

Writing and Brainstorming Assistance

GenAI excels in providing writing and brainstorming assistance, as acknowledged by students who appreciate its capabilities in offering guidance and suggestions throughout the writing process (Chan & Hu, 2023). This type of AI can assist with various aspects of academic writing, including grammar and vocabulary corrections, structural advice, and idea generation (Currie et al., 2023). Furthermore, ChatGPT, as one of the well-known uses of GenAI, has proven valuable in enhancing creativity, encouraging critical thinking, and enabling students to delve deeper into various topics (Iskender, 2023). It can also refine the phrasing and sentence structures in a student's work, significantly elevating the overall quality of written assignments. In addition to spelling, grammar, vocabulary, and punctuation corrections, ChatGPT ensures clarity, conciseness, and a smooth flow in the writing process. These capabilities make it a valuable tool for students seeking academic writing assistance.

Research and Analysis Capabilities

GenAI's potential for research and analysis support is indeed a remarkable revelation. The AI's capacity to gather, process, and analyze vast amounts of information streamlines the research process (Chan & Hu, 2023). It assists students in locating pertinent sources, summarizing findings, and deriving meaningful conclusions. This capability not only saves time but also cultivates critical thinking and analytical skills. By providing keywords and specific topics, it empowers students to streamline their research processes and pinpoint relevant sources for assignments and papers (Currie et al., 2023).

Visual and Audio Multimedia Support

In the digital age, multimedia elements play a pivotal role in education. Some of the GenAI implementations offer visual and audio support by generating interactive content and facilitating multimedia presentations (Chan & Hu, 2023). This aids in conveying complex concepts effectively and engaging students in a dynamic learning experience. At the time the study was conducted, ChatGPT could not deliver visual and audio output. However, OpenAI has since begun to roll out new voice and image capabilities for ChatGPT. The company has provided an interface that allows users to engage in voice conversations and upload images (OpenAI, 2023).

Administrative Support

Beyond academics, GenAI extends its utility to administrative tasks. Chan and Hu's (2023) study highlights that GenAI can assist in managing schedules, organizing assignments, and tracking progress. Hung and Chen (2023) also emphasize that GenAI aids students in creating schedules and establishing task priorities, ensuring they meet deadlines and maintain coursework efficiency. According to Currie et al. (2023), this support enhances the learning experience by allowing educators to focus more on teaching, while students can allocate more time to meaningful academic activities. This administrative assistance empowers students to manage their educational responsibilities and time better, promoting efficiency and self- directed learning (Kasneci et al., 2023; Keiper, 2023).

Mental Health Support

Importantly, ChatGPT extends its utility beyond academics. It acts as a resource hub for students grappling with stress and mental health issues. The system suggests coping strategies and provides access to essential resources, such as counseling services, fostering a supportive environment for students in need (Currie et al., 2023).

Enhancing Educational and Professional Opportunities with GenAI Systems

ChatGPT's human-like text generation has the potential to significantly enhance the educational experience in various ways. It can be a valuable tool for educators, helping to create engaging learning materials, answer students' questions, and even simulate conversations to facilitate language learning (Ellis & Slade, 2023; Uddin et al., 2023). This feature not only boosts teacher productivity but also ensures that students have access to high-quality resources for their studies (Hung & Chen, 2023). By harnessing the capabilities of ChatGPT, education can become more personalized, providing immediate assistance to students and benefiting both learners and educators alike.

Another key advantage of ChatGPT is its ability to assist students in producing coherent and well-structured content, thereby enhancing the quality of their work (Emenike & Emenike, 2023). This can offer students the advantage of swiftly seeking factual answers to their queries. In essence, ChatGPT can help correct spelling, grammar, vocabulary, and punctuation errors, making students' work more polished and professional. Therefore, ChatGPT's advanced text generation capabilities hold great promise for revolutionizing education. It can create engaging learning materials, answer questions, facilitate language learning, and improve the overall quality of student work, benefiting both students and educators (Ellis & Slade, 2023; Uddin et al., 2023; Hung & Chen, 2023; Emenike & Emenike, 2023). This innovation has the potential to make education more personalized and accessible while increasing productivity and resource quality in educational settings.

Empowering Faculty in Their Roles

Both students and faculty members can benefit from GenAI systems when it comes to drafting various documents. Faculty members can employ these systems not only for academic applications but also for other tasks, e.g., crafting letters of recommendation, award nominations, speeches, and promotional materials. This versatility enhances their productivity and ensures the creation of polished, effective documents (Emenike & Emenike, 2023). Additionally, faculty members can harness the power of these systems to swiftly access and synthesize vast amounts of data, facilitating informed decision-making and research (Emenike & Emenike, 2023). These models enable teachers to stay up-to-date with the latest

developments and techniques in education (Kasneci et al., 2023; You et al., 2023). By harnessing their capabilities, educators can generate innovative and engaging lesson plans tailored to their students' needs and curriculum requirements. This not only enhances the learning experience but also minimizes the time teachers spend on manual planning (Hung & Chen, 2023).

Automated Assignment Grading

ChatGPT's ability to grade written assignments efficiently can liberate instructors from the monotony of grading and allow them to focus on providing valuable feedback and conducting research (Iskender, 2023). It reduces the administrative burden of educators while ensuring consistent and unbiased assessment.

Promoting Machine Learning

Interacting with GPT can foster students' comprehension of natural language processing (NLP) and machine learning, both of which are in high demand in today's job market (Iskender, 2023). Engaging with GPT offers practical exposure to these domains and facilitates the development of pertinent skills (Kasneci et al., 2023). Furthermore, Kasneci et al. (2023) asserted that large language models enhance student engagement and participation.

Impact on Related Industries

Beyond higher education, GPT and similar machine learning models have the potential to revolutionize industries like hospitality, tourism, travel, events, and leisure services. They can provide personalized recommendations, improve customer service, and generate effective marketing materials (Iskender, 2023). These advancements can lead to enhanced customer experiences and industry growth.

Impact in Medical Field

In recent years, AI tools have proven to be invaluable assets within hospital settings, significantly reducing mortality rates among sepsis patients. By streamlining early detection and enabling swift, precise intervention, these AI systems have effectively lowered sepsis- related mortality rates by 20% (Šlapeta, 2023). This remarkable achievement underscores the immense potential of AI in revolutionizing healthcare. Šlapeta (2023) also noted that beyond healthcare, AI tools have demonstrated their prowess in accelerating the development of antiparasitic vaccines by accurately predicting structures with specific functions. These tools offer a promising avenue to expedite the creation of vital vaccines, crucial for combating various parasitic diseases afflicting populations worldwide. Therefore, AI tools have emerged as a powerful new weapon against society's most formidable challenges.

Impact on Identifying Construction Hazards

The study conducted by Uddin et al. in 2023 underscores the pivotal role played by ChatGPT in enhancing hazard recognition within the construction industry. By providing easily accessible information and guidance, ChatGPT empowers individuals to identify potential safety hazards more effectively. The discovery highlighted the versatility of ChatGPT, as it could adapt to diverse demographic backgrounds, including varying levels of experience and age among participants (Uddin et al., 2023). Consequently, ChatGPT emerged as a reliable tool that consistently offers substantial benefits to construction professionals in terms of hazard recognition.

b. Discussion on the Disadvantages of ChatGPT

Figure 1 shows not only the articles that outlined the advantages of ChatGPT, but also those that did not—a comparison list of researchers who have explored the disadvantages of utilizing ChatGPT among students. Notable contributions to this subject include the work of Ali et al. (2023), which conducted an examination of ChatGPT's capabilities and identified several noteworthy limitations. First and foremost, they found that ChatGPT was confined to answering text-based questions and lacked the capacity to process queries based on images. However, this limitation might have been eliminated, following the recently released capability for ChatGPT 4.0 to process audio files and images (OpenAI, 2023). Subsequently, Currie et al. (2023) also assessed ChatGPT's performance in the context of medical imaging content. Regrettably, they discovered that ChatGPT fell significantly short of expectations in this domain. Its inability to provide the required depth of insight and accuracy raises concerns not only regarding academic integrity but also its utility in situations where visual information is crucial (Ali et al., 2023; Currie et al., 2023).

Limitations in Responses to Sequential Questions (SEQs)

Ali et al. (2023) discovered that ChatGPT faces challenges in providing detailed information regarding clinical interventions and follow-up visits when responding to Sequential Questions (SEQs). This limitation results in comparatively lower scores when compared to other question formats. Currie et al. (2023) also questioned ChatGPT's ability to generate questions, offer assessment feedback, and create content for university-level independent learning. Emenike and Emenike (2023) reported a notable issue with ChatGPT, which is the inconsistency in responses to repeated prompts and in answering algorithmic questions. Specifically, ChatGPT struggles to generate accurate responses in chemistry-related questions, as exemplified by the case involving the paramagnetic or diamagnetic nature of phosphorus (Emenike & Emenike, 2023). Furthermore, it is important to note that ChatGPT's capacity to produce high-quality but superficial responses can potentially jeopardize academic integrity. At the same time, its deficiency in providing clinical details could hinder its efficacy in healthcare-related applications.

Issues in Written Assignments

In assisting students' written assignments, Currie et al. (2023) discovered significant issues with ChatGPT's performance. Notably, ChatGPT often provided answers without proper supporting evidence, relied on outdated sources, and, at times, even fabricated citations. Ali et al. (2023), in their work on generating reflective portfolio reports, also highlighted a noteworthy limitation of ChatGPT. Specifically, they found that it frequently omitted references to specific learning activities or events. Furthermore, ChatGPT's inability to adhere to professional writing standards and maintain an appropriate tone exacerbated its shortcomings when assisting students in doing their written assignments. This deficiency not only undermined the depth and relevance of the generated content but also diminished its overall educational value (Currie et al., 2023; Ali et al., 2023)

Challenges in Plagiarism Detection and Ethical Concerns

With ChatGPT's ability to generate original text, the effectiveness of routine plagiarism detection software applications comes into question. OpenAI has developed tools like AI Text Classifier, including DetectGPT and GPTZero, to address this challenge (Ali et al., 2023). However, Ali et al. cautioned that these tools might not always be entirely accurate and that misclassifications could occur (Ali et al., 2023). This raised significant concerns about potential academic dishonesty and cheating in academic works (Ali et al., 2023). Chan and Hu (2023) have also noted that students expressed their apprehension about the potential invasion of privacy, particularly as AI systems collect and process vast amounts of personal data. Ethical considerations include bias in AI algorithms and the responsible use of GenAI, highlighting the necessity for robust regulations and ethical guidelines (Chan & Hu, 2023; Kasneci et al., 2023).

Consequently, it is imperative to develop and implement more accurate and reliable detection mechanisms to uphold academic integrity.

Accuracy and Transparency

One pivotal concern raised by students in the study pertains to the accuracy and transparency of GenAI systems. As GenAI becomes more integrated into education and other sectors, ensuring the reliability and precision of generated content is paramount (Chan & Hu, 2023). GenAI-generated information must meet stringent standards, as errors or misinformation could have significant consequences. Hung and Chen's (2023) study revealed factual errors in ChatGPT-generated answers. ChatGPT's responses are generated based on the data it has been trained on, which may contain outdated or incorrect information. This can lead to misinformation and misunderstandings, hindering students' genuine learning experiences. To address this issue, it is crucial to continuously update and improve the training data used for GenAI systems to enhance their accuracy and reliability (Chan & Hu, 2023). This is to ensure that the generated information aligns with the current knowledge and minimizes the risk of spreading inaccuracies.

Academic Integrity and Learning Outcomes at Lower Taxonomies

Currie et al. emphasized that ChatGPT's impact on academic integrity and learning outcomes is most pronounced at lower cognitive levels. The ability of large language models to generate convincing yet unverified information can lead students to accept false or misleading data as fact, undermining the pursuit of knowledge (Kasneci et al., 2023). While it serves as a tool to enhance learning environments, it also poses a risk to academic integrity. Lower-level learning outcomes, such as basic comprehension and knowledge retention, are the areas where ChatGPT could excel by providing quick and accurate responses. Additionally, Chan and Hu's (2023) study revealed concerns about holistic competencies. While GenAI could assist in various tasks, students worry about its potential to diminish their overall skill development. Overreliance on ChatGPT may encourage a narrow view of knowledge, society, and culture, as students become accustomed to quick answers without depth and critical analysis (Hung & Chen, 2023; Iskender, 2023; Kasneci et al., 2023).

Human Values and Job Displacement

Chan and Hu's 2023 study highlighted the increasing importance of preserving human values in a world increasingly influenced by GenAI. The study raised questions about the potential erosion of societal values and the human touch in various fields, including education. As Iskender (2023) emphasized, an over-reliance on AI might reduce the need for human teachers, leading to concerns about the loss of personalization and essential human interaction within classrooms, which are crucial for effective learning. AI can be a valuable tool for teachers, but it should complement, not replace, their roles (Kasneci et al., 2023). Chan and Hu (2023) added that with GenAI automating certain tasks, individuals may encounter challenges in finding meaningful employment. Preparing students for a rapidly evolving job market, one that incorporates GenAI, becomes imperative for educational institutions. Balancing technological advancements with the preservation of core human values is a critical consideration for the future (Chan & Hu, 2023).

Accessibility Issues

AI adoption in education may inadvertently exacerbate existing educational inequalities. Students from low-income backgrounds may face challenges accessing AI-powered tools and resources, thereby widening the educational divide (Iskender, 2023). It is crucial to ensure equitable access for all students.

Bias Amplification

AI systems, including GPT, are susceptible to inheriting and exacerbating biases present in their training data. In education, this translates into the risk of perpetuating socio-economic and racial disparities (Iskender, 2023). For instance, if AI systems favor certain groups or fail to address the needs of marginalized students, it can exacerbate educational disparities (You et al., 2023). AI's algorithms can unintentionally discriminate against certain groups, reinforcing existing inequalities in the education system.

Data Privacy

A significant concern related to the use of AI tools in education revolves around data privacy and security. The collection and utilization of students' personal information without their consent can pose serious privacy risks (Iskender, 2023; Kasneci et al., 2023). Therefore, if proper safeguards are not in place, the indiscriminate gathering and processing of this information can lead to privacy violations (You et al., 2023). Another ethical challenge pertains to the lack of transparency and informed consent in data collection practices. Often, students, teachers, and parents have limited insight into how AI systems operate and how their data is employed (You et al., 2023). It is imperative for educational institutions to establish robust data protection measures and ensure transparent data usage practices.

Knowledge and Expertise Gap

Many educators and institutions may lack the knowledge and expertise required to effectively integrate these technologies into their teaching methods (Kasneci et al., 2023). Bridging this knowledge gap is crucial for maximizing the benefits of large language models in education.

Financial Burden

Kasneci et al. (2023) stated that the maintenance and deployment of large language models can impose substantial financial burdens on schools and institutions, particularly those with limited budgets. The costs associated with these technologies can strain already stretched resources.

Sustainable Usage

The computational demands of large language models lead to high energy consumption (Kasneci et al., 2023). To ensure environmentally sustainable operation, the adoption of energy- efficient hardware and renewable energy sources, such as cloud infrastructure, is imperative.

Lack of Adaptability

Large language models may struggle to adapt to the diverse needs of students and teachers, hindering the level of personalization essential for effective learning. Ongoing research may address this limitation (Kasneci et al., 2023).

Language Diversity and Access Equity

Most of the research in large language models primarily focuses on the English language, leaving a significant gap in support for other languages (Kasneci et al., 2023).

Significance and Contribution

Research significance and contribution can be attributed to the systematic literature review approach utilized in this study to explore the advantages and disadvantages of university students utilizing ChatGPT. The reason of this audit is to supply a comprehensive understanding of the subject by synthesizing existing inquire about and academic articles. Methodological thoroughness is included to the audit prepare by utilizing NVivo program for analyzing developing topics in chosen investigate papers. Researchers identified several key themes in their study, including the benefits and drawbacks of ChatGPT as well as the possibility of integrating it into university education. ChatGPT has been found to enhance university students' learning experiences, improve access to information, increase convenience, and provide personalized support. Student queries, study resources, and course selection guidance are all examples of how ChatGPT can assist students. Furthermore, it promotes self- directed learning and student engagement. Although the review acknowledges potential disadvantages, such as overreliance on AI technology, ethical concerns, and data security challenges, it also acknowledges potential advantages. As a result of addressing these concerns, the research establishes the need for further research and the establishment of best practices for integrating ChatGPT into university settings. The results of this research provide universities with insights into implementing and utilizing ChatGPT effectively to support their student's educational journeys. For successful implementation, it is important to consider both the benefits and challenges of ChatGPT access.

Recommendations:

1. Further research is needed to address the disadvantages identified in this review, such as overreliance on AI technology, ethical concerns, and privacy challenges. It is important to investigate these concerns thoroughly and propose solutions to mitigate any adverse effects.
2. ChatGPT should be integrated into university settings according to best practices. In addition, we discuss how AI technology must be balanced with human interaction and personalized support, as well as guidelines for using ChatGPT

appropriately. In order to ensure effective and responsible implementation of ChatGPT, universities should share these best practices.

3. Universities should consider ChatGPT to improve learning experiences, increase access to information, and provide personalized support to students. ChatGPT should, however, be used carefully to ensure that its goals and objectives are aligned with its limitations and challenges.
4. It's critical that colleges inform their student body of Chat GPT's possible advantages and drawbacks. AI technology has a place in education, and it is important to teach students how to utilize it properly. Programs for orientation, workshops, and instructional materials that explain ChatGPT and its ramifications can be used to do this.
5. To continuously enhance Chat GPT's features and resolve any shortcomings found in this review, universities should work with AI researchers and developers. Sustained research and development ought to concentrate on improving Chat GPT's comprehension of context, practical expertise, and precision in responding to domain-specific queries.
6. When using ChatGPT in academic settings, privacy and data security concerns ought to come first. Higher education institutions have an obligation to guarantee the security of student data and to implement suitable protocols to protect sensitive data. To address privacy concerns and safeguard student rights, it is important to develop clear standards and policies.
7. Collaboration and communication among stakeholders, including students, faculty, administrators, and AI developers, is essential to the successful integration of ChatGPT into university education systems. Feedback from students and faculty should be sought to resolve issues, improve the user experience, and ensure that the technology meets the needs of the university community. By following these guidelines, colleges and universities can determine how to implement and use ChatGPT to support their students and curriculum while addressing potential challenges and ensuring appropriate use of AI technology.

5. CONCLUSION

The integration of AI chatbots like ChatGPT into education holds significant promise, offering benefits such as personalized learning experiences, improved accessibility, and efficient resource allocation. These AI systems also aid educators in automating tasks and creating engaging materials, enhancing overall work efficiency. Moreover, they have potential applications in fields like occupational safety and medical education. Large language models, including GPT, have revolutionized education by enhancing student engagement, enabling automated grading, and preparing students for modern workforce demands. However, ethical considerations, such as data privacy, transparency, and algorithmic bias, must be addressed collaboratively among educational institutions, policymakers, and technology providers to ensure responsible AI integration. Additionally, the limitations of AI, including detectability, susceptibility to plagiarism detection, and factual errors, should be acknowledged, and AI should be used as a supplementary tool in educational assessments rather than a primary method.

While AI integration in education offers many advantages, it also raises ethical concerns that require careful attention. It is crucial to strike a balance between the benefits of AI and the preservation of essential human skills. The misuse of AI, such as ChatGPT, can compromise academic integrity, as it may be detectable by plagiarism detection software and contain factual errors. Therefore, educators and students should use AI judiciously, recognizing its limitations and potential negative consequences for genuine learning and intellectual development. In educational assessments, AI should be used cautiously, especially in specialized subjects, and educators and researchers must work together to refine its integration effectively, considering its inconsistency and potential for errors. Overall, responsible AI integration in education necessitates collaboration and proactive efforts to address ethical concerns while harnessing the technology's potential for enhancing learning experiences.

REFERENCES

Ali, K., Barhom, N., Tamimi, F., & Duggal, M. (2023). ChatGPT-A double-edged sword for healthcare education? Implications for assessments of dental students. *European Journal of Dental Education*, 00, 1–6. DOI: 10.1111/eje.12937 PMID: 37550893

Azman, N. A., & Maat, S. M. (2021). Integration of the History of Mathematics in Mathematics Education: A Systematic Literature Review. *International Journal of Academic Research in Business & Social Sciences*, 11(4). Advance online publication. DOI: 10.6007/IJARBSS/v11-i4/9768

Castro, R. A. G., Cachicatari, N. A. M., Aste, W. M. B., & Medina, M. P. L. (2024). Exploration of ChatGPT in basic education: Advantages, disadvantages, and its impact on school tasks. *Contemporary Educational Technology*, 16(3), ep511. DOI: 10.30935/cedtech/14615

Chan, C. K., & Hu, W. (2023). Students' Voices on Generative AI: Perceptions, Benefits, and Challenges in Higher Education. *International Journal of Educational Technology in Higher Education volume 20. Article*, (43), 1–8.

Chukwuere, J. E. (2024). The use of ChatGPT in higher education: The advantages and disadvantages. *arXiv preprint arXiv:2403.19245.*

Currie, G. M., Singh, C., Nelson, T., Nabasenja, C., Al-Hayek, Y., & Spuur, K. (2023). ChatGPT in medical imaging higher education. *Radiography*, 29(4), 792–799. DOI: 10.1016/j.radi.2023.05.011 PMID: 37271011

Ellis, A. R., & Slade, E. (2023). A New Era of Learning: Considerations for ChatGPT as a Tool to Enhance Statistics and Data Science Education. *Journal of Statistics and Data Science Education, 31:2, DOI:* , 128-133.DOI: 10.1080/26939169.2023.2223609

Emenike, M. E., & Emenike, B. U. (2023). Was This Title Generated by ChatGPT? Considerations for Artificial Intelligence Text-Generation Software Programs for Chemists and Chemistry Educators. *Journal of Chemical Education*, 100(4), 1413–1418. DOI: 10.1021/acs.jchemed.3c00063

Garcia Castro, R. A., Chura-Quispe, G., Velarde Molina, J. F., Espinoza Ramos, L. A., & Almonte Durand, C. A. (2024). Bibliometric review on teaching methods with artificial intelligence in education. *Online Journal of Communication and Media Technologies*, 14(2), e202419. DOI: 10.30935/ojcmt/14367

Hung, J., & Chen, J. (2023). The Benefits, Risks and Regulation of Using ChatGPT in Chinese Academia: A Content Analysis. *Social Sciences 12: 380. https://doi.org/* , 1-15.DOI: 10.3390/socsci12070380

Iskender, A. (2023). Holy or Unholy? Interview with Open AI's ChatGPT. *European Journal of Tourism Research, 34, 3414. https://doi.org/*, 1-11.DOI: 10.54055/ejtr.v34i.3169

Kasneci, E., Sessler, K., K¨uchemann, S., Bannert, M., Dementieva, D., Fischer, F., . . . Pfeffer, J. (2023). ChatGPT for Good? On Opportunities and Challenges of Large Language Models for Education. 1-13.

Keiper, M. (2023). ChatGPT in Practice: Increasing Event Planning Efficiency Through Artificial Intelligence. 33. . . *33*.. .DOI: 10.1016/j.jhlste.2023.100454

Mohamed Nor, H., & Sihes, A. J. (2021). Critical Thinking Skills in Education: A Systematic Literature Review. *International Journal of Academic Research in Business & Social Sciences*, 11(11). Advance online publication. DOI: 10.6007/IJARBSS/v11-i11/11529

Montenegro-Rueda, M., Fernández-Cerero, J., Fernández-Batanero, J. M., & López-Meneses, E. (2023). Impact of the Implementation of ChatGPT in Education: A Systematic Review. *Computers*, 12(8), 153. DOI: 10.3390/computers12080153

Open, A. I. (2023). ChatGPT can now see, hear, and speak. OpenAI. https://openai.com/blog/chatgpt-can-now-see-hear-and-speak (Access date: 19 October 2023).

Pahlevan Sharif, S., Mura, P., & Wijesinghe, S. N. R. (2019). Systematic Reviews in Asia: Introducing the "PRISMA" Protocol to Tourism and Hospitality Scholars. 13–33. DOI: 10.1007/978-981-13-2463-5_2

Ray, P. P. (2023). ChatGPT: A comprehensive review on background, applications, key challenges, bias, ethics, limitations and future scope. *Internet of Things and Cyber- Physical Systems Volume 3*, 121-154.

Roumeliotis, K. I., & Tselikas, N. D. (2023). ChatGPT and Open-AI Models: A Preliminary Review. *Future Internet*, 15(6), 192. DOI: 10.3390/fi15060192

Shahrol, S. J. M., Sulaiman, S., Samingan, M. R., & Mohamed, H. (2020). A systematic literature review on teaching and learning english using mobile technology. *International Journal of Information and Education Technology (IJIET)*, 10(9), 709–714. DOI: 10.18178/ijiet.2020.10.9.1447

Šlapeta, J. (2023). Are ChatGPT and other pretrained language models good parasitologists? *Trends in Parasitology*, 39(5), 314–316. DOI: 10.1016/j.pt.2023.02.006 PMID: 36872153

Stracke, C. M., & Trisolini, G. (2021). A systematic literature review on the quality of moocs. *Sustainability (Basel)*, 13(11), 1–26. DOI: 10.3390/su13115817

Uddin, S. M., Albert, A., Ovid, A., & Alsharef, A. (2023). Leveraging ChatGPT to Aid Construction Hazard Recognition and Support Safety Education and Training. *Sustainability, 15, 7121. https://doi.org/,* 1-22.DOI: 10.3390/su15097121

You, Y., Chen, Y., You, Y., Zhang, Q., & Cao, Q. (2023). Evolutionary Game Analysis of Artificial Intelligence Such as the Generative Pre-Trained Transformer in Future Education. *Sustainability, 15, 9355. https://doi.org/,* 1-12.DOI: 10.3390/su15129355

Young, J. C., & Shishido, M. (2023). nvestigating OpenAI's ChatGPT Potentials in Generating Chatbot's Dialogue for English as a Foreign Language Learning. *(IJACSA). International Journal of Advanced Computer Science and Applications*, 14(6), 65–72. DOI: 10.14569/IJACSA.2023.0140607

Chapter 3
AI-Powered ChatGPT in Education:
A Double-Edged Sword

Rahul Joshi
https://orcid.org/0000-0002-5467-5105
Manav Rachna International Institute of Research and Studies, Haryana, India

Suman Kumari
Manav Rachna International Institute of Research and Studies, Haryana, India

Krishna Pandey
https://orcid.org/0000-0001-8290-3681
Manav Rachna International Institute of Research and Studies, Haryana, India

ABSTRACT

The advent of AI in education provides new resources with the potential to revolutionize how we teach and learn. This chapter thoroughly summarizes artificial intelligence (AI) technologies, their possible uses in education, and the challenges that arise from using them. Researchers examine chatbots and similar algorithms that can mimic human interaction and create writing that reads naturally given linguistic input. While innovative chatbots like ChatGPT have many benefits, they also present significant ethical and practical concerns when used in the classroom. The chapter aims to promote AI's responsible and ethical use in education by providing insightful information on how AI may be effectively introduced into the educational context to benefit instructors and students.

DOI: 10.4018/979-8-3693-9811-1.ch003

1. INTRODUCTION

The educational practices of the 21st century have seen significant transformations driven mainly by technological advancements, particularly in artificial intelligence (Baidoo-Anu & Ansah, 2023). In recent decades, there has been a noteworthy surge in the development and use of artificial intelligence, leading to substantial transformations across all sectors of society. Artificial intelligence has emerged as a prominent study area across several academic domains, including health, robotics, education, and autonomous driving. In a general sense, artificial intelligence encompasses the capacity of a computer system to replicate the cognitive processes shown by the human brain. This process includes the acquisition of external data, the acquisition of knowledge via training, and the subsequent attainment of the intended objectives for which it was explicitly developed (Brazdil & Jorge, 2001).

The advancements and growth in machine learning have resulted in the developing of more advanced and novel technologies for generating digital content, such as generative artificial intelligence (Joshi et al., 2024). Generative modelling artificial intelligence (GAI) refers to a machine learning framework that operates unsupervised or partly supervised. It utilises statistical methods, probability, and other techniques to build artificial artefacts or relics (Jovanovic & Campbell, 2022). Generative artificial intelligence (AI) uses deep learning (DL) techniques to produce artificial artefacts by analysing pre-existing digital information, including but not limited to video, images/graphics, text, and audio. It involves studying training samples and understanding their distribution patterns (Abukmeil et al., 2021). The existing body of literature has recognised two prominent forms of generative artificial intelligence, namely the Generative Adversarial Network (GAN) and the Generative Pre-trained Transformer (GPT) (Gui et al., 2021). Generative Adversarial Networks are widely used as a prevalent approach in General Artificial Intelligence. The Generative Adversarial Network employs two neural networks: the generator and discriminator networks. The generator network produces synthetic data, such as a picture depicting a person's face. Conversely, the discriminator network evaluates the authenticity of the generated material, determining if it accurately represents an actual instance (e.g., verifying the authenticity of a human image). The verification process continues until the discriminator network reaches a point where it cannot differentiate between synthetic and genuine material, hence recognising the synthetic content as indistinguishable from actual content. The use of the Controller Area Network (CAN) is primarily focused on the creation of speech, graphics, and video. In contrast, Generative Pre-trained Transformer models leverage substantial quantities of publicly accessible digital content data, specifically in natural language processing (NLP), to comprehend and generate text that closely resembles human language. These models demonstrate the ability to exhibit creativity in their writing, ranging from a single paragraph to an entire research article. They can convincingly simulate various topics across multiple languages (Aydın & Karaarslan, 2023).

These models can interact with clients in a manner that closely resembles human communication, as seen in customer-service chatbots and fictitious characters inside video games (Pavlik, 2023). A more advanced Generative Pre-trained Transformer (GPT)-3 iteration has just been created. GPT-3, with its impressive 175 billion parameters, has been designed to improve task-agnostic capabilities and achieve competitive performance compared to previous fine-tuning methods (Mann et al., 2020). According to Brown et al. (2020), GPT-3 surpasses all prior non-sparse language models by a factor of 10. GPT-3 has emerged as the fundamental natural language processing engine powering ChatGPT. This newly created language model has garnered significant interest across several domains, such as education (Tate et al., 2023), economics, and finance (Terwiesch, 2023). The integration of artificial intelligence in the field of education has yielded noteworthy outcomes, as indicated by enhancements in the efficacy of the educational process, the facilitation of global learning, the customization of learning experiences, the development of more intelligent educational content, and the optimization of educational management in terms of both effectiveness and efficiency (Montenegro-Rueda et al., 2023). In the present day, emerging technologies have become essential in enhancing the educational process of teaching and learning. Education is not just confined to a commodified entity but rather encompasses a dynamic process whereby learning transcends the sheer accumulation of information. In this context, artificial intelligence is a novel technology with significant promise within education. AI-powered systems can facilitate individualized learning experiences by dynamically adjusting to each student's unique requirements and preferences.

In education, it is imperative to acknowledge that integrating artificial intelligence presents notable problems and ethical implications. Matters about data privacy, equitable access to education, and the implications for the role of educators require meticulous consideration. It is vital to conscientiously consider these facets to guarantee AI's advantageous and fair utilization for all individuals engaged in the learning process. It is also essential to establish an equilibrium between technology and the fundamental responsibility of educators to maintain the central emphasis on the comprehensive growth of students and their readiness for a dynamic global landscape. No comprehensive evaluations of the effects of ChatGPT on schooling have been identified so far, perhaps due to its recent introduction in late November 2022 (Neumann et al., 2023). Hence, the primary aim of this systematic review is to examine the extant scholarly works about utilizing ChatGPT in education to discern its effects, advantages, difficulties, and potential implementation domains within instruction and acquisition.

2. EVALUATION OF CHATGPT

The AI-driven conversational agent developed by OpenAI is referred to as ChatGPT. The phrase "Generative Pre-trained Transformer (GPT)" denotes a language processing model that has undergone extensive training on vast amounts of data to generate written content that closely mimics human-authored text. ChatGPT is an advanced natural language processing technique that leverages artificial intelligence to respond to inquiries. Consequently, this process conversationally facilitates information generation, acquiring knowledge from these interactions and enabling more tailored responses. ChatGPT exhibits human-like behaviour while delivering instructions to users and disseminating information. This technology can engage in many tasks, including generating poetic compositions, programming, responding to queries, composing electronic correspondence and written compositions, and facilitating the translation of textual materials. ChatGPT distinguishes itself from other chatbots by generating instantaneous responses, fostering more diverse and dynamic conversations over various subjects (Zhai, 2023; Kshetri et al., 2023). The utilization of ChatGPT is being examined as an educational instrument to enhance students' abilities in the areas of questioning and precise formulation, knowledge expansion through the provision of answers by ChatGPT, and the development of skills in evaluating the accuracy, reliability, and quality of ChatGPT's responses, as well as the ability to filter relevant information obtained from these responses. This versatile and adaptable technology makes it suitable for various applications since it can effectively accommodate varying climates and scenarios.

ChatGPT is a versatile tool that may be used in many applications within natural language processing. The system demonstrates a notable capacity to react to commands accurately and fluently, contingent upon possessing a comprehensive comprehension of the environment and the capacity to engage in human-like thinking (Halaweh, 2023). ChatGPT is an artificial intelligence application that facilitates engaging and authentic conversations with individuals. The system can understand textual input and respond to it by using artificial intelligence and machine learning techniques. Consequently, it may facilitate dialogues that bear more resemblance to interpersonal exchanges. Consequently, it is a proficient instrument for enterprises operating in customer support, marketing, and content production. ChatGPT is a virtual assistant capable of engaging in conversations and responding to questions resembling human interaction (Mitrović et al., 2023; Kitamura, 2023). ChatGPT employs deep learning algorithms to generate natural conversations between humans and machines.

3. CHATGPT'S IMPORTANCE FOR EDUCATIONAL PURPOSES

Among the facets of education that ChatGPT has the potential to influence are writing, instructional methodology, and teaching pedagogy. Writing has long been recognized as a fundamental tool for cultivating imaginative and analytical reasoning using structuring data and constructing stories. It remains an indispensable component of education despite the advent of artificial intelligence. Therefore, it should focus on providing AI with insights beyond its comprehension. The students' theses, essays, and assignments should be concise, reflective, and firmly rooted in a specific context. Education and AI are crucial subjects in discussions regarding the future of our society. ChatGPT is an invaluable resource for professionals who frequently generate text, including writers and marketers. It is utilized for chatbot responses, creating content for websites and social networking platforms, and marketing materials (Lund & Wang, 2023). The potential implementation of ChatGPT in higher education could lead to an automation revolution, reducing the demand for faculty members and potentially stifling chances for human interaction and interpersonal connections.

ChatGPT can examine the assignments for grammatical and structural errors to aid learners, enhance their writing proficiencies, and offer insightful feedback. Personalized feedback based on a student's writing style can assist the learner in recognizing and focusing on the specific areas that require improvement. ChatGPT enables computers to simulate human conversations (Perkins, 2023). This system can provide precise responses to user inquiries and facilitate personalization through its ability to understand user intent and context (Shahriar & Hayawi, 2023). ChatGPT could assist the students in various endeavours, including essay writing, computer program development, and mathematical problem-solving. ChatGPT is engineered to facilitate rapid connections and more cohesive responses, like conversational tools such as virtual assistants and bots. As a result of this interoperability, enterprises can enhance their existing products and services by creating unique AI-driven chatbots capable of promptly understanding and addressing customer inquiries. ChatGPT provides a private and secure environment for conducting conversations. It provides a secure environment without manipulation or intervention by identifying malicious content, spam, and censorship with artificial intelligence. Additionally, personal information is never stored or transmitted by ChatGPT to any third parties. To safeguard users' privacy, all communication is encrypted and stored locally (Cooper, 2023).

4. SWOT ANALYSIS FOR CHATGPT IN EDUCATION

Utilizing the SWOT framework and extensive research, this chapter comprehensively analyses ChatGPT's strengths. Its purpose is to identify the educational opportunities that ChatGPT can offer while gaining a thorough understanding of its limitations. Through this analysis, potential vulnerabilities that relevant stakeholders may encounter in the future are brought to light. Moreover, a personalized approach can be developed to address and mitigate these risks effectively. The SWOT analysis conducted on ChatGPT has identified a range of possible applications for this emerging AI technology within the field of education.

4.1 Strength involved in ChatGPT

1. ***Producing believable replies:*** ChatGPT is a sophisticated language model that employs the transformer architecture to effectively perform various natural language processing tasks, such as language production and comprehension (Xue et al., 2023). According to Li et al. (2019), this architectural design enables AI-driven chatbots to effectively capture the associations among words inside a phrase, maintaining contextual understanding and providing replies that exhibit coherence and relevance. The remarkable achievement of ChatGPT is credited to its access to an extensive repository of training data (Kasneci et al., 2023). This ample source of information empowers ChatGPT to grasp various linguistic nuances and correlations adeptly, leading to an in-depth comprehension of language and context (Lecler et al., 2023). These remarkable abilities enable ChatGPT to deliver notably authentic and credible responses compared to comparable AI solutions (Sobania et al., 2023).
2. ***Potential for self-improvement:*** One notable characteristic of ChatGPT is its capacity for self-improvement or autonomous learning. ChatGPT employs a linguistically intricate language processing model, generative pre-training (GPT), distinguishing it from other artificial intelligence chatbots. The GPT system is an artificial intelligence text generator that employs reinforcement learning techniques, namely using human input, to enhance its language model (Mann, 2023). The feature above allows ChatGPT to adapt and enhance its replies by incorporating feedback from human assessors (Shen et al., 2023).
3. ***Delivering individualized replies:*** ChatGPT can acquire knowledge via interactions with human users, enabling it to function as a flexible conversational agent. ChatGPT can retain and integrate past talks into its subsequent answers. This feature enables the system to retain context and engage in more fluid and coherent dialogues with users over an extended period. According to Haque et al. (2022), the extensive training of ChatGPT enables it to provide tailored

replies by considering the contextual information provided in a specific prompt (Gao et al., 2023). Furthermore, it should be noted that ChatGPT can produce replies using various tones and structures, which are tailored to accommodate the user's tastes and requirements (Aljanabi, 2023).

4.2 Weakness in ChatGPT

1. ***Insufficient awareness:*** According to Gao et al. (2023), ChatGPT needs more proficiency in comprehending the semantic nuances of the words it analyses. The system can identify patterns and provide reasonable replies (Bogost, 2022). However, it needs a comprehensive understanding of the underlying ideas conveyed by the words. According to a study, the outcomes of this approach may sometimes need more comprehensive analysis and profound understanding (Borji, 2023). Additionally, research suggests that receiving replies that deviate from the intended subject is possible (Gupta et al., 2023). According to Sinha et al. (2023), an empirical investigation revealed that ChatGPT showed proficiency in generating satisfactory solutions for intricate issues in Pathology (Rajain et al., 2023). However, it was observed that replies generated by ChatGPT needed a comprehensive comprehension of the underlying theoretical principles.
2. ***Assessing the efficacy of a given answer poses challenges:*** According to research conducted by Lecler et al. in 2023, there is room for improvement in ChatGPT's ability to assess the reliability and authenticity of its training data (Lecler et al., 2023). This limitation currently hinders the system's ability to accurately evaluate the content it generates, except for topics where there is widespread agreement, such as the 'flat-earth hypotheses.' It is worth noting that ChatGPT operates without internet access, and its knowledge of global events is limited to those occurring before 2021 (Stokel-Walker & Van Noorden, 2023).
3. ***The potential for biases and prejudices:*** ChatGPT can reinforce prejudices and contribute to discriminatory outcomes (Zhai, 2022). Many factors contribute to this phenomenon, including biases in the training data, the algorithms' design, and the broader cultural environment in which these technologies are developed and used. According to the established concept of 'garbage-in-garbage-out,' artificial intelligence algorithms can sustain prejudices using large datasets that embody these biases (Barocas & Selbst, 2016). Due to the above rationale, Amazon was compelled to terminate its AI-based recruitment tool in 2018 after realising that it exhibited bias against women, primarily due to the overrepresentation of male resumes within its training data. Artificial intelligence algorithms have the potential to perpetuate discriminatory practices due to their emphasis on goals such as maximising profitability and efficiency, sometimes neglecting to

account for unforeseen repercussions, such as reinforcing pre-existing prejudices (O'Neil, 2017).

4.3 Opportunities for Educational sector

1. ***The enhancement to data mobility:*** ChatGPT provides students and educators convenient access to information across several platforms, such as websites or smartphone applications, spanning diverse academic disciplines. Furthermore, this tool exhibits greater efficiency than conventional search engines because it provides comprehensive written responses instead of just lists of sources. According to Cascella et al. (2023), ChatGPT can locate and condense pertinent material, promptly facilitating students' access to detailed information. ChatGPT can potentially reduce students' time obtaining information, allowing them to spend more time reading and critically reflecting on the provided content. ChatGPT has the potential to aid educators in the identification and development of relevant instructional resources. Using this tool may also facilitate the development of lesson plans that adhere to certain limits and restrictions.
2. ***Enabling individualized instruction:*** ChatGPT can provide individualised assistance and evaluation to pupils across various degrees of intricacy. In higher education, argumentative essay writing is a significant learning task (Farrokhnia et al., 2022). In this study, we sought feedback from ChatGPT on an anonymous essay titled 'Video Games for Children' under three distinct scenarios. The feedback was evaluated based on three key aspects: constructive, effective, and critical. Using a praising prompt elicited a more significant amount of positive and emotive input from ChatGPT, while a critical prompt resulted in a higher incidence of critical feedback. Teachers should carefully consider the selection of question prompts to provide personalised feedback to students. It has been observed that when feedback consists solely of criticism without any positive elements, students tend to be less receptive to the feedback due to psychological and emotional factors.
3. ***Reducing the burden of teaching responsibilities:*** ChatGPT has significant promise for substantially reducing the burden of educators. For example, it may serve as a feedback instrument for evaluations on various academic projects, such as students' assignments, essays, and other similar undertakings (Qadir, 2023). Educators can request ChatGPT's assistance in generating diverse assessments, including open-ended inquiries, multiple-choice items, and even rubrics for grading students' tasks. ChatGPT has been proposed to automate the evaluation of assignments, particularly in the context of text-oriented educational courses (Cotton et al., 2024). Furthermore, it has been suggested by Mizumoto and

Eguchi (2023) that professors can efficiently provide comments to students on their writings within a limited timeframe (Mizumoto & Eguchi, 2023).

4.4 Threats to Education

1. ***Insufficient comprehension of the contextual factors:*** A sufficient comprehension of language's contextual nuances and underlying significance may give rise to several hazards, particularly in education. For instance, the use of ChatGPT in personalised learning scenarios may need to be improved in its comprehension of the curriculum, individual learning preferences, and the cultural milieu in which students are situated. Consequently, the content suggestions provided by ChatGPT may be excessively challenging or too simplistic for students. Another example is the use of ChatGPT in the essay evaluation process, whereby the absence of contextual information and requisite prior knowledge may impede its ability to assign appropriate grades to essays effectively.
2. ***Endangering the integrity of academia:*** Since the introduction of ChatGPT, numerous concerns have been expressed about the security of online assessments and the risk of cheating in online tests (Garg & Goel, 2022). The capacity of ChatGPT to produce writing that closely resembles human writing has been identified as a factor that could undermine the credibility of online assessments, particularly in higher education settings where such exams are becoming more prevalent (Susnjak & McIntosh, 2024). In a recent empirical investigation, Fijačko et al. (2023) revealed that ChatGPT's responses to university life support examinations were, on average, more relevant, accurate, and aligned with resuscitation guidelines than those from earlier studies using alternative AI technologies. These results underscore the significant risk that ChatGPT presents to academic integrity, especially within higher education.
3. ***Continuous educational prejudice:*** Concerns about the possibility of ChatGPT reinforcing prejudice and displaying biases in educational contexts exist. When the training data exhibits bias towards a specific group, it can result in unjust discrimination against other groups. For instance, if the AI is trained using data primarily from Western European students, its applicability and suitability may be less effective when utilized with groups from other parts of the world (Pedro et al., 2019).

5. ROLE OF AI IN EDUCATION

There are still several tasks to be accomplished to enhance the quality of teaching and learning. Educators seek technology-driven strategies for achieving these objectives that are secure, efficient, and adaptable. As technological advancements rapidly shape our daily lives, educators naturally inquire whether these developments can be harnessed to aid their goals. Like many individuals, educators integrate AI-powered tools into their daily routines, including home voice assistants, grammar correction tools, sentence construction aids, essay creation tools, and automated trip planning applications on their smartphones. Numerous instructors proactively explore the potential of AI capabilities as they become accessible to the broader public. Educators recognize the opportunity to utilize AI-driven capabilities, such as voice recognition, to enhance support for students with disabilities, multilingual learners, and those who could benefit from increased adaptability and customization in digital learning tools. They are exploring ways in which AI can contribute to the creation or enhancement of lessons and refining their approach to sourcing, selecting, and tailoring information for their classes. Educators are keenly aware of potential risks arising from integrating advanced technologies. The introduction of sophisticated features, while beneficial, may bring about new challenges in data privacy and security. They acknowledge that AI, while powerful, has the potential to generate inappropriate or inaccurate output independently. Concerns extend to the possibility of AI-generated connections or automation amplifying undesirable biases. Moreover, educators have observed novel methods employed by students to present others' work as their own. They value the nuanced handling of "teachable moments" and pedagogical strategies, which human instructors excel at but AI models may overlook or misinterpret. A central concern revolves around the fairness of suggestions put forth by algorithms (Schiff, 2022). Educators face a multitude of diverse challenges. Everyone in education is responsible for harnessing the good to fulfil educational aims while guarding against the risks that may occur due to AI integration in education.

5.1 Arguments for Immediate AI Education

First, AI could make it possible to accomplish educational aims more effectively, broadly, and affordably. It is a policy priority to address the many incomplete learning that pupils have due to the pandemic, and AI may make learning materials more adaptive to the requirements and skills of individual students. Enhancing teaching positions is a top goal, and AI can help instructors more by way of automated assistants or other technologies. When instructors run out of time, AI may also allow them to provide more help to specific kids. Priority should be given to creating resources considering the knowledge and experiences students bring to their education, cultural and communal assets. Artificial intelligence may make it possible for curriculum materials to be more locally tailored. AI can improve educational services, as shown by voice assistants, navigation tools, shopping suggestions, essay writing skills, and other commonplace uses.

Secondly, apprehension about possible future hazards and knowledge of system-level concerns give rise to urgency and significance. For instance, pupils can see more monitoring. While the Department vehemently denies that AI may replace teachers, some educators are concerned that they could be supplanted. People know algorithmic prejudice discrimination, such as voice recognition software that struggles to recognise accents from different regions or exam monitoring software that could unjustly flag some student groups for disciplinary action. AI may be used in unseen and infrastructure-related ways, raising questions about trust and transparency. AI often appears in new applications with a magical quality, but educators and procurement regulations demand that technology demonstrate its effectiveness. Artificial Intelligence may provide seemingly genuine knowledge that is, in fact, unreliable or unfounded. Most importantly, AI poses additional concerns beyond the well-known ones related to data security and privacy, such as the possibility of expanding automated processes and pattern detectors that lead to "algorithmic discrimination."

The importance of addressing AI lies in the potential for unforeseen or unintended outcomes. Educators may discover that AI-driven classroom decisions can have unintended consequences. For instance, if an AI system speeds up the curriculum for some students while slowing it down for others due to insufficient evidence, flawed theories, or biased assumptions about learning, it could result in more significant disparities in success. Additionally, the quality of data available may sometimes lead to unexpected results. While an AI-powered teacher selection system may appear more impartial than one that relies on manual assessments, it could inadvertently overlook candidates with unique abilities and diverse perspectives due to past data of poor quality.

6. POTENTIAL BENEFITS FOR STUDENTS

With its extensive prior knowledge, complex language production, and excellent natural language processing (NLP) capabilities, ChatGPT, created by OpenAI, offers several technological benefits to students. These advantages stem from its capacity to comprehend, interpret, and produce human-like writing, making it an effective tool for various educational uses (Firat, 2023).

1. ***Adaptive and Personalized Learning:*** The transformer design at the heart of ChatGPT allows for adaptable learning. Using adjustment and personalization, the model may be adapted to the unique learning patterns of each student, maximizing the dissemination of clarifications, illustrations, and evaluations to accommodate cognitive inclinations.
2. ***Dynamic Academic Support:*** The model's deployment as an on-demand resource makes dynamic academic assistance possible. Due to its round-the-clock accessibility, students may get real-time help and explanations, especially for academic projects with deadlines, like last-minute homework questions or test preparation.
3. ***Knowledge Augmentation:*** Since ChatGPT has a large amount of pre-existing information, it is an additional learning resource. The model can enhance existing curriculum by providing context, pertinent real-world applications, and insights not included in conventional instructional materials.
4. ***Language Generation for Enhanced Writing Skills:*** ChatGPT can produce logical and appropriate language for the situation because of the transformer design. This function is used to help students write better by offering structure, grammar, and style recommendations that result in written work that is more polished and eloquent.
5. ***Interactive Collaborative Learning:*** As ChatGPT is conversational, it may mimic collaborative, interactive learning settings. Participating in online dialogues with the model fosters critical thinking abilities by giving students a forum to express their thoughts, get helpful criticism, and improve their comprehension of complex subjects.
6. ***Natural Language Interaction for Language Learning:*** ChatGPT may be a conversation companion for language learners using its natural language interaction features. The approach facilitates language learning via realistic conversational simulations by offering contextualized language practice, vocabulary development, and grammatical support.
7. ***Accessibility and Customization:*** The design of ChatGPT allows modification to meet a range of learning requirements. The methodology ensures a more inclusive educational experience by improving accessibility for students with different learning styles by producing more examples, personalized help, and alternate explanations.
8. ***Research Assistance and Information Retrieval:*** The model is a valuable tool for research advice because it analyses and generates text. ChatGPT facilitates the process of information retrieval by helping students formulate research questions, find relevant sources, and organize their inquiry.

9. ***Programming Support and Code Assistance:*** ChatGPT's strengths in technical areas include helping students with coding problems. Because of its proficiency with programming languages, it may serve as a virtual instructor for computer science students by offering detailed explanations, samples of code, and help with debugging.
10. ***AI Technology Exposure and Understanding:*** Students are introduced to the real-world uses of artificial intelligence via interacting with ChatGPT. Students' comprehension of the area is deepened due to their increased grasp of the workings of AI models, the subtleties of language creation, and the potential and constraints of existing AI technologies.

When integrating ChatGPT into the education system, it is essential to approach the process with a perspective. While embracing the technological benefits it provides for students, ethical issues, model biases, and the proper use of AI in education should be prioritised. Educators and developers must work together to ensure that technology enhances traditional teaching methods and promotes a seamless combination of human knowledge and AI capabilities.

7. POTENTIAL BENEFITS FOR EDUCATORS

As a sophisticated language model created by OpenAI, ChatGPT offers educators a range of technological benefits. With its state-of-the-art natural language process features, extensive knowledge database, and interactive conversational abilities, it is an exceptional resource for enhancing teaching strategies, streamlining administrative tasks, and establishing customized and effective learning atmospheres in diverse domains (Whalen & Mouza, 2023).

1. ***Content Customization and Adaptation:*** The inherent transformer design of ChatGPT facilitates the process of fine-tuning and modification, empowering educators to adapt the model according to their pupils' requirements and preferences. The ability to customize learning materials allows for the development of adaptive resources tailored to match individuals' unique learning styles and cognitive skills.
2. ***Automated Administrative Assistance:*** Educators can use ChatGPT to streamline administrative responsibilities, like composing emails, creating announcements, and even automating replies to frequently asked questions. The capacity of the model to comprehend and produce coherent textual content enhances the efficiency of regular administrative procedures, enabling educators to dedicate more time to instructional preparation and student involvement.

3. ***Personalized Feedback and Assessment:*** The natural language-generating capabilities of ChatGPT facilitate the delivery of individualized feedback for assignments and exams. Educational practitioners can use this approach to give students a personalized and enhanced learning encounter by providing comprehensive insights, constructive feedback, and customized coaching.
4. ***Curriculum Augmentation and Enrichment:*** By integrating ChatGPT into the instructional toolbox, educators can supplement their current curriculum with more context, practical applications, and comprehensive explanations. The extensive knowledge bases the model possesses facilitate the enhancement of course materials, guaranteeing that students are provided with a thorough and current education.
5. ***Interactive Lesson Planning and Design:*** ChatGPT facilitates an interactive process of lesson preparation, allowing educators to exchange ideas and experiment with various methods of lesson design. Through this platform, they can receive timely feedback on the effectiveness of instructional tactics, which fosters a more collaborative and iterative planning process.
6. ***Intelligent Tutoring Systems:*** The conversational capabilities shown by the model render it well-suited for the creation of intelligent teaching systems. Educators can create interactive learning modules that effectively include students in dynamic discussions, offering prompt feedback, adaptable assistance, and customized coaching tailored to their achievements.
7. ***Natural Language Interface for Resource Discovery:*** ChatGPT can function as a natural language interface for resource finding. Educators can use the model to efficiently locate pertinent educational resources, scholarly publications, or other materials, simplifying collecting resources for lesson preparation and instructional development.
8. ***Facilitation of Collaborative Learning Environments:*** Educators can use ChatGPT to replicate collaborative learning settings inside virtual classrooms. The model's capacity to actively participate in interactive conversations promotes the development of critical thinking skills, stimulates participant engagement, and offers extra facilitation during group activities, thus cultivating a dynamic and captivating learning environment.
9. ***Enhanced Communication with Students:*** The model serves as a supplementary means of communication between instructors and students. Educators can optimise communication efficiency by integrating ChatGPT into messaging platforms, enabling prompt replies to student inquiries and fostering a dynamic and adaptable virtual classroom setting.
10. ***AI-Powered Assessment Tools:*** ChatGPT can be used to create evaluation tools driven by artificial intelligence. Educators can develop automated assessment systems that use the model's awareness of context and nuanced answers to test

students' comprehension, critical thinking, and problem-solving abilities in a more advanced way.

When examining the potential advantages of integrating ChatGPT into educational environments, it is crucial to consider ethical issues, privacy problems, and the appropriate use of artificial intelligence in educational contexts. Providing teachers with the authority to use these technologies correctly is essential, fostering a collaborative and seamless incorporation of technology within the educational domain.

8. ETHICAL ISSUES IN CHATGPT

The emergence of ChatGPT, a highly advanced language model created by OpenAI, has sparked a range of ethical concerns that reflect the far-reaching implications of artificial intelligence on society (Guleria et al., 2024). This discussion delves into the core ethical considerations linked to ChatGPT, examining its impact on various aspects of our lives and the potential consequences of its use.

8.1 Disparity of Opinion

Like several other artificial intelligence systems, ChatGPT has the potential to exhibit bias in the responses it generates. These biases have emerged due to many factors, including using machine learning algorithms for modelling purposes and incorporating specific data for training and fine-tuning. The presence of biases in data must be acknowledged since using human data labellers who adhere to ChatGPT's guidelines for training and fine-tuning datasets does not ensure the representation of varied ideas and perspectives. Moreover, the training data mainly originates from extensive online sources, which exhibit low variety and may include inherent biases. For instance, a study (Bender et al., 2021) revealed that these extensive datasets exhibit a notable bias towards younger users, particularly those from developed nations and those who speak English. The output of the model will reflect any biases that may be present in the data. Overcoming prejudice may be a challenging endeavour (Chan, 2023). OpenAI acknowledges this concern in its official blog article, highlighting that ChatGPT often exhibits verbosity and excessive reliance on specific terms, particularly in reiterating its identity as a language model developed by OpenAI. These flaws stem from inherent biases in the training data, as trainers tend to favour lengthier responses that appear more thorough. Additionally, there are well-documented concerns around over-optimization. The presence of biases among ChatGPT responses might lead to inequitable outcomes, especially for marginalized populations, which may be challenging to mitigate.

8.2 Confidentiality and Safety

The input that ChatGPT gets is used to inform the generation of its responses. These input-output pairs may also be used in the process of fine-tuning ChatGPT. These might divulge private information about users without their knowledge or consent. Tracking and profiling people may also be accomplished using ChatGPT interaction records of the persons in question. In addition, most of the databases that ChatGPT can use originate from the internet, including social media platforms like Twitter. It means that ChatGPT has the potential to learn content that does not undergo fact-checking and may leak the personal information of individuals. It may not only result in inaccurate or misleading information, but it may also lead to privacy concerns.

8.3 Openness and Honesty

Nevertheless, OpenAI needs to provide more details about the availability of ChatGPT. For instance, it is vital to inquire about the composition of the training data, the specifications of the training and testing datasets, their respective sizes, the model used in the study, the guidelines provided for the reviews, and the lack of transparency about the reviewers' identities. However, OpenAI placed significant emphasis on its proficiency in question answering. The internal mechanisms of ChatGPT remain obscure to users, posing challenges in comprehending the rationale behind its generated replies. The absence of openness in ChatGPT harms user confidence and their capacity to make well-informed choices about using the model. Moreover, the response generated by ChatGPT is determined by statistical models, which may be "random." The same query may provide varied responses in multiple inquiries using Chat GPT. The diminished level of understanding of the concept of randomization undermines the credibility of ChatGPT (Krügel et al., 2023).

8.4 Misuse and Abuse

The main objective of ChatGPT is to produce coherent and plausible text answers that resemble human-like communication via natural language. However, when subjected to reinforcement learning, the existing model lacks a reliable reference and fails to include a measure of its precision. The capacity to produce text that resembles human language has the potential to be exploited and misused, leading to many negative consequences, such as the dissemination of false information or the act of assuming someone else's identity. For instance, programmers are the first adopters and are more inclined to use generative AI techniques. According to a statement provided by Stack Overflow, a widely utilized platform for programmers seeking answers to their queries, the decision to prohibit the submission of answers generated by ChatGPT was made due to the site's observation that the average accuracy rate of such responses is insufficient.

Consequently, posting answers created by ChatGPT is deemed significantly detrimental to both the platform and the users who rely on it for accurate and reliable information (Andrews, 2023). Several prominent financial institutions have implemented a prohibition on the use of ChatGPT inside their operations, primarily due to concerns about its lack of accountability. In addition, using human-like text generated by ChatGPT may provide advantages to many forms of cyber deception, including phishing email scams, online job-seeking and dating scams, and the dissemination of political propaganda. In previous times, instances of cross-border fraudulent emails were often brought to light due to inadequate language translation. However, the advent of AI-driven translation and text production technologies raises concerns about the potential exacerbation of detection challenges. *Abuse* is a significant issue that warrants careful examination and analysis.

8.5 The Right to Authorship

Given the ability to create text closely resembling human language, ChatGPT can be used in various contexts requiring written content. For instance, students may use ChatGPT to complete their assignments or compose their academic papers. An increasing number of scholars affiliated with higher education institutions have raised awareness about the use of text reports produced by students via the employment of ChatGPT. Teachers need help distinguishing between student and AI authorship, raising worries about plagiarism. This problem hinders their ability to assess students' performance fairly. Moreover, ChatGPT has been used in artistic endeavours like creative writing and musical composition, raising issues about authorship and human creativity.

9. LIMITATIONS OF CHATGPT IN EDUCATION

The noticeable aspect of ChatGPT, namely its capacity to answer inquiries, gives rise to some apprehensions over the authenticity of educational sessions and homework tasks. A prevalent concern within the realm of education pertains to the potential utilisation of ChatGPT by students to complete their homework assignments, thereby enabling them to indiscriminately copy and paste the answers without the oversight or intervention of the instructor. Numerous educational institutions and universities have implemented regulations prohibiting the use of this technology for academic writing assignments since students have increasingly relied on it to complete their homework, essays, and theses. Determining the similarity or dissimilarity between AI-generated content and plagiarised text poses more challenges and entails less reliability. ChatGPT exhibits exceptional potency as a tool that caters to various tasks and academic domains. The use of AI-produced writing raises ethical concerns, particularly concerning the accuracy and reliability of ChatGPT's generated answers. The use of ChatGPT and other language models gives rise to significant ethical concerns about its societal impact.

ChatGPT has the potential to generate erroneous information and provide instructions that promote biased content on its website. When generating text, an automated conversational agent incorporates words with a high probability of following the preceding words. However, it does not undertake fact-checking or verifying the accuracy of the information provided. The ethical dilemma of using ChatGPT lies in the significant concern about potential biases inherent in the training data. The presence of biases within the training dataset of the chatbot is manifested in the output generated by the model, potentially resulting in the dissemination of inaccurate or hazardous information. Despite the presence of several precautions included in the opener to mitigate user abuse, the persistent problem of unfair, sexist, racist, and other unpleasant remarks remains prevalent. Considering the limited practical uses and incomplete understanding of the technology, institutions need assistance formulating regulations and protocols about ChatGPT.

The continuous flow of data from the cloud prevents its ability to choose to acquire information from a specific source. The strength of this system lies in its ability to integrate information from several sources and provide mostly novel responses to a given inquiry. ChatGPT may be compared to expert systems or mobile applications with restricted information access when it comes to relying on it for retrieving data from a specific source. Although this technology may address the issue of insufficient personnel by providing users with comprehensive information, it is incumbent upon each user to maximise the possibilities of this technology and exhibit caution in using the knowledge it offers. The system operates well when applied to a dataset of restricted scope, which must precisely represent the current situation. This phenomenon increases the probability of generating erroneous or imprecise data. Artificial intelligence in education raises a range of ethical and legal concerns, including but not limited to privacy, data security, and intellectual property. To ensure compliance with legal and ethical standards, a thorough assessment of these considerations will be required before implementing ChatGPT in the context of higher education.

ChatGPT users need to be aware of potential bias in their responses and take steps to mitigate it. Some worry that ChatGPT may negatively impact job opportunities for professions that rely heavily on written communication, such as writers and marketers. This concern arises from the possibility that machines and computers could replace human labour and lead to job displacement. ChatGPT developers use the tool to address development challenges effectively. The ChatGPT model is an artificial intelligence learning model that requires access to relevant data to function correctly. It is also essential to recognize that ChatGPT may provide inaccurate responses if its training data is incomplete or of poor quality. If the data used to train the model is not adequately anonymized or protected, the privacy of students and staff members may be compromised. Due to its network-based communication, malicious entities could intercept, access, or modify user data. ChatGPT could be used in social engineering attacks to gain unauthorized access to sensitive information, and attackers could use the platform to establish trust with users and collect data for malicious purposes, such as creating fraudulent credentials.

ChatGPT is a very efficient technology capable of generating textual responses to various inquiries, including even those with delicate or confidential content. The use of ChatGPT entails certain risks, such as the potential generation of inappropriate or offensive content, the possibility of privacy breaches, and susceptibility to manipulation by malicious actors. The answer generated by ChatGPT may vary based on the phrasing of the input or the frequency with which the same prompt is presented. The model's ability to generate accurate responses may vary, resulting in occasional incorrect or incomplete answers. The ChatGPT model may exhibit verbosity and abuse of specific terms due to biases in the training data and over-optimization.

10. AI IN OTHER AREAS

Just as ChatGPT has revolutionized communication, AI is changing the face of numerous sectors such as healthcare, tourism (Joshi, 2024), and smart cities (Joshi & Badola, 2024). The healthcare sector uses AI in diagnostics, personalized treatment/patient care and predictive analytics to enhance patient experience and improve operational efficiency. AI-powered platforms that can sift through enormous datasets of medical records to help get better at catching disease early, finding new drugs or optimizing treatment (Joshi et al., 2024). In tourism, AI revolutionizes the personalization of travel experiences, such as using chatbots for trip organization and recommendation engines/dynamic pricing, etc. This happens through smart buildings and public infrastructure; AI-enabled systems control energy usage based on real-time data to optimize the management of shared resources and services, which offers cities an efficient way of online urban living while reducing their ecological footprint (Joshi et al., 2024). Cities worldwide are safer, more efficient, and more sustainable because these technologies, using AI-driven sensors and camera algorithms, have optimized everything from street lighting on demand to prompt emergency response. Apart from these, AI is disruptive in education, finance, and agriculture. AI learning tools are changing how we learn: by creating a personalized approach, AI in education meets individual student's needs. AI detects fraud, automates trading, and improves risk management in finance AI. Enabled predictive analytics optimizes the crop yield and use of resources in agriculture. A conversational AI like ChatGPT fits into this more significant wave of technology, showing how we can use AI to accelerate communication, learning and decision-making in many other domains.

11. RECOMMENDATIONS FOR FUTURE RESEARCH

Future scholars can further explore the discoveries presented in this chapter via several avenues. To enhance the search approach, including non-Western sources would contribute to a more thorough global examination of the positioning of ChatGPT in various regions worldwide. Given the absence of student perspectives in most of the papers we analysed, it would be beneficial to use surveys and focus groups as additional methods to get deeper insights into how students use ChatGPT and other comparable generative AI technologies. Future researchers may also want to investigate the perspectives of academic personnel on ChatGPT, the degree to which it is used as an instructional aide, and the adaptations made to assessment assignments to minimize the potential for improper student utilization.

12. CONCLUSION

ChatGPT uses deep learning techniques and natural language processing algorithms to generate responses to textual inputs that exhibit human-like characteristics. ChatGPT has shown to be advantageous in education due to its varied applications, including language translation, interactive discourse, concise summarization, and proficient text generation. The technology is gaining popularity across several fields, such as research and education, because of its ability to analyze large datasets and provide accurate outcomes. The AI chatbot will significantly impact tutoring and customized learning in two crucial domains. Given its use of natural language processing for real-time interpretation and response to queries, ChatGPT can be employed for on-demand, live tutoring services catering to the needs of students. ChatGPT has the potential to function as a virtual teaching assistant by providing students with prompt feedback. Students may use ChatGPT to seek clarification on specific issues within their courses or to request repetitive explanations for better comprehension. ChatGPT can potentially support educators in developing educational resources, such as many iterations of assessments, evaluations of student learning, syllabi, rubrics, and other related materials. The use of this technology has the potential to promptly provide a satisfying solution to a given difficulty or assignment. However, revisions must be made to facilitate the application of students' knowledge and talents, enabling them to achieve the work properly. GPT models can generate natural language content that is both grammatically and structurally coherent because of their extensive training in large quantities of textual data. ChatGPT has undergone training to enhance its ability to generate text more suitable for conversational purposes in chatbot applications. In a chatbot context, the chatbot can initiate conversations, engage with user input, and provide users with information and assistance. As artificial intelligence progresses, ChatGPT exhibits enhanced cognitive abilities and an increased capacity to handle complex tasks.

REFERENCES

Abukmeil, M., Ferrari, S., Genovese, A., Piuri, V., & Scotti, F. (2021). A survey of unsupervised generative models for exploratory data analysis and representation learning. [csur]. *ACM Computing Surveys*, 54(5), 1–40. DOI: 10.1145/3450963

. Aljanabi, M. (2023). ChatGPT: Future directions and open possibilities. *Mesopotamian journal of Cybersecurity, 2023*, 16-17.

. Andrews, D. (2023). Human Evaluation of GPT-4's Answers to Stack Overflow Questions.

Aydın, Ö., & Karaarslan, E. (2023). Is ChatGPT leading generative AI? What is beyond expectations? *Academic Platform Journal of Engineering and Smart Systems*, 11(3), 118–134. DOI: 10.21541/apjess.1293702

Baidoo-Anu, D., & Ansah, L. O. (2023). Education in the era of generative artificial intelligence (AI): Understanding the potential benefits of ChatGPT in promoting teaching and learning. *Journal of AI*, 7(1), 52–62. DOI: 10.61969/jai.1337500

Barocas, S., & Selbst, A. D. (2016). Big data's disparate impact. *California Law Review*, 104, 671.

Bender, E. M., Gebru, T., McMillan-Major, A., & Shmitchell, S. (2021, March). On the dangers of stochastic parrots: Can language models be too big? In *Proceedings of the 2021 ACM conference on fairness, accountability, and transparency* (pp. 610-623). DOI: 10.1145/3442188.3445922

Bogost, I. (2022). ChatGPT is dumber than you think. *Atlantic (Boston, Mass.)*, •••, 7.

Borji, A. (2023). A categorical archive of chatgpt failures. *arXiv preprint arXiv:2302.03494*. DOI: 10.21203/rs.3.rs-2895792/v1

Brazdil, P., & Jorge, A. (Eds.). (2001). *Progress in Artificial Intelligence. Knowledge Extraction, Multi-agent Systems, Logic Programming, and Constraint Solving: 10th Portuguese Conference on Artificial Intelligence, EPIA 2001, Porto, Portugal, December 17-20, 2001. Proceedings* (Vol. 2258). Springer Science & Business Media.

Cascella, M., Montomoli, J., Bellini, V., & Bignami, E. (2023). Evaluating the feasibility of ChatGPT in healthcare: An analysis of multiple clinical and research scenarios. *Journal of Medical Systems*, 47(1), 33. DOI: 10.1007/s10916-023-01925-4 PMID: 36869927

Chan, A. (2023). GPT-3 and InstructGPT: Technological dystopianism, utopianism, and "Contextual" perspectives in AI ethics and industry. *AI and Ethics*, 3(1), 53–64. DOI: 10.1007/s43681-022-00148-6

Cooper, G. (2023). Examining science education in ChatGPT: An exploratory study of generative artificial intelligence. *Journal of Science Education and Technology*, 32(3), 444–452. DOI: 10.1007/s10956-023-10039-y

Cotton, D. R., Cotton, P. A., & Shipway, J. R. (2024). Chatting and cheating: Ensuring academic integrity in the era of ChatGPT. *Innovations in Education and Teaching International*, 61(2), 228–239. DOI: 10.1080/14703297.2023.2190148

Farrokhnia, M., Baggen, Y., Biemans, H., & Noroozi, O. (2022). Bridging the fields of entrepreneurship and education: The role of philosophical perspectives in fostering opportunity identification. *International Journal of Management Education*, 20(2), 100632. DOI: 10.1016/j.ijme.2022.100632

Fijačko, N., Gosak, L., Štiglic, G., Picard, C. T., & Douma, M. J. (2023). Can ChatGPT pass the life support exams without entering the American heart association course? *Resuscitation*, 185, 185. DOI: 10.1016/j.resuscitation.2023.109732 PMID: 36775020

Firat, M. (2023). What ChatGPT means for universities: Perceptions of scholars and students. *Journal of Applied Learning and Teaching*, 6(1), 57–63.

Gao, J., Zhao, H., Yu, C., & Xu, R. (2023). Exploring the feasibility of chatgpt for event extraction. *arXiv preprint arXiv:2303.03836.*

Garg, M., & Goel, A. (2022). A systematic literature review on online assessment security: Current challenges and integrity strategies. *Computers & Security*, 113, 102544. DOI: 10.1016/j.cose.2021.102544

Gui, J., Sun, Z., Wen, Y., Tao, D., & Ye, J. (2021). A review on generative adversarial networks: Algorithms, theory, and applications. *IEEE Transactions on Knowledge and Data Engineering*, 35(4), 3313–3332. DOI: 10.1109/TKDE.2021.3130191

Guleria, A., Krishan, K., Sharma, V., & Kanchan, T. (2024). ChatGPT: Forensic, legal, and ethical issues. *Medicine, Science, and the Law*, 64(2), 150–156. DOI: 10.1177/00258024231191829 PMID: 37528607

Gupta, P., Raturi, S., & Venkateswarlu, P. (2023). Chatgpt for designing course outlines: A boon or bane to modern technology. *Available atSSRN* 4386113. DOI: 10.2139/ssrn.4386113

. Halaweh, M. (2023). ChatGPT in education: Strategies for responsible implementation.

Joshi, R. (2024). Impact of AI in the Tourism Industry: A Double-Edged Sword. In *Technology and Luxury Hospitality* (pp. 135–150). Routledge. DOI: 10.4324/9781003488248-10

Joshi, R., & Badola, R. (2024). Digital Twin: A Transformative Tool for Smart Cities. In *Smart Cities: Innovations, Challenges and Future Perspectives* (pp. 177–198). Springer Nature Switzerland. DOI: 10.1007/978-3-031-59846-3_8

Joshi, R., Pandey, K., & Kumari, S. (2024). Artificial Intelligence for Advanced Sustainable Development Goals: A 360-Degree Approach. In *Preserving Health, Preserving Earth: The Path to Sustainable Healthcare* (pp. 281–303). Springer Nature Switzerland. DOI: 10.1007/978-3-031-60545-1_16

Joshi, R., Pandey, K., & Kumari, S. (2024, February). A Metaverse: Significance and Dimensions in Various Sectors. In *2024 11th International Conference on Computing for Sustainable Global Development (INDIACom)* (pp. 1781-1786). IEEE.

Joshi, R., Pandey, K., & Kumari, S. (2024, May). Exploring Various Government Health Schemes and the Use of AI in the Changing Dynamics of India. In *2024 5th International Conference for Emerging Technology (INCET)* (pp. 1-6). IEEE. DOI: 10.1109/INCET61516.2024.10643415

Jovanovic, M., & Campbell, M. (2022). Generative artificial intelligence: Trends and prospects. *Computer*, 55(10), 107–112. DOI: 10.1109/MC.2022.3192720

Kasneci, E., Seßler, K., Küchemann, S., Bannert, M., Dementieva, D., Fischer, F., Gasser, U., Groh, G., Günnemann, S., Hüllermeier, E., Krusche, S., Kutyniok, G., Michaeli, T., Nerdel, C., Pfeffer, J., Poquet, O., Sailer, M., Schmidt, A., Seidel, T., & Kasneci, G. (2023). ChatGPT for good? On opportunities and challenges of large language models for education. *Learning and Individual Differences*, 103, 102274. DOI: 10.1016/j.lindif.2023.102274

Kitamura, F. C. (2023). ChatGPT is shaping the future of medical writing but still requires human judgment. *Radiology*, 307(2), e230171. DOI: 10.1148/radiol.230171 PMID: 36728749

Krügel, S., Ostermaier, A., & Uhl, M. (2023). The moral authority of ChatGPT. *arXiv preprint arXiv:2301.07098.*

Kshetri, N., & Hughes, L. (2023). "So what if ChatGPT wrote it?" Multidisciplinary perspectives on opportunities, challenges and implications of generative conversational AI for research, practice and policy. *International Journal of Information Management*, 71, 102642. DOI: 10.1016/j.ijinfomgt.2023.102642

Lecler, A., Duron, L., & Soyer, P. (2023). Revolutionizing radiology with GPT-based models: Current applications, future possibilities and limitations of ChatGPT. *Diagnostic and Interventional Imaging*, 104(6), 269–274. DOI: 10.1016/j.diii.2023.02.003 PMID: 36858933

Lecler, A., Duron, L., & Soyer, P. (2023). Revolutionizing radiology with GPT-based models: Current applications, future possibilities and limitations of ChatGPT. *Diagnostic and Interventional Imaging*, 104(6), 269–274. DOI: 10.1016/j.diii.2023.02.003 PMID: 36858933

Li, Z., Niu, C., Meng, F., Feng, Y., Li, Q., & Zhou, J. (2019). Incremental transformer with deliberation decoder for document grounded conversations. *arXiv preprint arXiv:1907.08854*. DOI: 10.18653/v1/P19-1002

Lund, B. D., & Wang, T. (2023). Chatting about ChatGPT: How may AI and GPT impact academia and libraries? *Library Hi Tech News*, 40(3), 26–29. DOI: 10.1108/LHTN-01-2023-0009

Mann, B., Ryder, N., Subbiah, M., Kaplan, J., Dhariwal, P., Neelakantan, A., . . . Amodei, D. (2020). Language models are few-shot learners. *arXiv preprint arXiv:2005.14165, 1*.

Mann, D. L. (2023). Artificial intelligence discusses the role of artificial intelligence in translational medicine: a JACC: basic to translational science interview with ChatGPT. *JACC. Basic to Translational Science*, 8(2), 221–223. DOI: 10.1016/j.jacbts.2023.01.001 PMID: 36908674

Mitrović, S., Andreoletti, D., & Ayoub, O. (2023). Chatgpt or human? detect and explain. explaining decisions of machine learning model for detecting short chatgpt-generated text. *arXiv preprint arXiv:2301.13852*.

Mizumoto, A., & Eguchi, M. (2023). Exploring the potential of using an AI language model for automated essay scoring. *Research Methods in Applied Linguistics*, 2(2), 100050. DOI: 10.1016/j.rmal.2023.100050

Montenegro-Rueda, M., Fernández-Cerero, J., Fernández-Batanero, J. M., & López-Meneses, E. (2023). Impact of the implementation of ChatGPT in education: A systematic review. *Computers*, 12(8), 153. DOI: 10.3390/computers12080153

Neumann, M., Rauschenberger, M., & Schön, E. M. (2023, May). "We need to talk about ChatGPT": The future of AI and higher education. In *2023 IEEE/ACM 5th International Workshop on Software Engineering Education for the Next Generation (SEENG)* (pp. 29-32). IEEE.

O'neil, C. (2017). *Weapons of math destruction: How big data increases inequality and threatens democracy*. Crown.

. Pavlik, J. V. (2023). Collaborating with ChatGPT: Considering the implications of generative artificial intelligence for journalism and media education. *Journalism & mass communication educator, 78*(1), 84-93.

. Pedro, F., Subosa, M., Rivas, A., & Valverde, P. (2019). Artificial intelligence in education: Challenges and opportunities for sustainable development.

Perkins, M. (2023). Academic Integrity considerations of AI Large Language Models in the post-pandemic era: ChatGPT and beyond. *Journal of University Teaching & Learning Practice*, 20(2). Advance online publication. DOI: 10.53761/1.20.02.07

Qadir, J. (2023, May). Engineering education in the era of ChatGPT: Promise and pitfalls of generative AI for education. In *2023 IEEE Global Engineering Education Conference (EDUCON)* (pp. 1-9). IEEE. DOI: 10.1109/EDUCON54358.2023.10125121

Rajain, M., Bhatia, R., Tripathi, M., Kumar, N., & Mathur, R. (2023). Low-Frequency Repetitive Transcranial Magnetic Stimulation for Chronic Tension-Type Headache: A Randomized Controlled Study. *Cureus*, 15(2). Advance online publication. DOI: 10.7759/cureus.34922 PMID: 36938162

Schiff, D. (2022). Education for AI, not AI for education: The role of education and ethics in national AI policy strategies. *International Journal of Artificial Intelligence in Education*, 32(3), 527–563. DOI: 10.1007/s40593-021-00270-2

Shahriar, S., & Hayawi, K. (2023). Let's have a chat! A Conversation with ChatGPT: Technology, Applications, and Limitations. *arXiv preprint arXiv:2302.13817*.

Shen, Y., Heacock, L., Elias, J., Hentel, K. D., Reig, B., Shih, G., & Moy, L. (2023). ChatGPT and other large language models are double-edged swords. *Radiology*, 307(2), e230163. DOI: 10.1148/radiol.230163 PMID: 36700838

Sobania, D., Briesch, M., Hanna, C., & Petke, J. (2023, May). An analysis of the automatic bug fixing performance of chatgpt. In *2023 IEEE/ACM International Workshop on Automated Program Repair (APR)* (pp. 23-30). IEEE. DOI: 10.1109/APR59189.2023.00012

Stokel-Walker, C., & Van Noorden, R. (2023). What ChatGPT and generative AI mean for science. *Nature*, 614(7947), 214–216. DOI: 10.1038/d41586-023-00340-6 PMID: 36747115

Susnjak, T., & McIntosh, T. R. (2024). ChatGPT: The end of online exam integrity? *Education Sciences*, 14(6), 656. DOI: 10.3390/educsci14060656

Tate, T., Doroudi, S., Ritchie, D., Xu, Y., & Warschauer, M. (2023). Educational research and AI-generated writing: Confronting the coming tsunami. *EdArXiv. January, 10.* DOI: 10.35542/osf.io/4mec3

Terwiesch, C. (2023). Would Chat GPT3 get a Wharton MBA? A prediction based on its performance in the operations management course. *Mack Institute for Innovation Management at the Wharton School, University of Pennsylvania, 45.*

Whalen, J., & Mouza, C. (2023). ChatGPT: Challenges, opportunities, and implications for teacher education. *Contemporary Issues in Technology & Teacher Education*, 23(1), 1–23.

Xue, V. W., Lei, P., & Cho, W. C. (2023). The potential impact of ChatGPT in clinical and translational medicine. *Clinical and Translational Medicine*, 13(3), e1216. DOI: 10.1002/ctm2.1216 PMID: 36856370

Zhai, X. (2022). ChatGPT user experience: Implications for education. *Available atSSRN* 4312418.

Zhai, X. (2023). ChatGPT for next generation science learning. *XRDS: Crossroads. The ACM Magazine for Students*, 29(3), 42–46.

Chapter 4
Animal Species Explorer Scanimals: An AI Integrated Mobile Application for Young Learners

Anusha Achuthan
https://orcid.org/0000-0002-2015-2269
Universiti Sains Malaysia, Malaysia

Catherine Lam Pui Kiew
https://orcid.org/0009-0001-9770-5095
Universiti Sains Malaysia, Malaysia

Galib Muhammad Shahriar Himel
https://orcid.org/0000-0002-2257-6751
Universiti Sains Malaysia, Malaysia

ABSTRACT

Teaching children about animals is a fundamental aspect of early childhood education and is universally valued across different cultures and societies. Understanding animals fosters a sense of curiosity and wonder in young minds, encouraging them to explore the natural world and enhance their comprehension of the interdependence of life on our planet. Picture books, flashcards, stories, songs, and animal toys are some of the traditional ways that kids are introduced to animals. These resources make learning interesting and enjoyable for kids. However, if AI can be deployed for the enhancement of the current method, it can benefit the overall educational system. To make that a reality, "Scanimals" aims to introduce a novel and engaging educational experience for children through a mobile application focused on terrestrial animals. The main objective of this project is to establish an interactive learning

DOI: 10.4018/979-8-3693-9811-1.ch004

and image recognition-based digital platform to help young learners gain a greater grasp of the animal kingdom (terrestrial animal) in this digital age of technology.

1. INTRODUCTION

The human mind is an unimaginable system that can learn and understand the complex structures of the surroundings. It can question every single aspect of the w/h questions to learn more about the human anatomy, the environment, other animals, the galaxy, the ocean, and the skies which no other animal can. This intuition has been present since day one in every human child. Through the advancement of civilization, Human beings have developed many methods to teach children about their surroundings systematically. The first thing that the human child learns after being born is language which involves learning the number systems, names of the animals, names of plants, names of the objects, and also names of other people. However, not all of them, which they learn of are sentient beings. Learning about sentient beings is not only just a part of learning but also helps a child to grow empathy for living beings and makes them able to adjust to other animals that are part of the ecosystem.

In today's educational landscape, various methods cater to the diverse learning needs of children. Traditional classroom learning includes both teacher-centered approaches, where the teacher is the primary source of knowledge through lectures and demonstrations, and student-centered approaches, which focus on the student's needs, abilities, interests, and learning styles, encouraging active participation and collaboration (Team, 2021). The Montessori Method, developed by Dr. Maria Montessori, emphasizes self-directed activity, hands-on learning, and collaborative play, allowing children to make creative choices in their learning (Catherine et al., 2020). Waldorf Education, founded by Rudolf Steiner, integrates academics, arts, and practical skills for holistic development, emphasizing imagination and providing a nurturing environment (Randoll & Peters, 2015). The Reggio Emilia Approach, originating in Italy, is child-centered and encourages exploration and discovery, valuing the child's voice in the learning process and considering the environment as the third teacher (Hewett, 2001). Project-Based Learning (PBL) (Kokotsaki et al., 2016) engages students in real-world and personally meaningful projects, promoting critical thinking, problem-solving, and collaboration. The Flipped Classroom model (Akçayır & Akçayır, 2018) reverses traditional learning environments by having students watch lectures at home and engage in interactive activities in the classroom, allowing for more hands-on learning and personalized instruction. Blended Learning (Hockly, 2018) combines online digital media with traditional classroom methods, providing flexibility and personalized learning experiences. Game-Based Learning

(Tobias et al., 2013) uses games to enhance learning experiences, making learning fun and engaging, and helping students develop problem-solving skills and retain information better. Social Emotional Learning (SEL) (Hoffman, 2009) focuses on developing students' emotional intelligence, including self-awareness, self-regulation, social skills, empathy, and responsible decision-making. Inquiry-Based Learning (Panasan & Nuangchalerm, 2010)encourages students to ask questions and explore topics deeply, promoting curiosity, critical thinking, and independent learning. Experiential Learning involves learning through experience and reflection, including activities like field trips, experiments, and hands-on projects that help students connect theory with practice (Lewis & Williams, 1994). Personalized Learning (Shemshack & Spector, 2020) tailor's education to meet the individual needs, skills, and interests of each student, often involving adaptive learning technologies and customized learning plans. Collaborative Learning (Bruffee, 1987) involves students working together in groups to solve problems, complete tasks, or create projects, fostering teamwork, communication, and social skills. Differentiated Instruction (H.L., 2015) involves teachers modifying their instruction to meet the diverse needs of students, varying teaching methods, materials, and assessments to cater to different learning styles and abilities. Nature-Based Learning incorporates outdoor activities and nature exploration into the curriculum, promoting environmental awareness and physical activity. These methods reflect the dynamic nature of modern education, aiming to meet the varied needs and learning styles of children, ensuring a comprehensive and inclusive approach to learning (Miller et al., 2021).

Artificial Intelligence (AI) has the potential to revolutionize education by integrating various educational methods to enhance child psychology development and learning outcomes. Artificial Intelligence can create adaptive learning environments that mimic the Montessori method's (Catherine et al., 2020) emphasis on self-directed activity and hands-on learning. Intelligent tutoring systems can provide personalized activities and materials that align with each child's developmental stage and interests. It can support the Reggio Emilia approach (Hewett, 2001) by facilitating documentation and reflection. AI-powered tools can help teachers document children's learning processes through photos, videos, and notes, and analyze this data to tailor future activities that encourage exploration and discovery. AI can seamlessly integrate online and offline learning experiences. Adaptive learning platforms can provide personalized content and assessments, while AI-driven analytics can help teachers identify areas where students need additional support, ensuring a balanced blend of digital and traditional instruction. AI can enhance game-based learning (Tobias et al., 2013) by creating adaptive educational games that adjust difficulty levels based on the child's progress. AI algorithms can analyse gameplay data to provide real-time feedback and personalized learning experiences, making learning fun and engaging. Artificial Intelligence can facilitate inquiry-based learning

(Panasan & Nuangchalerm, 2010) by providing intelligent search tools and virtual assistants that help students explore topics deeply. AI can support experiential learning (Lewis & Williams, 1994) by simulating real-world scenarios through virtual and augmented reality. These immersive experiences allow students to apply theoretical knowledge in practical contexts, enhancing their understanding and retention of concepts. Artificial Intelligence excels in personalized learning (Shemshack & Spector, 2020) by analyzing individual learning patterns and preferences. AI-driven platforms can create customized learning paths, recommend resources, and provide targeted interventions, ensuring that each child receives a tailored educational experience. By integrating these methods, Artificial Intelligence can create a holistic and adaptive learning environment that caters to the unique needs of each child. This approach not only enhances academic performance but also supports the overall psychological development of children by fostering creativity, critical thinking, and emotional intelligence.

Since the kingdom Animalia contains a huge ocean of information, a systematic smart intelligent device can organize the data of animals in such a way that can help a child learn about animals in details in an effective way. Also, at the same time, they can easily go through the previous information very easily using the keyword search system which is not possible in the case of learning from real books. This will help the child save time and learn effectively about the terrestrial animals and foster empathy towards them and also develop a better understanding of the echo system to be able to contribute to the surrounding environment in a positive manner.

2. THE KINGDOM ANIMALIA

The Animal Kingdom, encompassing a vast array of species, is a testament to the incredible diversity of life on Earth. From the dense jungles of the Amazon to the icy expanses of Antarctica, animals have adapted to thrive in every conceivable environment. This essay explores the fascinating variety of animals across different regions of the world, highlighting their unique adaptations and ecological roles.

Africa is renowned for its rich biodiversity, hosting some of the most iconic animals on the planet. The African Elephant, the largest land animal, roams the savannas and forests, known for its intelligence and complex social structures. Lions, often called the "king of the jungle," live in prides and are apex predators in their habitats. Giraffes, with their long necks, are uniquely adapted to browse on tall trees, while cheetahs, the fastest land animals, rely on their speed to hunt. Gorillas, found in central African forests, exhibit remarkable strength and social behavior, making them one of our closest relatives in the animal kingdom (Bragg, 2024).

Asia's vast landscapes, from tropical rainforests to high mountain ranges, support a diverse array of wildlife. The Bengal Tiger, the largest tiger subspecies, prowls the forests of India and Bangladesh. Pandas, native to China, are known for their distinctive black and white fur and bamboo diet. The Asian Elephant, smaller than its African counterpart, inhabits forests and grasslands across the continent. The Komodo Dragon, the largest living lizard, is native to Indonesia, while the elusive Snow Leopard roams the mountains of Central and South Asia, perfectly adapted to cold, rugged terrains (WWF, 2019).

North America's varied climates and landscapes are home to a wide range of animals (BBC, 2024). The Bald Eagle, the national bird of the United States, is known for its powerful build and keen eyesight. Grizzly Bears, found in Alaska and western Canada, are formidable predators with a diverse diet. The American Bison, once roaming the Great Plains in vast herds, is the largest land mammal in North America. Alligators, predominantly found in the southeastern United States, thrive in freshwater environments, while Moose, the largest species in the deer family, inhabit northern forests.

South America's biodiversity is particularly rich in the Amazon rainforest, the world's largest tropical rainforest. Jaguars, the largest cats in the Americas, are powerful swimmers and apex predators. Llamas, domesticated in the Andes, are used as pack animals and are integral to Andean culture. Capybaras, the largest rodents in the world, live near water bodies and are highly social. Anacondas, among the largest snakes, inhabit the Amazon basin, while Sloths, known for their slow movement, are adapted to life in the treetops (Fracolli, 2020).

Europe's wildlife includes a mix of forest, mountain, and marine animals. Red Deer, one of the largest deer species, are found in forests and grasslands. The European Bison, or wisent, has been reintroduced to parts of Eastern Europe after near extinction. Brown Bears inhabit the forests and mountains, while the Eurasian Lynx, a medium-sized wildcat, roams European forests (National Geographic, 2015). The Atlantic Puffin, a seabird with a colorful beak, nests along the coasts and is known for its distinctive appearance.

Australia is famous for its unique and diverse wildlife, much of which is found nowhere else. Kangaroos, iconic marsupials, are known for their powerful hind legs and hopping movement. Koalas, tree-dwelling marsupials, feed primarily on eucalyptus leaves. The Platypus, an egg-laying mammal with a duck-bill and webbed feet, is one of the most unusual animals. The Tasmanian Devil, a carnivorous marsupial, is found in Tasmania, while the Emu, the second-largest bird in the world, is native to Australia (Anna, 2023).

Antarctica's harsh climate supports a limited but unique range of wildlife. The Emperor Penguin, the largest penguin species, is known for its breeding behavior in the extreme cold. Weddell Seals are adapted to the icy waters and ice floes. Krill, small crustaceans, form the basis of the Antarctic food web, supporting a variety of marine life. The Snow Petrel, a bird that breeds exclusively in Antarctica, and the Antarctic Fur Seal, found on sub-Antarctic islands, are key species in this frozen landscape (Nicklen, 2023).

The Arctic region supports a variety of animals adapted to cold environments. Polar Bears, the largest land carnivores, are found throughout the Arctic region. The Arctic Fox, with its thick fur, survives in extreme cold. Walruses, recognizable by their long tusks, inhabit Arctic waters, while Narwhals, known as the "unicorns of the sea," have long, spiral tusks. Caribou, also known as reindeer, migrate across the Arctic and sub-Arctic regions, playing a crucial role in the ecosystem (nsidc, 2017).

Learning about the animal kingdom can be significantly enhanced using Artificial Intelligence (AI) through various innovative methods. AI can create personalized learning experiences by adapting content to individual learning styles and paces. Platforms like Coursera and Khan Academy use AI to recommend courses, quizzes, and resources tailored to the learner's progress and interest (Compliance, 2024). AI-driven VR and AR applications can provide immersive experiences, allowing learners to explore different habitats and observe animals in their natural environments. This hands-on approach can make learning more engaging and memorable. AI can analyze vast amounts of footage to identify and highlight key behaviors and interactions among animals, leading to more informative and visually captivating documentaries that help viewers understand complex ecological relationships. AI tutors can provide real-time feedback and answer questions about the animal kingdom, simulating conversations that help learners explore topics in depth and clarify doubts instantly. Platforms like iNaturalist use AI to help users identify species from photos they upload, aiding in learning and contributing to scientific research by gathering data on biodiversity (Di Cecco et al., 2021). AI can process and analyze large datasets from wildlife studies, revealing patterns and insights that might be missed by human researchers, leading to new discoveries and a deeper understanding of animal behavior and ecology. Simulations driven by AI can model ecosystems and animal interactions, allowing learners to experiment with different scenarios and observe the outcomes, enhancing understanding of ecological dynamics and conservation strategies. Projects like the Earth Species Project use AI to decode animal communication, providing insights into how animals interact and express themselves (Parshley, 2023), deepening our understanding of animal behavior and social structures. Educational games powered by AI can make learning about the animal kingdom fun and interactive, adapting to the learner's level and providing challenges and rewards that keep them engaged. AI can create customized learning paths based on the learner's interests and progress, ensuring that each learner gets the most relevant and engaging content, making the learning process more efficient and enjoyable. By leveraging these AI-driven methods, learners can gain a comprehensive and engaging understanding of the animal kingdom, fostering a deeper appreciation for wildlife and conservation efforts.

To implement the AI-based learning module, a smart mobile device is the best choice as most of the people even the ones who belong to the lower class have access to a smartphone. To achieve this goal, this study focuses on developing a mobile application that will be interactive enough and at the same time user-friendly; following the different teaching techniques to effectively teach the children about terrestrial animals and assess their learning.

3. LITERATURE REVIEW

There are several existing applications that will be discussed later, such as iNaturalist and Google Lens. "Scanimals" is an enhancement project aimed at developing an original and cutting-edge learning environment. This project provides a learning platform for terrestrial animals suitable for kids, featuring interactive quizzes that offer a fun and engaging learning experience to spark children's curiosity.

iNaturalist (Van Horn et al., 2018) is a community and platform for citizen science that enables people to log and exchange observations of fungi, animals, plants, and other species. Users can post images and details about the animals they encounter in the field using the website and mobile apps. The community contributes to the expanding library of biodiversity data by assisting in the identification and verification of these observations. iNaturalist is available for Android and iOS devices, as well as on the web. It uses computer vision algorithms and functions as a crowdsourced species identification system.

Google developed a technique called Google Lens (Шаповалов et al., 2019) for image recognition. It analyzes photos, identifies objects, and utilizes artificial intelligence and machine learning to provide information or take actions. Initially available as a feature of Google Photos, it has since been integrated into various other Google services and products. Google Lens can identify and deliver details about flora, fauna, and other subjects, in addition to items and landmarks. It is also useful for activities like translating text from one language to another and storing information from business cards, as it can extract text from photos. Google Lens is available for Android and iOS devices.

EarthSnap (EarthSnap, 2016) allows users to identify every form of plant and animal species on Earth using their smartphone cameras. This app encourages people to take pictures and document the natural world around them. EarthSnap embodies citizen science in action: when users contribute images of plants or animals, the app uses a specially developed artificial intelligence machine learning system, which is awaiting a patent, to identify the subject and provide information about its habitat, population distribution worldwide, and Earth's known history.

The algorithms used by these existing applications are based on image recognition and machine learning. To identify a species from a photo, iNaturalist employs an image recognition and pattern matching method driven by machine learning models and Amazon Web Services. This method uses computer vision to recognize species based on input from previously taken photos. Similarly, Google Lens uses AI techniques and computer vision for image recognition. Google Lens's machine learning capabilities allow it to continuously learn and improve its recognition skills. The algorithm becomes better at detecting objects, reading text, and understanding context as more users engage with Google Lens and provide feedback.

Other than the well-established application mentioned above there are some other researches also going on in academic settings. For example, Aslan et al. (Aslan et al., 2022) developed Kid Space multimodal learning technology to actively engage children in physical movement and social collaboration through play-based learning. One significant contribution is the work by Jones, which evaluates automated image recognition apps for plant identification, emphasizing the potential of such technologies in ecological surveys. Jones argues that while automated systems can assist in identifying flora, they must adopt high confidence thresholds to ensure reliability in biodiversity assessments (Jones, 2020). This highlights the importance of accuracy in educational tools that rely on image recognition, a principle that could be applied to the development of "Scanimals" to ensure that children receive correct information about the animals they learn about. Moreover, the research by Gweon discusses inferential social learning, which emphasizes the cognitive foundations of how children learn from social interactions. This study suggests that incorporating social learning principles into educational applications can enhance children's cognitive development and their ability to engage with peers (Gweon, 2021). This insight could be instrumental in the educational experience. Anderson et al. (Anderson et al., 2018) also shows the performance study of Kid Space which is an augmented reality-based learning tool. Rasidah et al. (Rasidah binti Sardi, 2022) represent the Virtual Reality Kids Pedagogy Learning System which provides modern ways of doing educational exercises in a virtual world with 3D elements. One notable study by Aristeidou et al. explores how young volunteers contribute to citizen science projects, specifically focusing on iNaturalist. The authors emphasize the importance of training and support for volunteers to improve data quality and enhance the educational outreach of biodiversity research (Aristeidou et al., 2021). Furthermore, the work of Jiang on nature-based learning underscores the benefits of outdoor educational experiences in promoting children's physical, cognitive, and emotional development (Jiang & Hussain, 2023). This perspective could inform the design of "Scanimals" by incorporating outdoor exploration features, where children can learn about animals in their natural habitats, thus bridging the gap between digital learning and real-world experiences. Sheard discusses the role of emerging technologies in citizen science, particularly for insect monitoring. The study highlights how technology can facilitate participation and enhance public engagement in scientific research (Sheard et al., 2024). Javed et al. (Sobia Tariq Javed et al., 2024) introduced a social cognitive learning-based metaheuristic called Kids Learning Optimizer (KLO), which leverages the concept of children's learning through social interactions within families and communities. Additionally, the findings of Salum indicate that creative learning approaches, such as engaging games and activities, significantly contribute to children's holistic development, including their cognitive and emotional skills (Salum et al., 2024). The work of Ping et al. presents a mobile learning model tailored for children with special learning needs, emphasizing the importance of personalized learning experiences (Ping et al., 2022). The research by Zedda highlights the significance of transformative biodiversity education, particularly in urban environments. The study advocates for the integration of digital media and participatory approaches in biodiversity education, which can enhance children's understanding and appreciation of local ecosystems (Zedda, 2023). Karimkhanlooei and Seifiniya (Karimkhanlooei & Seifiniya, 2015) found that the phonics method can be more effective in teaching English to children. Hsi (Hsi, 2007) demonstrates the increased interaction of youth with digital technologies and presents a framework for exploring digital fluency—the skills, innovative representational practices, design awareness, sense of ownership, and strategic expertise that learners acquire or display through the use of digital tools to collect, design, assess, critique, synthesize, and create digital media artifacts, communication messages, or other electronic expressions. Ayman et al. (Reham Ayman et al., 2018) developed "MiniColon" is a learning game designed for children aged 8-9, which teaches three fundamental programming concepts: sequential blocks, conditional blocks, and iterative blocks. Patiño-Escarcina et al. (Patiño-Escarcina et al., 2021) proposed a new collection of methods along with guidelines and strategies for implementing the educational robotics standard curriculum for children, called EDUROSC-Kids. Semeraro et al. (Semeraro et al., 2024) created a blended learning program for primary and secondary students (Kids Save Lives – KSL) that includes short lectures, hands-on training with mannequins, and virtual reality experiences. Lastly, the systematic review by Antezana et al. on young people's engagement with well-being apps reveals that user engagement is influenced by design elements and motivational factors (Antezana et al., 2022). This finding underscores the necessity of creating an engaging and user-friendly interface for "Scanimals," which can enhance children's interaction with the app and promote sustained interest in learning about terrestrial animals.

4. METHODOLOGY

Developing an advanced AI-based app to help children learn about various animals around the world needs to involve a comprehensive approach that integrates effective data collection, educational methodologies, and robust software development practices. Combining the Dynamic Systems Development Method (DSDM) and Scrum methodologies can create a robust framework for mobile application development that leverages the strengths of both approaches. This hybrid method can enhance flexibility, stakeholder engagement, and iterative development, which are crucial in the fast-paced mobile app environment.

DSDM

The Dynamic Systems Development Method (DSDM) is an Agile framework focused on delivering business value through iterative development across the entire project lifecycle. It emphasizes active user involvement, fostering collaboration between developers and users to enhance understanding of requirements, improve user satisfaction, and reduce project failure risks..DSDM also promotes frequent delivery of functional increments, allowing for regular feedback and adjustments, which is crucial in adapting to changing market conditions. Additionally, DSDM emphasizes the importance of quality throughout the development process, integrating testing and validation at every stage to ensure that the product meets the required standards. By prioritizing business needs, fostering collaboration, and embracing flexibility, DSDM enables teams to deliver high-quality software solutions efficiently and effectively, making it particularly suitable for projects with dynamic requirements and tight deadlines (Margini et al., 2017).

Scrum

Scrum is an Agile framework that supports incremental and iterative software development, enabling teams to deliver high-quality products while adapting to changing requirements. It includes defined roles: the Scrum Master facilitates the process, the Product Owner represents stakeholder interests, and the Development Team executes the work. Scrum operates in time-boxed iterations called sprints, lasting two to four weeks, during which a potentially shippable product increment is developed. The framework emphasizes collaboration, transparency, and continuous feedback, enhancing team productivity and engagement through accountability and self-organization. Regular ceremonies, such as Sprint Planning, Daily Stand-ups, Sprint Reviews, and Retrospectives, help keep the team aligned and focused on delivering value. Additionally, Scrum promotes early and frequent delivery of functional software, which not only increases customer satisfaction but also reduces the risks associated with late-stage project changes. The empirical process control inherent in Scrum allows teams to inspect and adapt their processes and products continuously, leading to improved quality and efficiency over time. Furthermore, Scrum's flexibility makes it suitable for a wide range of projects, from small start-ups to large enterprises, as it can be tailored to fit various organizational contexts and project complexities Schwaber & Sutherland (2012)Moe & Dingsøyr, 2008).

Hybrid Approach

Combining the Dynamic Systems Development Method (DSDM) and Scrum for mobile application development offers a synergistic approach that enhances flexibility, stakeholder engagement, and the ability to incorporate continuous new data and information. DSDM provides a structured framework that emphasizes the importance of business needs and active user involvement throughout the project lifecycle, ensuring that the final product aligns with user expectations and market demands. By integrating Scrum's iterative sprints, teams can deliver functional increments of the application regularly, allowing for rapid feedback and adjustments based on user input. This iterative process is particularly beneficial in mobile app development, where user preferences and technological advancements can change swiftly. The combination of DSDM's focus on quality and Scrum's emphasis on collaboration fosters a culture of continuous improvement, enabling teams to adapt to new data and insights effectively. Furthermore, this hybrid approach promotes a shared understanding among team members and stakeholders, enhancing communication and reducing the risks associated with misaligned objectives. As new information becomes available, teams can quickly pivot their development efforts, ensuring that the mobile application remains relevant and competitive in a dynamic market. Overall, the integration of DSDM and Scrum not only streamlines the development process but also enhances the quality and responsiveness of mobile applications, making them better suited to meet the evolving needs of users.

- **Enhanced Flexibility**: The hybrid approach allows teams to respond quickly to changing requirements and market conditions, ensuring that the application remains relevant and user-focused. In this research, this method helps develop the app in a flexible manner.
- **Active User Involvement**: DSDM's emphasis on user engagement ensures that stakeholders are actively involved throughout the development process, leading to higher satisfaction and better alignment with user needs. During the app development, the prototype was tested by multiple users to ensure user satisfaction and a good user experience.

- **Iterative Development**: Scrum's sprint-based structure facilitates the delivery of functional increments, allowing for regular feedback and adjustments based on user input and new data which iteratively develop the project by continuous adjustment.
- **Quality Assurance**: The integration of DSDM's quality principles ensures that testing and validation are embedded throughout the development process, leading to a higher-quality final product.
- **Improved Communication**: The combination fosters better communication and collaboration among team members and stakeholders, reducing the risks associated with misunderstandings and misaligned objectives.
- **Continuous Improvement**: The iterative nature of both methodologies promotes a culture of continuous improvement, enabling teams to learn from each sprint and adapt their processes accordingly.
- **Rapid Feedback Loops**: Regular reviews and retrospectives allow teams to incorporate new information and insights quickly, ensuring that the application evolves in line with user expectations.
- **Streamlined Processes**: The structured framework of DSDM combined with Scrum's agility streamlines the development process, making it more efficient and effective.
- **Adaptability to New Data**: The hybrid approach allows teams to incorporate continuous new data and information seamlessly, ensuring that the mobile application can adapt to emerging trends and user preferences. Since this was an academic project, ample data was not used during the preliminary stage. However following the hybrid approach, new data can be incorporated within the AI model continuously to improve the generalizability of the ML model used by the mobile app.
- **Increased Stakeholder Satisfaction**: In this research project, educators, children, and other experts were involved to make sure the app was suitable for the users.

To achieve the goal of developing an effective mobile app for children to educate them on terrestrial animals, this research divided the tasks into three categories.

4.1. Data Collection

The first step is to gather accurate and extensive data on animals. This can be achieved through partnerships with wildlife organizations, access to scientific databases, and crowdsourced contributions from platforms like iNaturalist. The data should include detailed information on animal species, habitats, behaviors, and conservation status. AI algorithms can be employed to continuously update and verify this data, ensuring it remains current and reliable.

Since this is an academic work, a less rigorous data collection method was followed. The total number of images collected for this research is 30,000. The number of animal classes is 150; each containing 200 images. All the images are collected from platforms like Google, Pexels (https://www.pexels.com/), and Pinterest (https://www.pinterest.com/). A web scrapping method was implemented to form this dataset. All the images are of 100 × 100 pixels. **Table 1** shows the list of animal classes.

Table 1. List of Animal Classes

1. Aardvark	**31. Buffalo**	**61. Guinea Pig**	**91. Orangutan**	**121. Sea Otter**
2. Aardwolf	32. Camel	62. Hamster	92. Oryx	122. Sheep
3. Abyssinian	33. Caracal	63. Hedgehog	93. Ostrich	123. Skunk
4. Addax	34. Caribou	64. Hippopotamus	94. Otter	124. Sloth
5. Afghan Hound	35. Cat	65. Honey Badger	95. Panda	125. Snail
6. African Wild Dog	36. Cheetah	66. Horse	96. Pangolins	126. Snake
7. Ainu	37. Chihuahua	67. Hyena	97. Penguin	127. Spectacled Bear
8. Airedale Terrier	38. Chimpanzee	68. Iguana	98. Pig	128. Sphynx Cat
9. Alabai	39. Chipmunk	69. Indri	99. Pika	129. Spider Monkey
10. Alaskan Husky	40. Coyote	70. Japanese Macaque	100. Pine Marten	130. Squirrel
11. Alpaca	41. Crocodile	71. Jerboa	101. Polar Bear	131. Tapir
12. Anteater	42. Crow	72. Kangaroo	102. Pomchi	132. Tarsier
13. Antelope	43. Deer	73. Kaoola	103. Pomeranian	133. Tasmanian Devil
14. Appenzeller Dog	44. Dog	74. Koala	104. Poodle	134. Thorny Devil
15. Arctic Fox	45. Donkey	75. Komodo Dragon	105. Porcupine	135. Tiger
16. Arctic Hare	46. Dragon-Li	76. Lemur	106. Proboscis Monkey	136. Toad
17. Arctic Wolf	47. Elephant	77. Leopard	107. Quokka	137. Tortoise
18. Armadillo	48. Fennec Fox	78. Lion	108. Quoll	138. Tree Kangaroo

continued on following page

Table 1. Continued

1. Aardvark	31. Buffalo	61. Guinea Pig	91. Orangutan	121. Sea Otter
19. Aussiedoodle	49. Fox	79. Lynx	109. Rabbit	139. Vervet Monkey
20. Australian Cattle Dog	50. Frilled Lizard	80. Maine Coon	110. Raccoon	140. Walrus
21. Australian Terrier	51. Frog	81. Mandrill	111. Rat	141. Warthog
22. Aye-aye	52. Gazelle	82. Marmoset	112. Red Fox	142. Weasel
23. Badger	53. Gecko	83. Meerkat	113. Red Panda	143. White Tiger
24. Bat-Eared Fox	54. Gelada	84. Mouse	114. Rhinoceros	144. Wild Boar
25. Beaver	55. Giraffe	85. Musk Ox	115. Rockhopper Penguin	145. Wildebeest
26. Bison	56. Goat	86. Naked Mole Rat	116. Saiga	146. Wolf
27. Black Bear	57. Gorilla	87. Name	117. Saluki	147. Wombat
28. Black Panther	58. Greater Flamingos	88. Nutria	118. Samoyed	148. Woolly Mammoth
29. Black-footed Ferret	59. Greater Kudu	89. Okapi	119. Scorpion	149. Zebra
30. Brown Bear	60. Groundhog	90. Opossum	120. Sea Lion	150. Zonkey

4.2. Strategy for Educational Framework

To develop the app, the contents should be designed following the advanced teaching frameworks to cater to various learning styles and developmental stages. Incorporating methods from Montessori, Reggio Emilia, and game-based learning can make the app interactive and engaging. The goal of the system is to contribute to the educational landscape by leveraging technology to make learning about animals informative and entertaining. Integrating interactive quizzes, image classification, and information retrieval modules into the app lets kids learn about the amazing world of terrestrial animals. When the user uploads a terrestrial animal's image, the uploaded image is sent to the backend of the system, where the machine learning model is deployed. The ML model will generate the results of image classification and will be displayed. The following modules ensure that the app follows the Montessori,

i. Scan Image Module

- Able to recognize and classify the terrestrial animals from uploaded images (Visual Learning- Reggio Emilia approach)
- Able to support real-time/near real-time image processing

ii. Animal Information Module

- Able to allow user to access the information on terrestrial animal
- Able to allow the user to search for their interested terrestrial animal by using their specific name (Enquiry-based Learning)

iii. Quizzes Module

- Able to provide interactive quizzes to engage the user, especially kids (Interactive Learning)
- Able to shuffle/randomize the quiz's questions and answers
- Able to design a scoring system to improve the learning experience (Experimental Learning)

iv. Account Module

- Normal users are able to view their quizzes history (Personalized Learning)
- Normal users are able to view their scanning history (Personalized Learning)
- Admin able to edit/add the terrestrial animal's information
- Admin able to add/edit the quizzes (questions & answers)

The whole module is based on Montessori Learning which enables auto-education, independence, intrinsic motivation, freedom of movement of choice, and individualized learning. The environment of the app is built in such a way that will be able to attract the kid's minds and keep them busy for a longer period of time.

4.3. Software Development Strategy

The development of the application is done following agile methodology to ensure flexibility and continuous improvement. The software architecture should be scalable and capable of handling large datasets and real-time updates in the future. AI and machine learning models are made as an integral part of the app as they will enable features like species identification, personalized learning recommendations, and interactive simulations. User experience (UX) design is crucial in this research as it needs to be user-friendly for a specific group of age. Regular testing and feedback loops with educators, parents, and children will help refine the app and ensure it meets educational goals. Security and privacy measures are also implemented to protect user data, especially given the app's target audience of children.

By integrating these elements, the AI-based app can provide a comprehensive and engaging platform for children to learn about the animal kingdom, fostering a deeper appreciation for wildlife and supporting their overall educational development. **Table 2** represents the functional and non-functional requirements of our app.

Table 2. Functional and Non-Functional Requirements of Different Modules

Module	Functional Requirements	Non-functional Requirements
Scan Image Module	• Able to upload an image for identification • Able to open the camera to snap a photo • Able to determine whether the animal is a terrestrial animal or not based on the image uploaded • Able to store the scanning history	i. Able to have a higher accuracy for animal recognition ii. Able to perform the image recognition in real-time or near real-time
Animal Information Module	• Able to search using the terrestrial animal's name for their information • Able to display detailed information such as their behavior, habitats, diet, and so on	i. Able to provide a well-designed user interface so that kids can use it easily ii. Able to provide the accurate and verified information of the terrestrial animals
Quizzes Module	• Able to provide a randomize quiz questions and answer • Able to implement a scoring system to improve the gaming experience	i. Able to keep the kids interested and curious about animals.
Account Module	**User** • Able to sign in / signup account • Able to edit the user profile • Able to display the quiz history	i. Able to design a user-friendly interface for login/register ii. Able to ensure the privacy of user's personal information
	Admin • Able to add/edit the information of terrestrial animal • Admin able to add /edit the quizzes (questions & answers)	

The Animal Species Explorer System contains three main modules and two other modules. In the main modules, the Scan Image module allows users to upload images or snap a photo for animal classification, the Animal Information module allows users to search the animal's related information and the Quiz module allows user to answer the quizzes provided and earn points. In the other module, users are able to view their scanning history and quizzes history and manage their profile, and the system admin has the permission to add, edit, or delete the animal's information and quizzes. **Figure 1** illustrates the overall system modules.

Figure 1. Overall System Modules Diagram

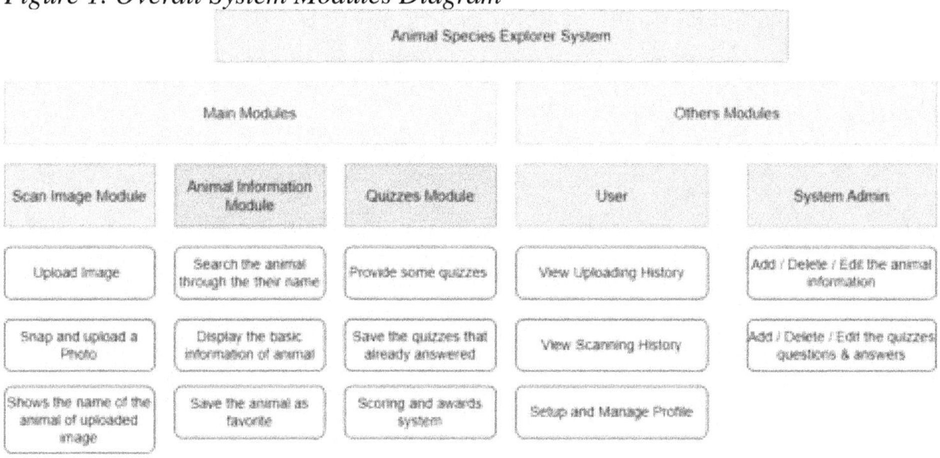

Figure 2 illustrates the use case diagram for our system.

Figure 2. Use Case Model Diagram of the Developed System

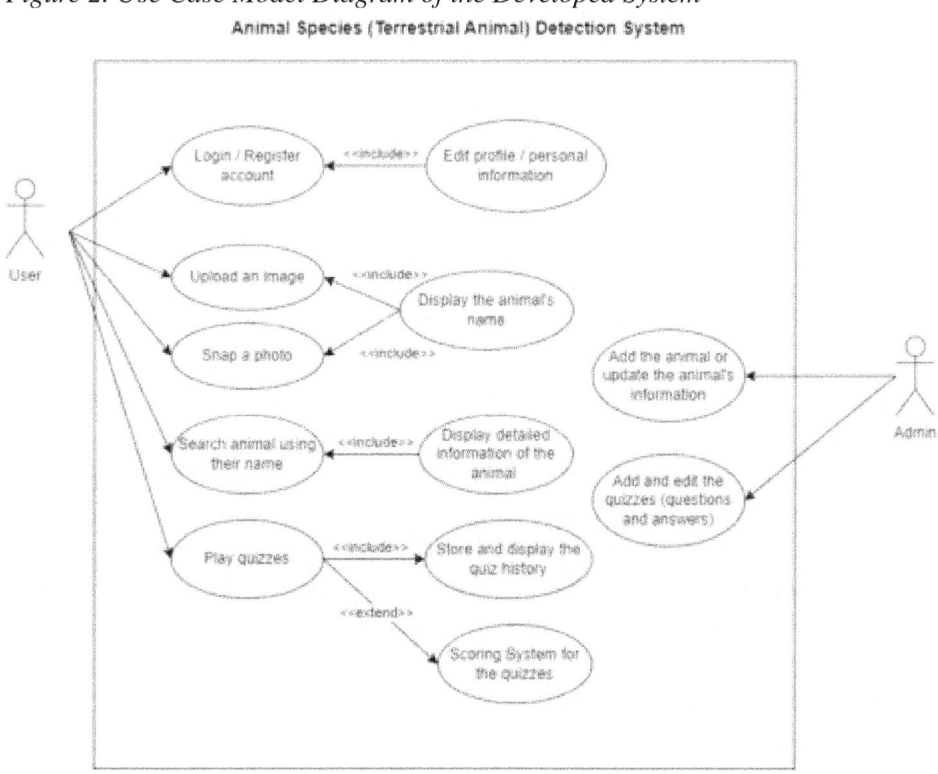

Figure 3 shows the process flow of four pages (Account Page, Quizzes Page, Animal Information Page, and Scan Image Page)

Figure 3. System Flow Chart Diagram

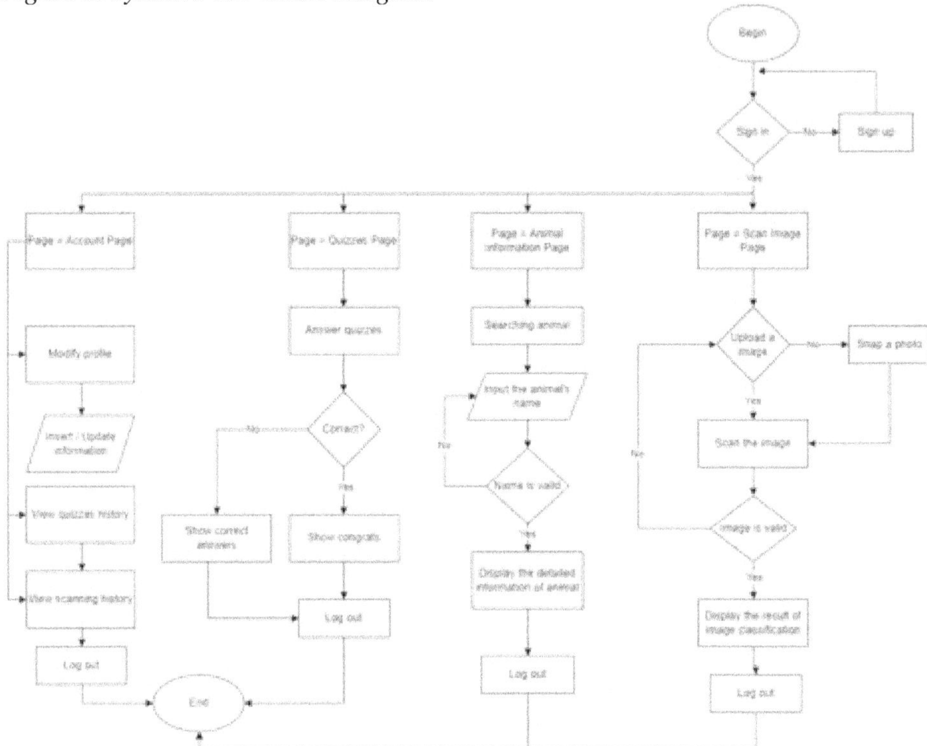

5. APPLICATION DEVELOPMENT, ANALYSIS AND EVALUATION

For mobile development, Kivy is selected as the framework for building the application using the Python language. Kivy stands out as an open-source framework, providing free access for anyone to utilize, modify, and distribute. It offers a diverse array of customizable widgets and layouts, enabling the creation of dynamic user interfaces. These widgets range from buttons and labels to text inputs and images, empowering developers to craft visually appealing and highly interactive interfaces tailored to their application's needs.

As for the backend, SQLite is chosen as the database management system. SQLite is preferred due to its lightweight nature and self-contained design, making integration into applications seamless without the need for a separate server process.

Operating as a serverless database, SQLite simplifies deployment and management, eliminating the overhead of managing a separate server process. **Figure 4** demonstrates the overall system architecture.

Figure 4. System Architecture

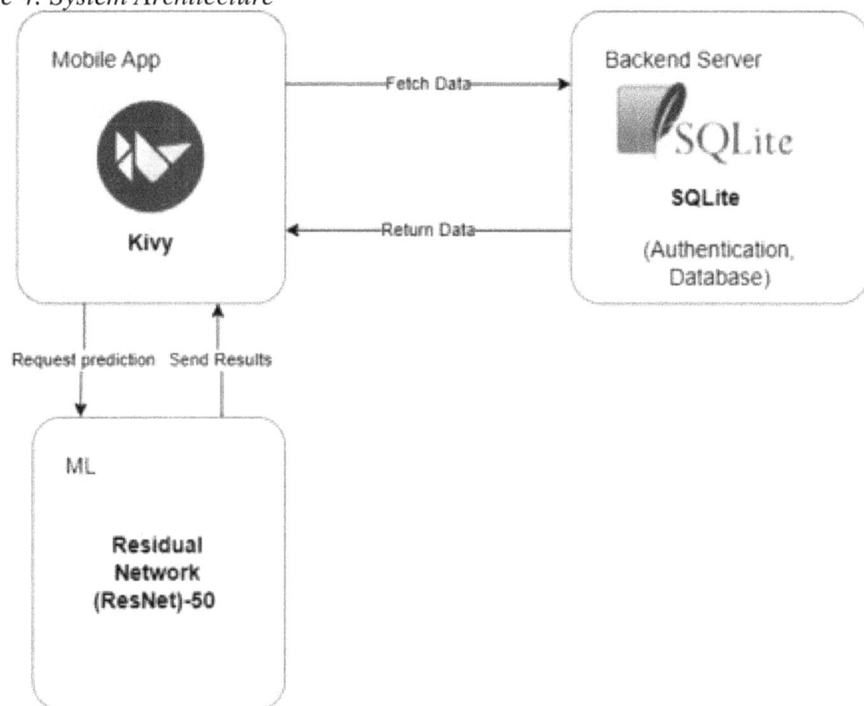

5.1. Tools and Technologies

Table 3 represents the hardware specifications for developing the project.

Table 3. Hardware Specification

Component	**Minimum**
Personal Laptop	ASUS Vivobook S15 X530UN
Mobile Device	Vivo X60
Operating System	Windows 11 (Laptop)Android 13.0 (Mobile Device)

Table 4 represents the software and software specifications of our system development.

Table 4. Software Specification

Component	Software
IDE	Pycharm
Database	SQLite
Model Build	TensorFlow
Version Control	Github
Framework	Kivy
Programming Language	Python, MySQL
Algorithm	Residual Network (ResNet-50) (Deep Learning for image classification)

5.2. Database Development

In **Figure 5**, 5 database tables which are User, ScanImage, AnimalDetail, Quiz, and QuizHistory are demonstrated. One user has one account with one email address. In the ScanImage table, it stores scan_id, the BLOB / Binary data of the image, the address of the scanned image, and the predicted result. In the AnimalDetail table, it stores the animal_id, animal information which includes name, scientific name, locations, habitats, prey, lifespans, and size, is_favourite is used to determine whether this animal is set as favorite or not. Users can set many animals as favorites. In the Quiz table, it stores quiz_id, question, 4 choices, and the answer. When the user answers, the quiz database will save it to the QuizHistory table. In QuizHistory table, it stores history_id, score_points of the quiz (Correct – 5 points, Wrong – 0 poitns), user selected_choice, user_id and quiz_id. Users can answer one or more quizzes.

Figure 5. Database Schema Diagram

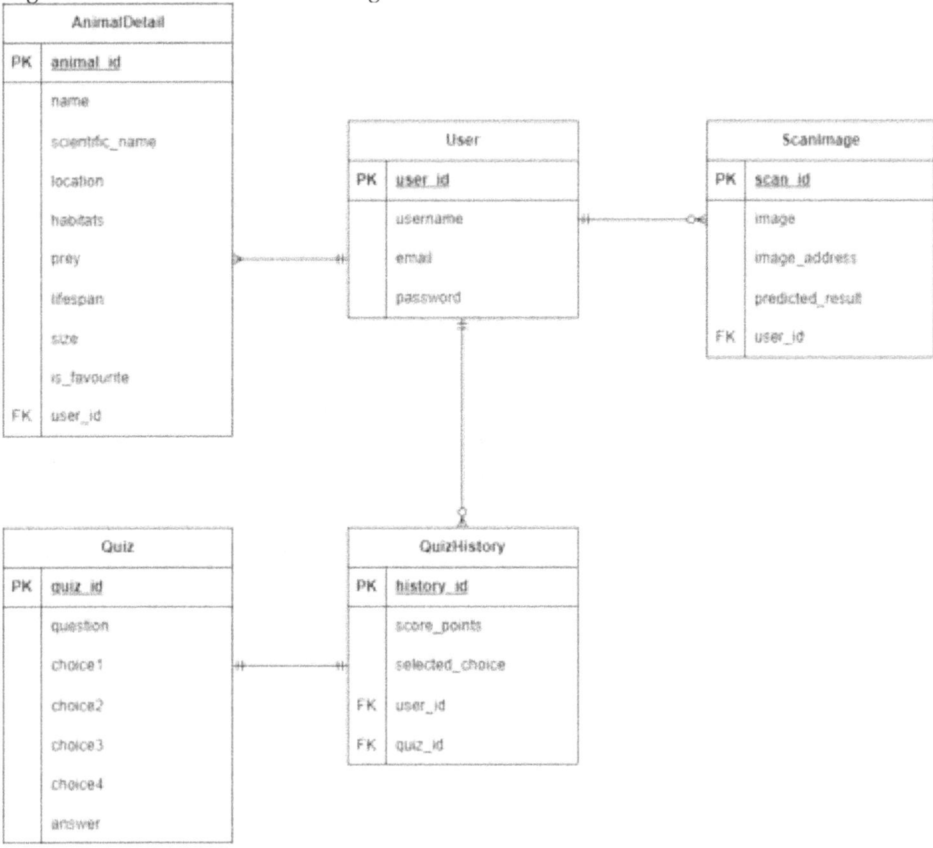

5.3. Intelligent Model

When it comes to machine learning, the Residual Network (ResNet-50) model is employed for image classification tasks. Leveraging pre-trained weights on datasets like ImageNet provides a robust starting point, significantly reducing the data and computational resources required for effective model training. By utilizing ResNet-50, the application can accurately classify images with minimal computational overhead, enhancing performance and user experience. **Figure 6** shows the base architecture for the ResNet-50 model.

Figure 6. ResNet-50 architecture

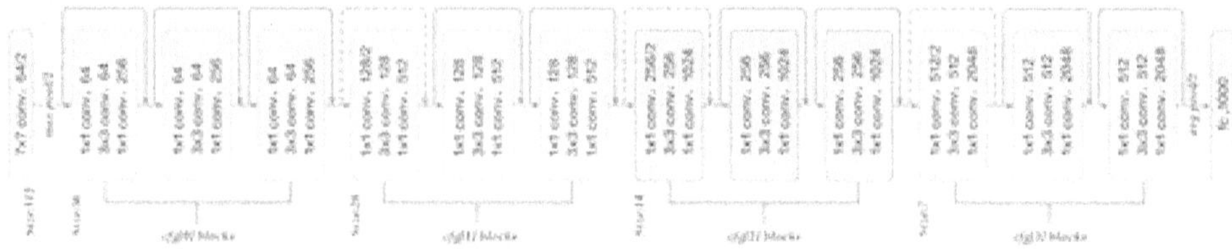

5.2.1. ResNet-50 Model Evaluation

The ResNet-50 model was evaluated under three hyper-parameter tuning. For parameter tuning, the method discussed by Thomas et al. (Thomas et al., 2023) and Krishnamoorthy et al. (Krishnamoorthy et al., 2020) were followed. **Figure 7,Figure 8,** and **Figure 9** illustrate different accuracy graphs for three hyper-parameter tuning. At first, the hyperparameter used for this model is learning rate: Adam (0.01), dropout rate: 0.5, and batch size (64). The validation accuracy of this model is 0.72. The accuracy of this model is 0.69 while the accuracy loss is 0.18. Based on the graph below, we can see that the highest accuracy of the model is around 0.65 but the accuracy decreases when the number of epochs increases. This situation occurs due to the model overfitting.

Figure 7. Accuracy graph of 1st Hyperparameter Tuning

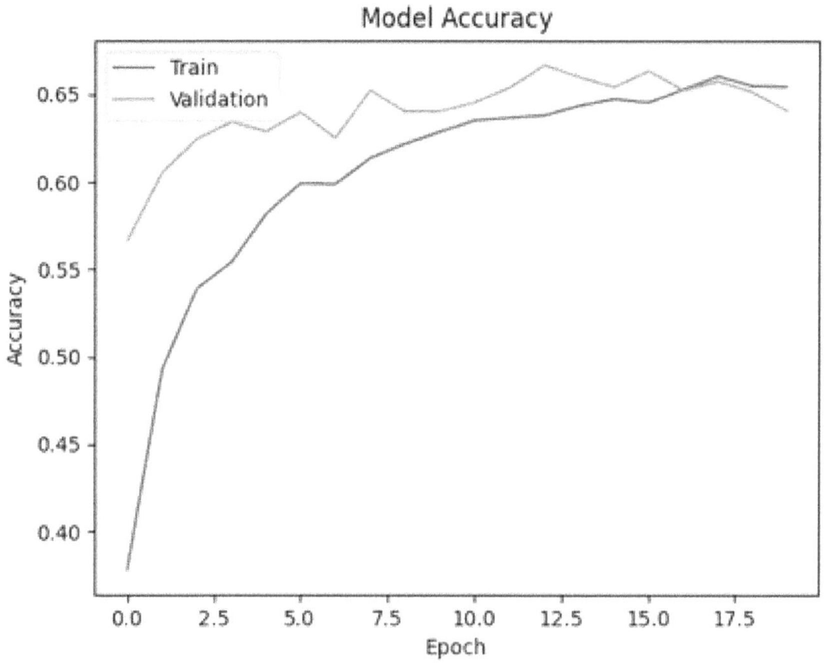

For this experiment, multiple LR values were tested starting from a higher value to a lower (0.1, 0.01. 0.001, 0.0001). It was observed that, for higher values of LR, the accuracy was low and less stable while the training was done faster. When the value was reduced the accuracy was higher while training took a longer amount of time. However, after the 0.001 value, the accuracy did not increase; only the time taken for training increased. By editing the hyperparameters of the Learning Rate = 0.001, Dropout rate = 0.6, and Batch size = 32, the accuracy increases but still inconsistency. Moreover, the model loss of train loss and validation loss increases when epochs increase, although the accuracy increases compared to Experiment 1, however, the overfitting issue still exists.

Figure 8. Accuracy graph of 2nd Hyperparameter Tuning

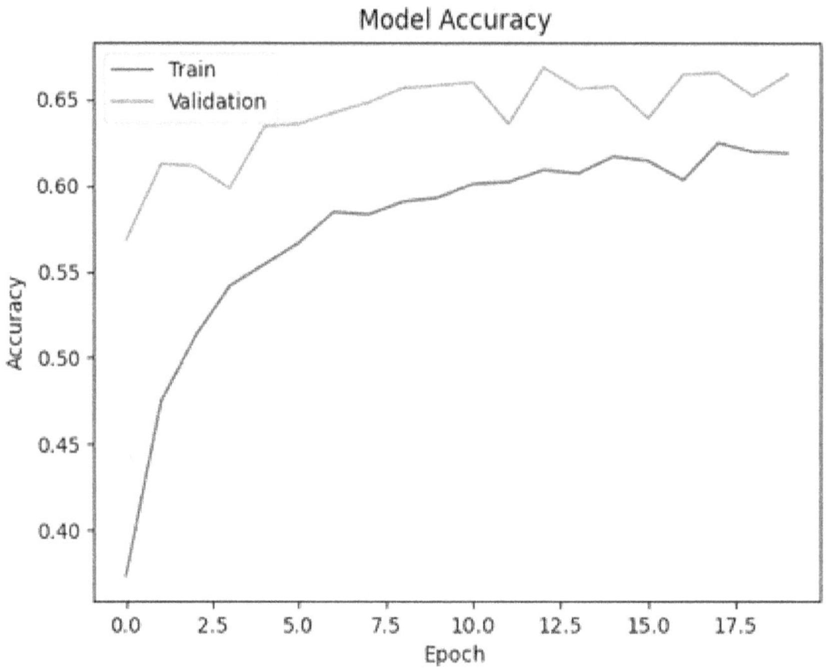

In order to avoid overfitting, early stopping, and learning rate scheduling are added. The function of early stop is added to halt training when the validation loss stops improving while the learning rate scheduling is added a learning rate scheduler to reduce the learning rate when the validation loss plateaus. Since the model is fine-tuned with early stopping and learning rate scheduling the model fitting is avoided. The model loss is decreased per epoch for train loss and validation loss. The hyperparameter for this model is Learning rate = 0.001, Batch size = 32, Epochs =30, and Dropout Rate = 0.6. The accuracy of this model is around 0.52.

Figure 9. Accuracy graph of 3rd Hyperparameter Tuning

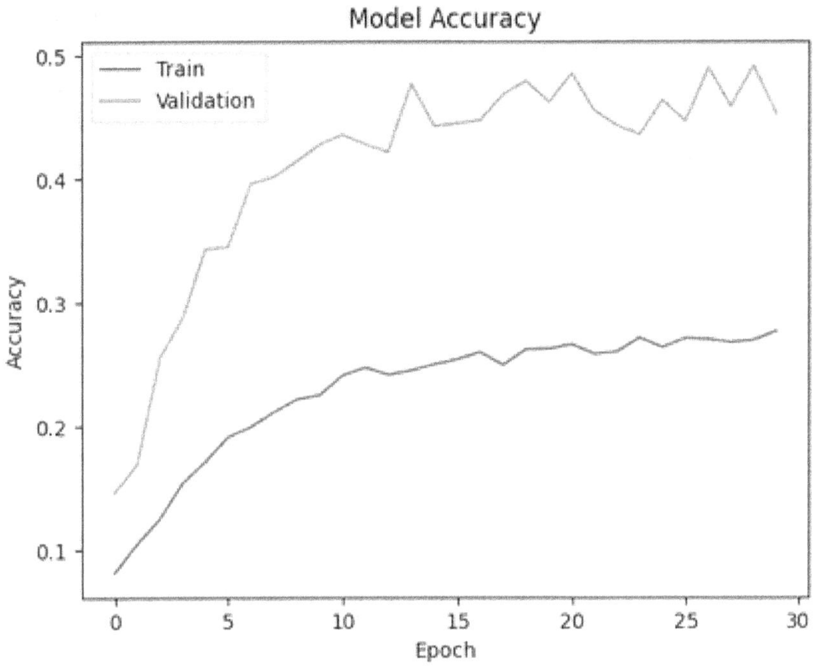

5.4. User Interface

When a new user opens the app, the first screen will ask the user to log in to an account or register a new account. If the user already has an account, when clicking on the 'LOGIN ACCOUNT' button will go to the login page, the user needs to enter the email address and correct password to log in to their account. If the user doesn't have an account yet, click on the 'REGISTER ACCOUNT' button, which will go to the register page, the user needs to enter their username, email address, and password to register a new account. When logging into an account, if the user inputs an invalid account such as a wrong email, wrong password, or email that does not exist, the system will pop up a wrong message ("Invalid Account!"). When registering a new account, the user needs to input the username, email, and password. If the user's input of username, and password's number of characters is less than 3, and the email address format does not contain '@' and '.', the system will pop up a wrong message ("Invalid Input!"). On the main page, it shows the name of this application. On the top app bar, it has a navigation menu which contain three function, scan image, search animal and quiz function. When clicking on 'Scan' will go to the Scan Page, 'Animal' will go to the Animal Page, 'Quiz' will go to the Quiz page, and 'Profile' will go to the Profile Page. There are two options for user which are upload an image from directory, and open a camera to snap a photo. When the user clicks on the 'UPLOAD IMAGE' button, a dialog will be displayed to allow the user to choose the image they want for animal classification. When the user clicks on the 'OPEN CAMERA' button, will go to the snap a photo page, which allows the user to open the camera to snap a photo. Once the user uploads an image or snap a photo, the system will do the prediction based on the uploaded image and give the results. A pop-up window will display shows the image of user and the predicted animal class name. When the user clicks on the text "View History" on the Scan Page, the User will navigate to the Scan History Page that shows the scanning history made by the user. The image and the predicted result are shown. On the Animal Page, there is a search bar that allows users to search for the animal they are interested in based on their names. Besides, there is a list of animals that is stored in the database system, the user can click on the specific animal to see the animal's details. In Animal Detail Page, will show the animal details information with an image of the animal. In the Quiz page, it displays the quiz with one question and 4 choices. If the user selects the wrong answer, it will show "Incorrect!" and give the correct answer. If the user selects the correct answer, will show "Correct! Congrats! Earn 5 points!!". By clicking the View History on the Quiz Page, it will navigate to the Quiz History Page. This page will display the quizzes that are answered by the user. On the profile page, it shows the username and the email address of the user account. When click on the "Edit Profile" button, user is allowed to edit the username, email address and password. When the user login as admin ("user_id == 1), it will navigate to the Admin Page that shows two options which are Animal Management and Quiz Management When the user clicks on the "ANIMAL MANAGEMENT" button on the admin page, it will navigate to the Animal Management Page (Search Page). In the Search Page, there is a '+' icon that allows the system admin to add new animal and their information. In the Animal Detail Page, there is an edit button on the top app bar that allows the admin to edit the information of the animal and a 'DELETE ANIMAL' button that allows the admin to delete the animal. When the user click on the "QUIZ MANAGEMENT" button on the admin page, it will navigate to the Quiz Management Page. In the Search Page, there is a '+' icon allow system admin to add new quiz and the information (choices, answer). In the Quiz Detail Page, there is an edit button on the top app bar that allows the admin to edit the information of the quiz and a 'DELETE QUIZ' button that allows the admin to delete the quiz. **Figure 10** shows some example interfaces of our app.

Figure 10. User Interfaces Example

5.5. Expert's Evaluation and Recommendation

After analyzing the mobile application, experts have suggested some crucial points and according to those points, some adjustments were made. Those recommendations and adjustments are mentioned in detail as follows:

1. ***User Engagement and Retention Issues***
   * Problem: Users may not be returning to the app regularly, leading to low engagement levels.
   * Solution: Introduce gamification elements such as rewards, badges, and leaderboards to motivate children to engage with the app consistently.
2. ***Content Relevance and Freshness***
   * Problem: The educational content may become outdated or irrelevant, failing to capture children's interest.
   * Solution: The establishment of a content update schedule to regularly add new information about terrestrial animals, including recent discoveries and conservation efforts has been done to solve that while incorporating user-generated content features where children can submit their observations or facts.
3. ***Difficulty Level of Educational Material***
   * Problem: The app may present information that is either too complex or too simplistic for the target age group.
   * Solution: Utilization of machine learning algorithms has been done to analyze user performance and adapt the difficulty level of quizzes and educational content dynamically.
4. ***Limited Interactivity***
   * Problem: The app may lack interactive features that engage children effectively.
   * Solution: Integration of augmented reality (AR) features that allow children to visualize animals in their natural habitats will solve this. This can enhance interactivity and provide a more immersive learning experience. However, this feature will be added later.
5. ***Feedback Mechanisms***
   * Problem: The app may not provide adequate feedback to users on their learning progress.
   * Solution: Implementation of a robust feedback system that gives children immediate responses after quizzes and activities, highlighting areas of strength and suggesting areas for improvement has been incorporated .
6. ***Accessibility Concerns***

* Problem: The app may not be fully accessible to children with different learning abilities.
* Solution: It should be ensured that the app includes features such as text-to-speech, adjustable font sizes, and color contrast options to accommodate children with visual impairments or learning disabilities which is done properly.

7. ***Data Privacy and Security***
   * Problem: Concerns regarding the collection and use of personal data from children may arise.
   * Solution: Implementation of strict data privacy policies is present in the app in compliance with regulations such as COPPA (Children's Online Privacy Protection Act).
8. ***Technical Performance Issues***
   * Problem: Users may experience slow loading times or crashes, leading to frustration.
   * Solution: Optimization of the app's performance by refining the code, reducing the size of images and videos, and ensuring that the machine learning models are efficient has been done to resolve this.
9. ***Limited Social Interaction Features***
   * Problem: The app may not facilitate social learning or peer interaction among users.
   * Solution: The introduction of features that allow children to share their achievements or discoveries with friends and family is done. There are also discussion forums and collaborative challenges where users can work together to learn more about terrestrial animals.
10. ***Lack of Parental Involvement***
    * Problem: Parents may not have a way to engage with their children's learning experience through the app.
    * Solution: A parent dashboard is created that provides insights into their child's progress, achievements, and areas needing improvement and offers suggestions for offline activities that parents can do with their children to reinforce learning.

After adjusting all of the above-mentioned things the app was cross-checked with the experts again thus validated to make sure that the adjustments were done properly. However, since, this is an academic project, so some things will have to be left for future improvement such as augmented reality. Augmented reality with real-time 3D effect will make the app even more interactive which will be developed later.

6. DISCUSSION AND CONCLUSION

In conclusion, the primary goal of this project is to develop an Animal Species Explorer System specifically for terrestrial animals for educational purposes. This project is aimed at creating an engaging and educational mobile application for children. The application seeks to attract kids in a lively and educational manner by incorporating multiple modules, such as Scan Image, Animal Information, Quizzes, and Account Management. The machine learning algorithm ResNet-50 has been chosen for the image classification of terrestrial animals. The benefits and uniqueness of this proposed solution lie in its innovative fusion of technology and education, transforming passive learning experiences into active and interactive journeys for children. The inclusion of elements, such as quizzes, aims to stimulate children's curiosity, encouraging continuous exploration and learning. This study includes use case descriptions that clearly outline the essential use cases required for system implementation. Since this system is newly developed, there are several potential improvements that can be made in the future. One is implementing a cloud database for better database management and deploying the mobile application online instead of as a standalone application. Additionally, communication features could be added, allowing users to share their scanning history with others. Furthermore, more categories can be added as there are a lot more species in this beautiful world, and also at the same time, for every category, a lot more images can be deployed to train the ML model. The main categories can be even divided into birds, fish, lizards, etc. which will contain a lot of sub-categories. Besides increasing the category and images in each category, other machine learning algorithms can also be used for training; even model customization can be done to increase the accuracy of the ML model for this specific task. To collect more credible data, wildlife conservation organizations can be contacted. At the same time after getting feedback from the current users, new modules can also be developed such as voice commands with re-enforcement learning along with upgradations in the existing features of the app.

REFERENCES

Шаповалов, Є., Білик, Ж., Атамась, А., & Учитель, О. (2019). The Potential of Using Google Expeditions and Google Lens Tools under STEM-education in Ukraine. *Педагогіка вищої та середньої школи, 51*, 90–101. https://doi.org/DOI: 10.31812/pedag.v51i0.3659

Akçayır, G., & Akçayır, M. (2018). The flipped classroom: A review of its advantages and challenges. *Computers & Education*, 126(1), 334–345. DOI: 10.1016/j.compedu.2018.07.021

Anderson, G. D., Panneer Selvakumar, Shi, M., Marshall, C. S., Agrawal, A., Rebecca, C., Raffa, G., Sherry, J. F., Loi, D., & Durham, L. M. (2018). Kid Space. *Proceedings of the Group Interaction Frontiers in Technology*. DOI: 10.1145/3279981.3279986

Anna. (2023, April 24). *Australian Wildlife: 35+ Animals You Can See in Australia*. Really Wildlife. https://www.reallywildlife.com/animals-in-australia/

Antezana, G., Venning, A., Smith, D., & Bidargaddi, N. (2022). Understanding what we know so far about young people's engagement with wellbeing apps. A scoping review and narrative synthesis. *Digital Health*, 8, 205520762211441. DOI: 10.1177/20552076221144104 PMID: 36532119

Aristeidou, M., Herodotou, C., Ballard, H. L., Higgins, L., Johnson, R. F., Miller, A. E., Young, A. N., & Robinson, L. D. (2021). How do young community and citizen science volunteers support scientific research on biodiversity? The case of iNaturalist. *Diversity*, 13(7), 318. DOI: 10.3390/d13070318 PMID: 35873351

Article detail - International Journal of Advanced Research. (2020, June 5). International Journal of Advanced Research. https://www.journalijar.com/article/18752/comparison-of-different-agile-methodologies-and-fit-assessment-in-an-industrial-context./

Aslan, S., Agrawal, A., Alyuz, N., Chierichetti, R., Durham, L. M., Manuvinakurike, R., Okur, E., Sahay, S., Sharma, S., Sherry, J., Raffa, G., & Nachman, L. (2022). Exploring Kid Space in the wild: A preliminary study of multimodal and immersive collaborative play-based learning experiences. *Educational Technology Research and Development*, 70(1), 205–230. DOI: 10.1007/s11423-021-10072-x PMID: 35035182

Ayman, R., Sharaf, N., Ahmed, G., & Abdennadher, S. (2018). MiniColon; Teaching Kids Computational Thinking Using an Interactive Serious Game. *Lecture Notes in Computer Science*, 11243, 79–90. DOI: 10.1007/978-3-030-02762-9_9

BBC. (2024). *North America: A land of pioneering animals*. BBC Bitesize. https://www.bbc.co.uk/bitesize/articles/zntnt39

Bragg, C. (2024, January 2). *Parenting, Wahlberg's eagle style: making financial brokers proud*. Africa Geographic. https://africageographic.com/stories/parenting-wahlbergs-eagle-style/

Bruffee, K. A. (1987). The Art of Collaborative Learning. *Change*, 19(2), 42–47. DOI: 10.1080/00091383.1987.9939136

Catherine, L., Javier, B., & Francisco, G. (2020). Four Pillars of the Montessori Method and Their Support by Current Neuroscience. *Mind, Brain and Education : the Official Journal of the International Mind, Brain, and Education Society*, 14(4), 322–334. DOI: 10.1111/mbe.12262

Compliance. (2024, June 3). *AI in Online Learning Platforms: Personalized Learning Paths*. Redresscompliance.com. https://redresscompliance.com/ai-online-learning-platform/

Deep learning techniques and optimization strategies in big data analytics. (2019). In Advances in systems analysis, software engineering, and high performance computing book series. https://doi.org/DOI: 10.4018/978-1-7998-1192-3

Di Cecco, G. J., Barve, V., Belitz, M. W., Stucky, B. J., Guralnick, R. P., & Hurlbert, A. H. (2021). Observing the Observers: How Participants Contribute Data to iNaturalist and Implications for Biodiversity Science. *Bioscience*, 71(11), 1179–1188. DOI: 10.1093/biosci/biab093

EarthSnap. (2016). *EarthSnap - Nature Identifier * Earth.com*. Www.earth.com. https://www.earth.com/earthsnap/

Fracolli, B. (2020, August 7). *30 Amazon Rainforest Animals to Spot in the Wild*. Peru for Less. https://www.peruforless.com/blog/amazon-rainforest-animals

Gweon, H. (2021). Inferential social learning: Cognitive foundations of human social learning and teaching. *Trends in Cognitive Sciences*, 25(10), 896–910. DOI: 10.1016/j.tics.2021.07.008 PMID: 34417094

Hewett, V. M. (2001). Examining the Reggio Emilia Approach to Early Childhood Education. *Early Childhood Education Journal*, 29(2), 95–100. DOI: 10.1023/A:1012520828095

H.L. P. (2015). *Differentiated Instruction And The Need To Integrate Teaching And Practice - ProQuest*. Www.proquest.com. https://www.proquest.com/openview/718ee130f5c18bd09e667e85b304b6ce/1?pq-origsite=gscholar&cbl=2026890

Hockly, N. (2018). Blended Learning. *ELT Journal*, 72(1), 97–101. DOI: 10.1093/elt/ccx058

Hoffman, D. M. (2009). Reflecting on Social Emotional Learning: A Critical Perspective on Trends in the United States. *Review of Educational Research*, 79(2), 533–556. DOI: 10.3102/0034654308325184

Hsi, S. (2007). Conceptualizing Learning from the Everyday Activities of Digital Kids. *International Journal of Science Education*, 29(12), 1509–1529. DOI: 10.1080/09500690701494076

Javed, S. T., Zafar, K., & Younas, I.Sobia Tariq Javed. (2024, October). Zafar, K., & Younas, I. (2024). Kids Learning Optimizer: Social evolution and cognitive learning-based optimization algorithm. *Neural Computing & Applications*, 36(28), 17417–17465. Advance online publication. DOI: 10.1007/s00521-024-10009-4

Jiang, Z., & Hussain, Y. (2023). THE EFFECTIVENESS OF NATURE-BASED LEARNING IN PROMOTING PHYSICAL, COGNITIVE, AND EMOTIONAL DEVELOPMENT IN YOUNG CHILDREN: A CASE STUDY IN CHINA. *International Journal of Education & Technology*, 1(3). Advance online publication. DOI: 10.59021/ijetech.v1i3.24

Jones, H. G. (2020). What plant is that? Tests of automated image recognition apps for plant identification on plants from the British flora. *AoB Plants*, 12(6), plaa052. Advance online publication. DOI: 10.1093/aobpla/plaa052 PMID: 33173573

Karimkhanlooei, G., & Seifiniya, H. (2015). Teaching Alphabet, Reading and Writing for Kids between 3-6 Years Old as a Second Language. *Procedia: Social and Behavioral Sciences*, 192, 769–777. DOI: 10.1016/j.sbspro.2015.06.090

Kokotsaki, D., Menzies, V., & Wiggins, A. (2016). Project-based learning: A review of the literature. *Improving Schools*, 19(3), 267–277. DOI: 10.1177/1365480216659733

Lewis, L. H., & Williams, C. J. (1994). Experiential Learning: Past and Present. *New Directions for Adult and Continuing Education*. https://eric.ed.gov/?id=EJ492388

Miller, N. C., Kumar, S., Pearce, K. L., & Baldock, K. L. (2021). The outcomes of nature-based learning for primary school aged children: A systematic review of quantitative research. *Environmental Education Research*, 27(8), 1–26. DOI: 10.1080/13504622.2021.1921117

Moe, N. B., & Dingsøyr, T. (2008). Scrum and team effectiveness: theory and practice. In Lecture notes in business information processing (pp. 11–20). https://doi.org/DOI: 10.1007/978-3-540-68255-4_2

National Geographic. (2015). *Europe: Physical Geography | National Geographic Society*. Education.nationalgeographic.org. https://education.nationalgeographic.org/resource/europe-physical-geography/

Nicklen, P. (2023, September 14). *Top 10 facts about Emperor penguins*. WWF. https://www.wwf.org.uk/learn/fascinating-facts/emperor-penguins

nsidc. (2017). *Why Sea Ice Matters*. National Snow and Ice Data Center. https://nsidc.org/learn/parts-cryosphere/sea-ice/why-sea-ice-matters

Panasan, M., & Nuangchalerm, P. (2010). Learning Outcomes of Project-Based and Inquiry-Based Learning Activities. In *ERIC* (Vol. 6). https://eric.ed.gov/?id=ED509723

Parshley, L. (2023, October 1). *Artificial Intelligence Could Finally Let Us Talk with Animals*. Scientific American. https://www.scientificamerican.com/article/artificial-intelligence-could-finally-let-us-talk-with-animals/

Patiño-Escarcina, R. E., Barrios-Aranibar, D., Bernedo-Flores, L. S., Alsina, P. J., & Gonçalves, L. M. G. (2021). A Methodological Approach to the Learning of Robotics with EDUROSC-Kids. *Journal of Intelligent & Robotic Systems*, 102(2), 34. Advance online publication. DOI: 10.1007/s10846-021-01400-7 PMID: 34025033

Ping, T. P., Dennis, P. S., Julaihi, A. A., & Ling, M. G. B. (2022). Mobile Learning Model for Children with Special Learning Needs. *European Journal of Teacher Education*, 4(3), 1–12. DOI: 10.33422/ejte.v4i3.815

Randoll, D., & Peters, J. (2015). Empirical research on Waldorf education. *Educar em Revista*, 56(56), 33–47. DOI: 10.1590/0104-4060.41416

Rasidah binti Sardi. (2022). Vr Kidos – Virtual Reality Kids Pedagogy Learning System. *European Proceedings of Multidisciplinary Sciences*. DOI: 10.15405/epms.2022.10.62

Salum, R. N., Khadijah, K., & Nasution, N. L. (2024). Optimization of early childhood social emotional development through exercise methods and learning videos. [JIES]. *Journal of Islamic Education Students*, 4(1), 163. DOI: 10.31958/jies.v4i1.12281

Scalable and distributed machine learning and deep learning patterns. (2023). In Advances in computational intelligence and robotics book series. https://doi.org/ DOI: 10.4018/978-1-6684-9804-0

Semeraro, F., Imbriaco, G., Del Giudice, D., Antognoli, M., Celin, D., Cuttitta, M., Lo Guasto, V., Giulini, G., Gnudi, T., Monesi, A., Nava, E., Tucci, R., Carenzio, A., Lo Jacono, S., Gordini, G., Gamberini, L., Coniglio, C., Baldazzi, M., Landini, C., & Tortolani, D. (2024). Empowering the next Generation: An innovative "Kids Save Lives" blended learning programme for schoolchildren training. *Resuscitation*, 194, 110088. DOI: 10.1016/j.resuscitation.2023.110088 PMID: 38101506

Sheard, J. K., Adriaens, T., Bowler, D. E., Büermann, A., Callaghan, C. T., Camprasse, E. C. M., Chowdhury, S., Engel, T., Finch, E. A., Von Gönner, J., Hsing, P., Mikula, P., Oh, R. Y. R., Peters, B., Phartyal, S. S., Pocock, M. J. O., Wäldchen, J., & Bonn, A. (2024). Emerging technologies in citizen science and potential for insect monitoring. *Philosophical Transactions of the Royal Society B Biological Sciences, 379*(1904). DOI: 10.1098/rstb.2023.0106

Shemshack, A., & Spector, J. M. (2020). A systematic literature review of personalized learning terms. *Smart Learning Environments*, 7(1), 33. Advance online publication. DOI: 10.1186/s40561-020-00140-9

The Scrum Guide. (2012). *Software in 30 Days*, 133–152. https://doi.org/DOI: 10.1002/9781119203278.app2

Tobias, S., Fletcher, J. D., & Wind, A. P. (2013). Game-Based Learning. *Handbook of Research on Educational Communications and Technology, 1*, 485–503. DOI: 10.1007/978-1-4614-3185-5_38

Van Horn, G., Mac Aodha, O., Song, Y., Cui, Y., Sun, C., Shepard, A., Adam, H., Perona, P., & Belongie, S. (2018). *The INaturalist Species Classification and Detection Dataset*. Openaccess.thecvf.com. https://openaccess.thecvf.com/content_cvpr_2018/html/Van_Horn_The_INaturalist_Species_CVPR_2018_paper.html

WWF. (2019). *Where do tigers live? And other tiger facts*. World Wildlife Fund. https://www.worldwildlife.org/stories/where-do-tigers-live-and-other-tiger-facts

Zedda, L. (2023). The importance of a transformative biodiversity education for perceiving, appreciating and supporting lichen diversity in German urban environments. *Lichenologist (London, England)*, 55(5), 161–168. DOI: 10.1017/S0024282923000312

Chapter 5
Challenges and Acceptance of AI Integration in Classroom Learning: A Preliminary Study

Nurul Maziah Mohd Barkhaya
https://orcid.org/0009-0002-2192-0706
Universiti Sains Malaysia, Malaysia

ABSTRACT

The potential of artificial intelligence (AI) in fostering inclusive education, especially in classroom learning, is well-documented by current educational trends. This study aims to explore students' acceptance of AI in higher education as a preliminary investigation. It identifies the challenges faced from the students' perspectives. Data will be collected through interviews with three students enrolled in instructional technology course, focusing on their experiences and acceptance levels. Thematic analysis will be used to uncover recurring themes and patterns in the interview transcripts. Expected findings include a diverse range of student attitudes, with some highlighting the enhanced learning opportunities and personalization provided by AI. Overall, this research addresses fostering the inclusivity of AI among students to create an effective learning environment.

DOI: 10.4018/979-8-3693-9811-1.ch005

INTRODUCTION

Artificial Intelligence (AI) has rapidly emerged as a transformative force across various sectors, including education. The integration of AI technologies in classroom learning promises to revolutionize the educational landscape by enhancing personalized learning, providing real-time feedback, and automating administrative tasks. Despite these promising benefits, the acceptance and successful implementation of AI in higher education hinge significantly on the perceptions and readiness of the students who are expected to use these technologies (Jaiswal & Arun, 2021). Understanding students' acceptance and the challenges they face in integrating AI into their learning environments is crucial for maximizing the potential benefits of these advanced technologies.

In recent years, numerous studies have explored the potential advantages of AI in education, highlighting its capacity to support personalized learning experiences, foster active learning, and improve overall educational outcomes (Huang, Saleh, & Liu, 2021). However, the successful integration of AI in higher education is not without its obstacles. Challenges such as inadequate training, limited access to AI tools, and ethical concerns are significant barriers that need to be addressed to facilitate effective AI adoption in educational settings (Chan & Zary, 2019; Akgun & Greenhow, 2022). As AI technologies continue to evolve, there is a growing need to investigate these challenges from the students' perspective to develop strategies that can support their successful integration.

Therefore, this study aims to explore students' acceptance of AI in higher education as a preliminary investigation. It seeks to identify the challenges faced by students in integrating AI into their learning processes and to understand their overall perceptions of AI technologies. By examining the factors that influence students' acceptance and the barriers they encounter, this research provides valuable insights that can inform the development of targeted interventions and support mechanisms to enhance AI integration in educational contexts (Labadze, Grigolia, & Machaidze, 2023). This preliminary study sets the foundation for more extensive research on AI in education, aiming to foster a more inclusive and effective learning environment.

BACKGROUND

The integration of AI in education has garnered significant attention from researchers and educators alike, driven by the potential of these technologies to transform traditional teaching and learning methods. AI applications in education range from intelligent tutoring systems and adaptive learning platforms to AI-driven administrative tools that streamline various educational processes. These advancements have the potential to enhance educational efficiency, provide personalized learning experiences, and support students in achieving better academic outcomes (Cui, Xue, & Thai, 2019).

Challenges of AI integration in classroom learning

Despite the growing interest and investment in AI for education, several challenges impede its widespread adoption and effective implementation. One of the primary challenges is the lack of adequate training for both students and educators on how to effectively use AI tools. Chan and Zary (2019) highlighted that insufficient training from AI experts can hinder the implementation of AI in educational settings, particularly in specialized fields such as medical education. Moreover, ethical concerns and the implications of AI systems are often overlooked, raising questions about privacy, bias, and the overall impact of AI on the educational experience (Borenstein & Howard, 2021). The addressing of these challenges to ensure the successful integration of AI in higher education, it is essential to understand the perspectives and experiences of the students who are expected to use these technologies.

In classroom learning, educators encounter challenges when integrating AI technology due to difficulties fitting it into an already crowded curriculum. Lee and Perret (2022) highlight that standardized assessments do not yet test content knowledge and skills related to AI, resulting in these areas being undervalued and deprioritized in the school day. Furthermore, there is a reported lack of professional development for teachers in AI, particularly through effective pedagogical approaches (Eguchi, 2021; Sanusi & Oyelere, 2020).

Acceptance of AI in classroom learning

The acceptance of AI integration Previous research has shown that students' perceptions of AI, particularly regarding ease of use and usefulness, significantly influence their willingness to adopt AI-based tools (Kim et al., 2020). Through this study, the delve deeper into these perceptions, examining the specific challenges students face and their overall acceptance of AI in their learning environments will be investigated. By identifying the factors that influence students' acceptance and the barriers they encounter, this research seeks to inform the development of strategies and interventions that can support the effective integration of AI in higher education (Lo, 2023).

METHODOLOGY

Thematic analysis will be employed to analyze the interview data in this study. As been stated by Kiger and Varpio (2020), thematic analysis as a powerful and flexible method in qualitative method for identifying, analyzing, and reporting patterns or themes within data. This approach will be used to uncover recurring themes and patterns in the interview transcripts of three students enrolled in an instructional technology course, focusing on their experiences and acceptance of AI integration in the classroom. Additionally, the insights gathered from these discussions will help inform and influence the implementation of AI in higher education learning environments.

RESULTS

The analysis of the interview transcripts from three higher education students revealed two key themes: "challenges of AI in classroom learning" and "acceptance of AI in classroom learning". These themes reflect the students' overall attitudes toward the integration of AI technologies in their educational experience. The "challenges" theme highlights the difficulties and concerns the students face when interacting with AI tools in a classroom setting, while the "acceptance" theme encompasses the perceived benefits and positive impacts AI has had on their learning processes. These findings provide valuable insights into both the advantages and potential barriers of adopting AI in higher education environments.

Challenges of AI in classroom learning

Based on the interview session, three students who are the participants of this study were asked about the challenges they encountered when using AI tools in classroom learning, how AI integration impacted their interactions with instructors and peers, and the difficulties they face in trusting AI to provide accurate or unbiased information. The responses have been transcribed as follows.

Table 1. Challenges faced by students in AI integration in classroom learning

No.	Role	Questions and Response
1	Researcher	What challenges have you encountered when using AI tools in your classroom learning experience?
	Student 1	*"I've noticed that some AI tools don't always fit well with the course content. They seem more general and aren't customized to the specific curriculum, which can make it harder to align with what the lecturer is teaching."*
	Student 2	*"One of the biggest challenges for me is relying too much on AI for quick answers, which sometimes prevents me from fully understanding the material. I feel like it reduces my critical thinking skills when I depend on it too much."*
	Student 3	*"There have been instances where the AI gave me incorrect or incomplete information, and if I hadn't double-checked, it could have affected my assignments. It's frustrating when you can't fully trust the tool."*
2	Researcher	How has AI integration impacted your interaction with instructors and peers?
	Student 1	*"With AI being used more often, I feel like there are fewer opportunities to ask my educator questions directly. It's like AI is replacing some of the instructor's role, and that makes it harder to have meaningful discussions in class."*
	Student 2	*"Sometimes AI tools are helpful for collaboration, like group projects, but I've also noticed that it makes some students more isolated. We tend to rely on the AI instead of discussing problems with each other."*
	Student 3	*"I think it's made communication less personal. If I use AI for feedback or assistance, I don't feel like I'm getting the same level of guidance that I would from a lecturer or classmate."*
3	Researcher	What difficulties do you face in trusting AI to provide accurate or unbiased information?
	Student 1	*"I find it hard to trust the AI when it comes to nuanced topics. Sometimes, it seems biased or doesn't fully understand the context of the question, which leads to skewed information."*
	Student 2	*"The lack of transparency about how AI generates answers bothers me. I don't always know where the information comes from or how reliable it is, so I have to verify it with other sources."*
	Student 3	*"AI sometimes simplifies information too much, and I worry that it misses out on important details. I feel like I have to cross-check everything with my lecture notes, reference book or lecturer, which makes me hesitant to trust it fully."*

According to Table 1, the research findings highlight several challenges faced by higher education students in integrating AI into classroom learning. One of the key challenges is the misalignment between AI tools and specific course content, as students found that these tools often lack the customization necessary to fully complement the curriculum. This disconnect sometimes led to difficulties in understanding course materials. Additionally, students expressed concern that over-reliance on AI for quick answers can hinder the development of critical thinking skills. There were also frustrations regarding the reliability of AI tools, with students reporting instances of incorrect or incomplete information, underscoring the need to verify AI-generated content.

Another significant challenge relates to the impact of AI integration on student interactions with instructors and peers. Students felt that AI has reduced direct engagement with lecturers, diminishing opportunities for meaningful discussions and personalized feedback. While AI can facilitate collaboration, some students noted that it can also foster isolation, as reliance on AI tools limits peer-to-peer problem-solving and communication. Furthermore, trust in AI was a recurring concern, with students expressing doubt on the accuracy and bias of AI-generated information. Many felt the need to cross-check AI responses with other sources due to a lack of transparency in how AI systems produce answers, making it difficult to fully trust the tools in their learning process.

Acceptance of AI in classroom learning

Regarding students' acceptance of AI integration in classroom learning, the participants were asked to identify the aspects of AI they find most beneficial in the field of education. Additionally, they were asked to share their opinions on whether the benefits of AI outweigh its potential drawbacks in educational settings, as well as their trust in AI tools to provide accurate and reliable information. The details of the questions and responses are provided below.

Table 2. Students' acceptance of AI in classroom learning

No.	Role	Questions
1	Researcher	What aspects of AI in education do you find most advantageous?
	Student 1	*"I really appreciate how AI helps personalize my learning experience. It adapts to my pace, focusing more on the areas where I struggle, and that's really helpful in understanding complex topics."*
	Student 2	*"The AI tools we use provide immediate feedback on assignments and quizzes, which is incredibly useful. I don't have to wait for a lecturer to correct my work, so I can learn from my mistakes right away."*
	Student 3	*"I like how AI tools help in organizing study materials and automating tasks like generating study guides or summaries of readings. It saves time, allowing me to focus on preparing for exams and completing assignments."*
2	Researcher	In your opinion, do the benefits of AI outweigh its potential drawbacks in educational settings?
	Student 1	*"Yes, the benefits definitely outweigh the drawbacks. AI has made learning more flexible, and the personalized assistance I get is really effective in improving my performance."*
	Student 2	*"I'm on the fence. AI has made studying more efficient, but I worry about becoming too dependent on it, and sometimes it lacks the human insight that I get from my lecturers."*
	Student 3	*"Not entirely. While AI offers a lot of convenience, I'm concerned about the accuracy of the content and how it can sometimes oversimplify concepts that require deeper understanding."*

continued on following page

Table 2. Continued

No.	Role	Questions
3	Researcher	Do you trust AI tools to deliver accurate and reliable information?
	Student 1	*"Generally, I trust AI tools for basic tasks like summarizing articles or fact-checking, but for complex topics, I prefer to verify the information with additional resources or ask my lecturers."*
	Student 2	*"I trust AI for quick answers and help with assignments, but I always double-check the information, especially when it comes to important concepts or projects."*
	Student 3	*"Not fully. AI can be useful, but there have been times when the information it provided wasn't entirely accurate or didn't cover the full depth of what I was looking for, so I still rely heavily on traditional sources."*

By referring to Table 2, the results reveal that students perceive various advantages of AI in education, particularly in how it personalizes the learning experience. Students appreciate that AI adapts to their individual learning pace, helping them focus on areas where they struggle, which enhances their understanding of complex topics. Another major benefit is the immediate feedback provided by AI tools on assignments and quizzes, allowing students to learn from their mistakes in real time. Additionally, AI helps in organizing study materials and automating tasks like creating study guides, which saves time and enables students to focus more on exam preparation and project completion.

However, students express mixed opinions on whether the benefits of AI outweigh its potential drawbacks. While some view AI's personalized assistance as significantly improving their academic performance, others worry about becoming overly dependent on it, noting that AI lacks the human insight that professors provide. Concerns also arise regarding the accuracy and depth of AI-generated information, with students often feeling the need to verify content with other sources. Although AI is generally trusted for basic tasks like summarizing or fact-checking, students remain cautious, especially with more complex subjects, preferring to rely on traditional sources and their instructors or lecturers for deeper understanding.

DISCUSSION

Challenges of AI in Classroom Learning

The integration of artificial intelligence (AI) in classroom learning presents several challenges that can impact both student learning experiences and the overall educational process. One of the major issues is the lack of customization and alignment of AI tools with specific course content. According to Luckin (2019), many AI tools used in education are developed with a general framework, leading to challenges in adapting them to individual course requirements and the diverse needs of students. This lack of specificity can create a disconnect between AI-generated content and what instructors are teaching, ultimately affecting student comprehension. Furthermore, students have reported that relying too much on AI for quick solutions can undermine critical thinking and problem-solving skills. This aligns with findings from Holmes et al. (2020), which suggest that AI, while efficient, can inadvertently reduce deep engagement with material when overused.

Another significant challenge is the trustworthiness and accuracy of AI-generated information. Students often face difficulties in trusting AI tools due to the lack of transparency in how data is processed and the potential for biased or incomplete information (Zawacki-Richter et al., 2020). This problem is compounded by AI's tendency to oversimplify complex topics, as noted by Alhumaid (2022), leading to concerns about the depth of understanding it offers. Additionally, the impersonal nature of AI feedback can reduce meaningful interaction between students and instructors, creating a barrier to deeper learning and communication (Selwyn, 2021). This detachment from human instruction can obstruct collaborative learning, which is essential for developing soft skills such as communication and teamwork, as identified by Heffernan (2023).

Acceptance of AI in Classroom Learning

Apart from challenges, in classroom learning, AI has shown clear advantages, particularly in personalizing educational experiences. Students highlighted how AI tailors their learning, offering personalized assistance by adapting to their pace and focusing on areas of difficulty. This aligns with findings by Ayeni et al. (2024), who noted that AI enables real-time data analysis, allowing educators to gain valuable insights into student engagement, understanding, and areas where they may be struggling. Immediate feedback from AI tools was also considered beneficial by participants, echoing research from Bui and Stracke (2021), which emphasizes that real-time feedback promotes self-regulation and enhances academic performance. Moreover, the ability of AI to automate tasks like organizing study materials or generating summaries contributes to better time management, allowing students to concentrate on more critical learning tasks.

While the advantages of AI in education are widely acknowledged, there are concerns about over-reliance on these technologies and the potential drawbacks. One participant pointed out that while AI has made learning more efficient, it can sometimes lack the depth of human insight provided by professors. This concern is supported by Selwyn (2020), who discusses the risk of AI reducing complex subjects to oversimplified solutions, potentially limiting critical thinking skills. Other students expressed concerns about dependency on AI tools, mirroring similar anxieties discussed by Luckin et al. (2021), where excessive dependence on AI systems could lead to diminished problem-solving capabilities in students. Hence, while AI provides flexibility and convenience, it must be implemented alongside human mentorship to maintain educational integrity.

Trust in AI tools remain an important consideration for students. Although some trust AI for basic tasks like summarizing articles and fact-checking, others raised concerns about its reliability, particularly for complex topics. This scepticism is consistent with broader discussions in AI ethics, where issues of transparency and accuracy in AI-driven educational tools have been explored (Zawacki-Richter et al., 2020). As students note, cross-referencing AI-generated content with traditional resources remains crucial to ensure a comprehensive understanding of academic material. Thus, while AI can serve as a valuable supplement to learning, students continue to value human input and traditional sources for deeper insights.

CONCLUSION AND SUGGESTIONS

In conclusion, the integration of AI in classroom learning presents both significant opportunities and notable challenges. The acceptance of AI among students is largely influenced by its ability to personalize learning experiences, provide instant feedback, and automate routine tasks, which enhances overall efficiency. However, concerns regarding over-reliance on AI, the accuracy of information provided, and the potential for AI to oversimplify complex subjects remain prevalent. These concerns highlight the importance of balancing AI's use with traditional learning methods to ensure a comprehensive and deep understanding of educational content. While students recognize the benefits of AI, their trust in these tools is not absolute, particularly in areas requiring critical analysis and human insight.

To address the challenges and enhance AI's effectiveness in education, it is essential for educational institutions to invest in training both students and educators on the best practices for utilizing AI tools. Additionally, continuous improvements in AI technology are needed to ensure accuracy and depth in the information it provides. Encouraging collaboration between AI developers, educators, and students can help create AI systems that are better tailored to the unique demands of the educational environment. Finally, a hybrid model that combines the strengths of AI with the irreplaceable value of human interaction may provide the most effective and balanced approach to AI integration in the classroom.

REFERENCES

Akgun, B., & Greenhow, C. (2022). Addressing ethical implications of AI in educational contexts. *Journal of Educational Technology*, 55(3), 245–260.

Alhumaid, K. (2022). The role of AI in higher education: Opportunities and challenges. *Journal of Educational Technology*, 33(2), 105–120.

Ayeni, O. O., Al Hamad, N. M., Chisom, O. N., Osawaru, B., & Adewusi, O. E. (2024). AI in education: A review of personalized learning and educational technology. *GSC Advanced Research and Reviews*, 18(2), 261–271. DOI: 10.30574/gscarr.2024.18.2.0062

Borenstein, J., & Howard, A. (2021). The ethics of AI in education: A critical review. *AI & Society*, 36(1), 1–12.

Bui, A., & Stracke, C. M. (2021). The impact of immediate feedback on student engagement in AI-based learning environments. *Journal of Educational Technology Research*, 42(2), 225–240.

Chan, K. S., & Zary, N. (2019). Challenges of implementing AI in medical education: A review. *Medical Education*, 53(8), 775–785.

Cui, X., Xue, F., & Thai, T. (2019). The impact of AI-driven learning technologies on educational outcomes. *Computers & Education*, 141, 103620.

Eguchi, A., Okada, H., & Muto, Y. (2021). Contextualizing AI education for K-12 students to enhance their learning of AI literacy through culturally responsive approaches. *Kunstliche Intelligenz*, 35(2), 153–161. DOI: 10.1007/s13218-021-00737-3 PMID: 34376922

Heffernan, N. (2023). AI in education: Bridging the gap between innovation and application. *Educational Research and Innovation*, 27(1), 88–102.

Holmes, W., Bialik, M., & Fadel, C. (2020). *Artificial intelligence in education: Promises and implications for teaching and learning*. Routledge.

Huang, Y., Saleh, M., & Liu, Y. (2021). AI in education: Enhancing learning experiences and outcomes. *Journal of Educational Computing Research*, 59(1), 56–75.

Jaiswal, A., & Arun, C. (2021). Transforming education with AI: Potential and challenges. *Educational Technology Research and Development*, 69(2), 1–20.

Kiger, M. E., & Varpio, L. (2020). Thematic analysis of qualitative data: AMEE Guide No. 131. *Medical Teacher*, 42(8), 846–854. DOI: 10.1080/0142159X.2020.1755030 PMID: 32356468

Kim, J., Park, H., & Lee, C. (2020). Students' perceptions of AI in education: Implications for adoption. *International Journal of Educational Technology in Higher Education*, 17(1), 48.

Labadze, M., Grigolia, M., & Machaidze, R. (2023). AI chatbots in education: Real-time feedback and efficiency. *Interactive Learning Environments*, 31(2), 217–232.

Lee, I., & Perret, B. (2022). Preparing high school teachers to integrate AI methods into STEM classrooms. *Proceedings of the AAAI Conference on Artificial Intelligence*, 36(11), 12783–12791. DOI: 10.1609/aaai.v36i11.21557

Lo, H. (2023). The potential of ChatGPT in enhancing active learning in higher education. *AI in Education Journal*, 5(1), 34–45.

Luckin, R. (2019). The impact of AI on student learning outcomes: What we know and what we need to know. *AI and Education Journal*, 19(3), 45–60.

Luckin, R., Holmes, W., Griffiths, M., & Forcier, L. B. (2021). AI in Education: A Systemic Perspective. *AIED Journal*, 31(4), 239–248.

Sanusi, I. T., & Oyelere, S. S. (2020). *Pedagogies of machine learning in K-12 context. In 2020 IEEE frontiers in education conference (FIE)*. IEEE.

Selwyn, N. (2020). *Should Robots Replace Teachers? AI and the Future of Education*. Polity Press.

Selwyn, N. (2021). *AI in education: The challenges of classroom integration*. Oxford University Press.

Zawacki-Richter, O., Marín, V. I., Bond, M., & Gouverneur, F. (2020). Systematic review of research on artificial intelligence applications in higher education. *International Journal of Educational Technology in Higher Education*, 17(1), 1–27.

Chapter 6
Creating Immersive Learning Experiences in Education Domain Through Virtual and Augmented Reality

Manpreet Kaur Riyat
https://orcid.org/0000-0002-1190-0819
Lovely Professional University, India

Ruchi K. Kakkar
Lovely Professional University, India

Amit Kakkar
Lovely Professional University, India

ABSTRACT

This chapter examines the transformation of education with Virtual and Augmented Reality (VR and AR), highlighting their remarkable capability to create immersive learning experiences that significantly enhance student engagement and knowledge retention. AR and VR technologies, with their interactive and dynamic interfaces, overcome traditional educational challenges, thereby enhancing the inclusivity and personalisation of learning. The chapter analyses the current state of the education system, presents use cases of AR and VR in education, and discusses the steps to incorporate AR and VR into educational settings for immersive learning experiences. As AR and VR progress, their integration into education offers substantial prospects to connect theoretical knowledge with practical implementation, establishing these

DOI: 10.4018/979-8-3693-9811-1.ch006

technologies as crucial catalysts for the future of learning.

1. INTRODUCTION

Technological innovations have significantly impacted the education sector, transforming traditional teaching methods into engaging and interactive learning experiences (AlGerafi et al., 2023; Almufarreh & Arshad, 2023). Growing technological developments have facilitated the widespread use of technology-enhanced learning in education and sparked a surge in research on this subject matter. This growing body of research underscores the increasing importance of technology in education (Lampropoulos et al., 2024). Transformative technologies such as Augmented Reality (AR) and Virtual Reality (VR) are not just tools, but they are leading the way in a new era of education. Their transformative nature indicates the potential to impact how students learn and engage with their subjects significantly (AlGerafi et al., 2023). These technologies can transform educational environments by providing immersive and interactive learning experiences that surpass the constraints of traditional classroom settings. VR generates a fully simulated environment that accurately replicates real-life situations or generates entirely novel ones, enabling students to investigate and engage with their subjects within a three-dimensional realm. AR superimposes digital information onto the environment, enriching real-world encounters with supplementary data, visuals, and multimedia.

Incorporating AR and VR into education amplifies student engagement and motivation and addresses students' various learning styles and demands. Visual and kinesthetic learners can derive advantages from the tactile involvement offered by these technologies, resulting in enhanced recall and comprehension of intricate concepts (Liu et al., 2024). Moreover, AR and VR can enhance experiential learning by providing students with a safe and regulated environment to practice skills and apply knowledge (Asad et al., 2021). As technology advances, educators and researchers investigate ways to integrate AR and VR into curricula. Research has demonstrated that these technologies have the potential to boost spatial awareness, promote collaboration, and offer immediate feedback, thus facilitating a more dynamic and participatory learning experience (Martín-Gutiérrez et al., 2015; Molina-Carmona et al., 2018).

Virtual Reality (VR) and Augmented Reality (AR) have established themselves as effective teaching aids, providing engaging and dynamic learning environments that are frequently absent from conventional approaches. Students can examine 3D human body models and see organs and systems up close in courses like biology or anatomy by donning virtual reality headsets. Augmented reality applications can improve language acquisition by superimposing translations and providing real-time interactive vocabulary instruction in the student's surroundings. Through virtual reality (VR) simulations, students can practice languages in authentic environments without leaving their classrooms, making communication skills immersive and valuable. Students can use AR apps that bring historical personalities to life with stories and interactive questions, or they can use VR recreations of historical places to "visit" them instead of reading about them. This method improves recollection and emotional connection by making historical situations more exciting and memorable. Students can conduct chemistry experiments or investigate physics simulations in secure, interactive virtual labs based on virtual reality technology. This fosters a more profound understanding, conserves resources, and lowers safety hazards compared to actual labs. Using gamified augmented reality applications, students can work together to tackle challenging, scenario-based coding, math, or critical thinking tasks. Students gain problem-solving abilities faster when they engage with real-time data and outcomes than when they engage in static case studies. Through customised simulations that provide visual, aural, and kinesthetic signals, AR and VR technologies can meet various learning demands, including those of people with disabilities, enhancing learning outcomes, engagement, and confidence. Virtual field trips to museums, other nations, or cultural sites are made possible by AR and VR, which enable immersive learning experiences that are not possible with regular classroom instruction. Engaging in such realistic simulations increases students' likelihood of retaining and applying what they have learned.

As society progresses in the digital technology era, AR and VR in education will continue to expand. Ongoing technological breakthroughs and improved availability will support this growth. Institutions worldwide are increasingly acknowledging the worth of these immersive technologies and allocating resources towards pilot projects and research activities to explore their capabilities thoroughly. The future of education may involve the conversion of classrooms into interactive, virtual environments where individuals from different countries can connect and work together in real-time without being limited by geographical constraints. By using AR and VR, educators may improve the learning experience and equip students for a future when digital literacy and technological aptitude are crucial. This book chapter will further discuss the present state of the education system, challenges and issues presented in the education system, different technological advancements and the role of AR and VR in creating immersive experiences. It will offer valuable knowledge and practical advice for educators, policymakers, and researchers who want to utilise these technologies to develop immersive learning experiences in education.

1.1. Present Education System

Global education varies due to geographic, economic, and social considerations. Digital transformation in the education domain is one of the phenomena that has changed how education is disseminated around the globe. Because of digital transformation, online learning systems have grown, especially after the COVID-19 epidemic, and the same has forced a quick change, bringing remote schooling into the picture to educate individuals. This transition led to blended learning approaches integrating face-to-face and online training. As educators attempt to prepare kids for a technologically driven society, digital literacy is becoming more critical.

Nowadays, researchers and practitioners often use “Blended learning”, although its definition is still unclear. Oliver and Trigwell (2005) stated that the blended learning approach combines two or more different education approaches; this may involve web-based technologies, pedagogical approaches, adaptive learning, and instructional technologies. Thus, the present education system is a blend of traditional practices and modern innovation, providing a flexible approach to meeting the evolving expectations of learners and society (Kumar et al., 2021). The primary objective is to give students an extensive base of information and skills, equipping them for achievement in both personal and professional realms. Traditional components, such as well-organized curricula, in-person teaching, and standardised assessments, remain crucial in maintaining educational benchmarks and responsibility (Vallée et al., 2020).

The rapid advancement of technology has significantly influenced everyday life and the establishment of a digitalised society (Puncreobutr, 2016). The emerging digital culture has a fundamentally distinct perception of time and space, together with an abundance of knowledge that seems almost limitless (Mukul & Büyüközkan, 2023). At the same time, the education system has progressively integrated to incorporate technology to augment learning experiences and adapt to current demands. The present education system is undergoing a significant change by digitisation, fundamentally changing how education is provided and experienced. Classrooms are becoming more technologically advanced by integrating digital resources like interactive whiteboards, tablets, gamification, and online learning platforms (Barneva et al., 2017). This transformation provides students with convenient access to a wide variety of information and resources, adaptive learning, and, most importantly, allows for flexible, remote education that was previously impossible (Dumitru, 2023). Online learning platforms, interactive software, and educational apps are now widely used digital tools and resources. These advancements promote learning preferences, enhance understanding, and allow customised learning and teaching. Technology is playing a more significant part in India's education system. The government and business sectors are allocating resources towards digital transformations, e-learning, and technology advancements such as AR and VR in education and online courses to improve educational outcomes and create immersive learning experiences.

2. ISSUES IN THE PRESENT EDUCATION SYSTEM

Despite these advances, access and equity remain issues. Access to technology and the Internet is still a significant issue for low-income and rural students. The promotion of inclusive education aims to incorporate disabled and marginalised students. However, there are significant differences in education quality and access between wealthy and developing countries, underscoring the need to address global imbalances. The education system is changing rapidly due to technological advances and a greater awareness of equality and inclusivity. However, securing equal access to quality education for all students, regardless of socioeconomic background or geographic location, remains challenging. Innovation, investment, and a commitment to an effective and fair education system are needed to address these difficulties.

Figure 1. Issues in the Education System

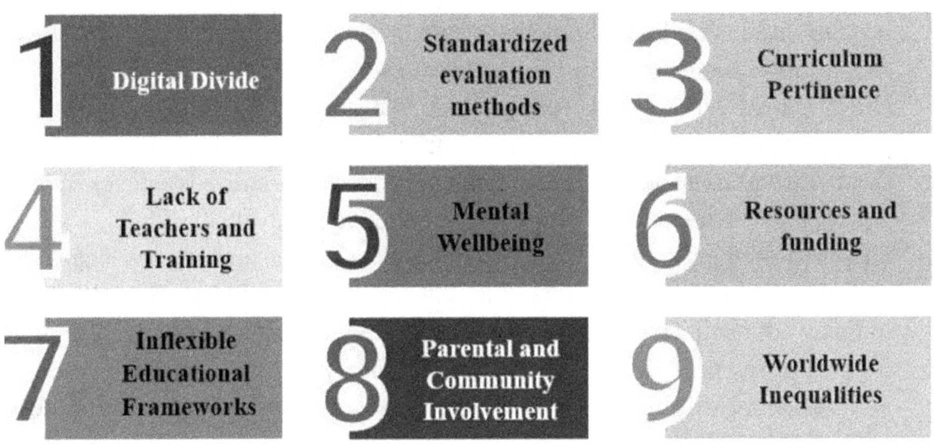

a) **Digital Divide:** Due to events like Covid 19, a significant shift in the educational sector can be seen from manual to digital learning platforms. It has led to reducing the "digital divide" among learners, referred to as the gap between people who have access to modern information and communication technology and those who don't (Cheshmehzangi et al., 2023). But, even today, many factors lead to the digital divide among the learners. Many learners from remote or low-income households do not have access to laptops, tablets, or cell phones, which are essential technologies for online learning (Lembani et al., 2020), making the students struggle to participate fully in online learning environments. Fast-speed internet is required for successful and dependable online learning. Nonetheless, poor connectivity might make it difficult for students in marginalised areas to participate in online courses, prepare homework assignments, or access online resources. The digital divide exacerbates disparities in academic achievement. Its existence exacerbates success gaps; students with little exposure to technology may lag behind those with more effective technological resources.

b) **Standardized Evaluation Methods:** Although standardised evaluation has been a mainstay of educational evaluation for decades, it has been controversial for several reasons. Standardised evaluations frequently concentrate on knowledge domains and topics such as arithmetic and comprehension of texts. However, other crucial components of learning, such as innovative thinking, problem-solving ability, and mental agility, might not be sufficiently assessed using these evaluation parameters. Similarly, teachers may prioritise "teach to the test" rather than a more comprehensive and all-encompassing approach to education to achieve standardised test results (Correia, 2020). It may restrict kids from being exposed to balanced curricula and impede their capacity to acquire various abilities. Students may experience severe anxiety and stress as a result of high stakes for standardised testing (Kubiszyn & Borich, 2024). Due to parents, society, and peer pressure, the stress of excelling in exams can harm learners' mental wellness and their entire educational journey. Students from diverse ethnic and socioeconomic groups may face disadvantages if their learning styles or experiences are not reflected in the evaluations.

c) **Curriculum Pertinence:** Rapid technological advancements and changing societal needs cannot maintain traditional school curricula. Traditional curricula may not adequately prepare students for the job market's needs, placing a higher priority on technological proficiency, analytical skills, problem-solving abilities, and flexibility. Analysing data, innovative problem-solving, and developing technology are becoming essential abilities that may not be adequately addressed in conventional education systems (Mauki et al., 2020). On the side, many colleges and universities still strongly emphasise mechanical learning and traditional academic courses, which may not necessarily incorporate application in the real world and practical abilities. A slow-moving curriculum may omit essential elements like media competence, virtual citizenship, and the capacity to learn about and adjust to novel information sources and technology. Modifying the curriculum to reflect current knowledge and abilities takes time and money (Ghafar, 2020). Moreover, strict assessments and learning standards can be challenging for schools to balance incorporating new and developing topics and covering old academic curricula. As there are diverse learners, it becomes imperative for educators to ensure the academic program's appropriateness for learners with a range of passions and past experiences (Guangul et al., 2020). Education systems must be adaptable and accessible, providing possibilities and career trajectories that accommodate a range of learning styles and objectives (Vreuls et al., 2022).

d) **Lack of Teachers and Training**: The lack of professional development for teachers and the shortage of teachers, especially in the fields of Science, Technology, Engineering, and Mathematics (STEM), are two of the main reasons for slow changes in the education domain. Many regions have a dearth of competent STEM educators, negatively impacting students' access to high-quality instruction in critical subject areas. Low pay, stress, and a lack of financial motivation for professionals to enter or remain in the teaching industry are reasons for this scarcity. There is a lack of opportunities for the continuous professional development of teachers, which is essential for keeping up with new pedagogies, educational technologies, and changing curricular standards (Sancar et al., 2021). They require resources, support, and formal training to successfully adopt new methods and technologies. Inadequate assistance may result in exhaustion and a lack of commitment when using novel teaching methods. Lack of professional development and non-addressing teacher shortages can also lower teachers' motivation and retention (Alexander et al., 2020). Offering possibilities for career progression, acknowledgement, and support can create a more dependable and productive teaching workforce.

e) **Mental Wellbeing:** It is becoming more widely acknowledged that student's mental wellness is essential to their overall achievement in school. Students' levels of stress and anxiety are on the rise due to a variety of factors, including societal demands, stress from their studies, and worries about their futures (Jiang et al., 2021). These problems can be made worse by students' modern education requirements, outside influences like networking sites, and families' expectations. Many schools are underequipped to deal with mental health issues effectively as they have insufficient availability of mental health services, experts in psychology, and educational counsellors (Oke & Fernandes, 2020). Due to a lack of funds and resources, schools may find it challenging to incorporate mental health support into their regular operations. Mental health concerns can significantly impact students' capacity to concentrate, participate, and achieve academic success. Addressing mental health issues have become crucial for students' overall growth, academic success, and general well-being. Having thorough mental health support systems in schools is critical. It entails implementing school-wide mental health programs, training teachers to identify and handle mental health concerns, and ensuring that children can get professional assistance (McGorry et al., 2022).

f) **Resources and Funding:** Due to lack of funding in Education, many educational institutions, especially those in underprivileged or economically disadvantaged areas, struggle financially, which affects their capacity to deliver high-quality education (Tyner, 2023). Older or insufficient textbooks and educational resources frequently result from inadequate funding. It may be difficult for educators to give pupils the most recent materials needed for successful learning, which might negatively impact students' educational experiences. Underfunded schools may have restricted access to computers, software, and internet connectivity. Due to the lack of technology, students may struggle to interact with digital learning resources and acquire critical digital skills. Inadequate financing can result in schools having deteriorating physical spaces that do not support a secure and effective learning environment. Problems like packed classrooms, poor upkeep, and dangerous infrastructure can impact employees and children. Teacher workloads and school enrolments are increasing day by day (Farrie, 2022). Overworked but less paid instructors may struggle to give each student the personalised care and assistance they need. Excessive class sizes may also affect student participation and the general standard of instruction. Budgetary restrictions may make it more difficult for schools to provide a wide range of recreational activities and program options, negatively affecting students' overall educational experiences and possibilities for personal development (Ortagus et al., 2020).

g) **Inflexible Educational Frameworks:** The established structures and practices of many education systems can make them resist change, which makes it difficult for them to adopt novel approaches to teaching and technological advancements. It is possible that educational materials and methods of instruction from the past are outdated and do not consider contemporary teaching strategies that emphasise solving problems, innovation, and analytical thinking (Whalley et al., 2021). Strict adherence to old-fashioned methods might stifle creativity and restrict students' capacity to acquire necessary skills. Certain educational systems move slowly when new technologies are incorporated into the classroom. A conservative approach to educational change, a lack of resources, or inadequate teacher preparation may cause this resistance. Schools may struggle to adapt frameworks to their students' varied requirements and interests because of their inflexible behaviour towards adopting the changes (Arán Filippetti & Krumm, 2020). Educational institutions might not fully assist all students in realising their full potential if they do not adjust to various learning styles or the unique needs of each student. Due to an inflexible educational system, students may find it challenging to discover and develop their abilities and interests. Creativity in teaching and learning is crucial to engaging students and preparing them for a rapidly changing world.

h) **Parental and Community Involvement:** Improving educational results requires strong collaboration across communities, schools, and parents. Children are more likely to be motivated and encouraged when community members and parents actively participate in their educational process (Kelty & Wakabayashi, 2020). Engagement can result in improved behaviour, enhanced academic performance, and a more profound feeling of community inside the school. Time constraints, a lack of resources, and cultural differences can make it difficult for parents and the community to get involved in their children's education. Establishing strong relationships between schools and local businesses may be complex, which could affect student support services. Schools may better utilise their resources, knowledge, and assistance by collaborating with their communities to improve learning possibilities. It can involve collaborations with nearby companies, associations, and volunteers who support the learning process. One way to increase parental and community engagement is to create chances for meaningful involvement, such as parent-teacher conferences, community events, and cooperative programs. Schools might also offer tools and assistance to encourage parents to participate more actively in their children's education.

i) **Worldwide Inequalities:** Global education systems differ significantly in accessibility and quality. These worldwide differences can have significant ramifications and reflect more considerable socioeconomic disparities. In certain nations, educational systems offer top-notch instructions supported by ample resources, qualified instructors, and state-of-the-art facilities. On the other hand, other countries struggle with issues like low literacy rates, poor infrastructure, and restricted access to educational resources. Global educational disparities are frequently associated with more generalised socioeconomic inequality. Differing regions may have differing levels of accessibility and the standard of education due to factors including poverty, political unpredictability, and a lack of infrastructure. Many international organisations and projects aim to solve global educational inequities by giving underprivileged areas money, resources, and support. Initiatives include expanding educational infrastructure, strengthening teacher preparation programs, and expanding educational access. International cooperation and funding are necessary to address the inequities in education worldwide. To improve education globally, it is imperative to exchange best practices, fund educational development initiatives, and advocate for laws that deal with the underlying causes of inequality.

3. DIFFERENT TECHNOLOGICAL ADVANCEMENT IN EDUCATION SYSTEM

Figure 2. Technological Advancements in Education

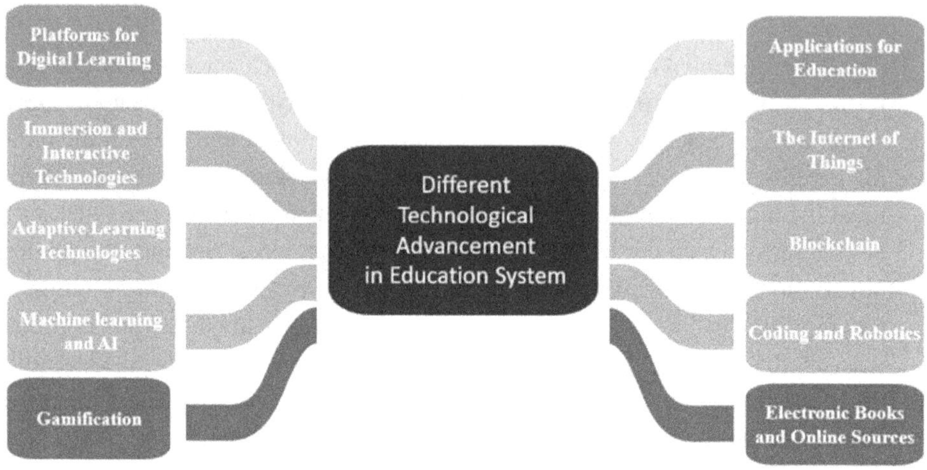

The education system has evolved significantly in the last decade, but the advent of COVID-19 has accelerated the pace of change. The following section discusses the recent technological advancements in the education system.

a) **Platforms for Digital Learning:** Many platforms like Moodle, Canvas, and Blackboard have been conceptualised and developed that offer a centralised location for assignments, grades, and student-teacher communication (Bujang et al., 2020). They support blended learning, which combines traditional classroom instruction with internet resources for students to access. Apps like Mathway for problem-solving math problems, Khan Academy for various courses, Duolingo for language learning, and many others provide learners with interactive and customised learning experiences. These resources accommodate various learning requirements and styles. Platforms like Zoom, Microsoft Teams, and Google Meet make possible real-time virtual classrooms that facilitate distance learning and collaboration with learners who have joined the platforms from far-off places. These platforms facilitate collaborative projects, live lectures, and conversations, increasing accessibility to education regardless of geography. This has also helped reduce the cost of space and physical facilities in offline classrooms, helping even low-funding institutions increase student footfall via online classes.

b) **Immersion and Interactive Technologies:** Apps with Augmented Reality (AR) capabilities improve learners' learning experiences by superimposing digital data over the actual world. For example, students can use augmented reality apps to engage with 3D models of anatomical structures, visualise complex scientific concepts, and tour historical sites (Binytska et al., 2020). Similarly, Virtual Reality (VR) produces immersive settings for hands-on education. Virtual reality (VR) solutions such as Google Expeditions and Oculus Education enable interactive scenarios, simulations, and virtual field trips that offer experiential learning opportunities in a safe and controlled virtual environment.

c) **Adaptive Learning Technologies:** Data analytics can be used to develop adaptive learning systems that help customise course materials to meet the requirements and skill levels of specific students (Morze et al., 2021)Learning problems and gaps can be addressed with programs such as DreamBox and Smart Sparrow, which modify the type and level of content according to student performance. Interactive tutoring systems use student interactions to deliver tailored advice and feedback. Such systems understand the students' online queries and respond in real time. Carnegie Learning and the Knewton platform are examples of how algorithms customise education and offer focused support.

d) **Machine learning and artificial intelligence (AI)**: Algorithms and data are used in machine learning and artificial intelligence (AI) to improve educational procedures. AI is capable of intelligent teaching, individualised learning experiences, and task automation, such as AI chatbots, which can respond to students' questions. Squirrel AI and other AI-based systems offer individualised learning programs, promoting students' overall development and mental growth. Machine learning algorithms examine student data to find patterns, forecast results, personalise learning experiences, modify content according to students' performance and learning preferences, and provide tailor-made solutions as per the student's requirements.

e) **Gamification**: Gamification is the process of adding elements of game design to educational sessions to boost motivation and engagement (Pakinee & Puritat, 2021). Learning games, like Prodigy Math and Kahoot!, use game elements to impart knowledge engagingly and dynamically (Fuchs, 2022). These games frequently have challenges, levels, and awards to encourage pupils. Apps such as Class Craft incorporate gaming components into classroom management, giving students points and incentives for good behaviour and academic performance. This method increases the interaction and engagement of learning. Incorporating gamification elements like leaderboards, rewards, progress bars, competitions, badges, and other elements helps increase the engagement levels of learners with educational platforms.

f) **Learning Management Systems (LMS):** Learning Management Systems (LMS) are software platforms that track students' progress and oversee the management and delivery of instructional materials. These platforms lower the manual work of teachers and help them maintain proper digital records of students. LMSs allow teachers to design and oversee classes, assign reading materials, and support distance learning. They also offer resources to monitor students' advancement, organise homework, and produce reports on their scholastic achievements. Forums, chat rooms, and virtual classrooms are a few tools being frequently included in LMS platforms to facilitate communication between students and teachers.

g) **Applications for Education**: Educational apps are called smartphones or web applications that assist teaching and learning and are being developed with different functionalities. The Khan Academy app provides interactive lessons, practice questions, and tutorials for various subjects, and the Duolingo app helps in language learning (Nami, 2019). Apps like Code.org and Tynker teach coding and computational thinking through engaging activities and games. Similarly, Flashcard apps like Anki and Quizlet help students review and memorise information using digital flashcards and practice tests.

h) **The Internet of Things:** "Internet of Things" (IoT) describes a network of interconnected devices that exchange data and converse online (Rana et al., 2021). By creating teaching spaces that are more pleasant and interactive, IoT-enabled gadgets like lighting controls, temperature controls, and smart whiteboards can improve the student learning environment (Al-Emran et al., 2020). Wearable technology, such as fitness trackers, can track students' physical and mental wellness and provide information that can be used to encourage well-being and participation in physical education. Similarly, Internet of Things (IoT) instruments such as sensors and smart lab equipment can access real-time data and interactive science experiments, enhancing student engagement.

i) **Blockchain**: Blockchain is a technology for decentralised ledgers that can safely store and validate data and transactions (Ali et al., 2023). It can produce tamper-proof records of certifications, degrees, and other educational credentials and securely and transparently manage and share student records and achievements across institutions.

j) **Electronic Books and Online Sources**: In today's ever-evolving world, updated books and curricula are the need of the hour as hard books cannot be printed and updated frequently (Haleem et al., 2022). Digital resources and e-books are electronic versions of teaching aids and additional resources that can be easily updated periodically. Textbooks and other works of literature are available digitally on devices like tablets and e-readers. E-books have interactive elements like multimedia, search capabilities, and hyperlinks, improving learners' engagement levels. These technological developments improve education by increasing accessibility, personalisation, and interaction in the classroom. They assist in addressing various issues that students and teachers encounter and promote various educational objectives.

4. ROLE OF AUGMENTED REALITY AND VIRTUAL REALITY IN EDUCATION DOMAIN

In recent years, the adoption of emerging technologies in education, such as augmented reality (AR) and virtual reality (VR), has gained significant popularity. AlGerafi et al. (2023) and Liberati (2016) opined that AR integrates digital information into the actual environment, enabling users to interact with virtual aspects seamlessly without focusing on a device's screen. Whereas within the realm of education, VR pertains to using immersive digital environments and simulations to augment teaching and learning experiences. It enables students to actively participate in educational material through interactive and immersive means, frequently surpassing conventional instructional approaches (AlGerafi et al., 2023). AR/VR technologies have experienced notable progress in recent years and hold the potential to transform multiple domains, such as engineering, medicine, tourism, industry, entertainment and gaming, and education (Alvarez-Marin & Velazquez-Iturbide, 2021; Barteit et al., 2021; Dede et al., 2017; Yung & Khoo-Lattimore, 2019).

These technologies have transformed learning methods by providing immersive digital experiences, interactive environments, simulations, and increased engagement (Al-Ansi et al., 2023). They provide students with immersive digital experiences that cannot be replicated using conventional methods of learning (Phakamach et al., 2022). However, these technologies are still in the development phase and necessitate substantial investment and extensive customization to fulfil the significant demand in the field of education. Still, technologies hold the potential to improve the learning experience for students (Cevikbas et al., 2023; Graafland, 2018). AR/VR provides virtual field trips and simulations, offering immersive experiences without the necessity of actual transportation. Moreover, they serve as a connection between traditional classroom teaching and practical implementation in real-life scenarios, facilitating the growth of professional skills and equipping students for careers in technology-driven industries (Al-Ansi et al., 2023).

AR in Education and its Used Cases

AR is a recent advancement that is gaining significant attention (Koumpouros, 2024). AR overlays computer-generated graphics onto real-world situations in real time, effectively integrating the physical and digital realms (K. Lee, 2012). It is a complex collection of technologies that enable users to see a significant aspect of the natural world by utilizing technological devices that gather data from the surrounding environment, such as smartphones and tablets (Li et al., 2018). Also, it often overlays images, sounds, films, and graphics onto a specific scene (Martina Tritthart, 2021).

AR seamlessly combines the digital world with the real world, allowing students to perceive both digital and physical surroundings concurrently through devices such as smartphones and screens. AR improves real-world experiences by offering additional contextual information. Students can engage with tangible things enhanced with digital annotations or examine three-dimensional models integrated into their natural environment. AR enhances practical experiences by serving as an instructive layer, making it especially helpful for guided tours, interactive visualizations, and the incorporation of real-time data (AlGerafi et al., 2023). The implementation of augmented reality in education has benefitted by improving learning methods within the classroom (Moreno-Guerrero et al., 2020). For instance, AR has garnered significant attention in geometry education because it can explain fundamental operations, such as perimeters, areas, and diameters. Augmented reality revolutionizes education by enriching textbooks with interactive three-dimensional models and films, fostering captivating classroom experiences through interactive courses, and facilitating virtual laboratories for secure experimentation (Selvan et al., 2024).

The real-world application of AR extends beyond simply chasing Pokémon around the city or applying photo effects on Instagram and Facebook. Qulver Masks merges augmented reality coloring pages with face-tracking technology, enabling children to design interactive AR masks. Children can utilize their creativity and recount thrilling narratives by employing masks. The application provides an educational platform that allows educators and students to enhance their understanding and visualization of different learning topics. Another AR-based application called Catchy Words AR is an enjoyable augmented reality application designed for children to enhance their spelling abilities (Pereira et al., 2020). Through the application, children will be able to observe letters floating through the space surrounding them. Students must move their phones or tablets and adjust their positions to capture the letters and solve the word. This enhances their spelling proficiency and cognitive recall. Leometry is an educational game that aims to instruct elementary school pupils about fundamental geometric forms by immersing them in an adventure story in the African savannah (Laine, 2018). The narrative revolves around a mother leopard and her cub escaping from poachers, as players assist them in moving through challenges such as rivers infested with crocodiles and traps set by the poachers. The role of the dung beetle is to aid players by offering constructive comments and helpful tips on geometry problems. The game is centered around the Van Hiele model of geometry education, with a specific emphasis on the initial two levels: recognising fundamental shapes by their visual characteristics (Level 0) and examining objects by their attributes, such as sides and angles (Level 1). Leometry can be played individually or in pairs.

VR in Education and its Used Cases

Virtual Reality (VR) involves the use of immersive digital environments and simulations to enrich teaching and learning experiences (Beck, 2019). This technology enables students to interact with educational content more engagingly and experientially, often surpassing traditional instructional methods. Unlike AR, VR operates within a completely artificial environment, where users are either fully immersed or partially involved in the simulated world (AlGerafi et al., 2023). For instance, participants can interact with and manipulate computer-generated objects through haptic interfaces while fully engaged in the virtual environment. VR transforms education by establishing immersive learning environments that enable students to fully immerse themselves in educational content, such as navigating through ancient Rome or conducting in-depth exploration of the human body (Pellas et al., 2021). It also enables students to engage in practical training for areas such as medicine, engineering, and aviation by allowing them to simulate procedures or construct structures in virtual environments (Potkonjak et al., 2016). VR improves language acquisition by enabling students to engage in immersive experiences that resemble real-life situations (Natale et al., 2020). In addition, it facilitates special education by creating customised settings for learners with impairments, enabling them to engage in social interactions and encounter sensory-stimulating situations that conventional classrooms cannot provide (Monahan et al., 2008).

VR has various novel applications in education that greatly enhance the learning experience. Google Expeditions permits students to explore diverse worldwide locations, encompassing ancient ruins, submerged ecosystems and celestial realms (Parmaxi et al., 2017). These expeditions offer immersive experiences that augment conventional classroom learning. For instance, by utilising Google Expeditions, students can virtually explore ancient Rome and visit prominent historical sites such as the Colosseum, the Roman Forum, and the Pantheon. This interactive encounter enables students to vividly perceive and enhance their comprehension of Roman architecture, culture, and history, surpassing the limitations of relying just on textbooks. Similarly, Meta Quest is a collection of VR headsets that enable users to navigate, interact, and engage in gaming activities within virtual environments (Ko,

2024). Within an educational setting, the Meta Quest VR headsets provide profound learning opportunities, allowing students to interact with content in an immersive environment (Sadler & Thrasher, 2023). This allows students to participate in immersive learning experiences that can profoundly impact their understanding and engagement with educational content. These autonomous devices enable users to navigate virtual environments independently without requiring a linked personal computer, allowing users to engage in various activities, such as virtual field trips, interactive simulations, and collaborative projects (Šašinka et al., 2018). Unimersiv provides a wide variety of educational opportunities, enabling students to delve into intricate topics more captivating and immersively (Lee & Kim, 2023). The software provides immersive experiences in human anatomy, allowing students to engage with 3D representations of organs and systems. It also offers virtual tours of past civilisations such as Greece and Rome, making history and science more vivid and tangible. The design of this educational tool is to accommodate students of various age groups, enhancing the learning experience through dynamic and memorable activities that enable direct engagement with the material in a virtual setting. Moreover, Labster is a Denmark-based company that creates interactive laboratory simulators facilitating unrestricted studies. It provides virtual laboratory simulations encompassing a wide range of scientific fields, including biology, chemistry, and mathematics (Tsirulnikov et al., 2023). These simulations enable students to perform experiments and investigate scientific concepts within a secure and regulated virtual setting (Navarro et al., 2024).

Combined AR and VR Experiences in Education

Integrating AR and VR allows hybrid learning models to be implemented, which provide students with a holistic educational experience (Richards, 2023). For instance, students can utilise AR to identify and name various components of an actual plant, while VR can be used to immerse them in a virtual rainforest for exploration. In addition, AR and VR enhance collaborative learning by enabling students to collaborate in virtual environments, fostering cooperation and worldwide connections (Tan et al., 2022). As an illustration, students from various regions across the globe might convene in a digital environment to engage in collaborative endeavours, enhancing their educational journey through shared encounters and varied viewpoints.

5. CREATING IMMERSIVE EXPERIENCE

An immersive experience is characterised by an individual's deep engagement and complete absorption in an area or activity, frequently to the point that they perceive themselves as an integral part of it (Han et al., 2024). This type of experience is intended to engage the senses, blur the boundaries between the actual world and the simulated or augmented environment, and create a sense of immersion for the participant rather than just passive observation (Dincelli & Yayla, 2022). Regarding technology and education, immersive experiences are commonly generated using AR and VR technologies. These technologies facilitate consumers' engagement with digital content in a manner that is intuitive and realistic (Tang et al., 2022). Using AR and VR together to achieve an immersive educational experience requires a careful and methodical strategy. Figure 3 provides a detailed explanation of creating a learning environment that uses technology to fully engage students and improve their comprehension.

Figure 3. Immersion Process

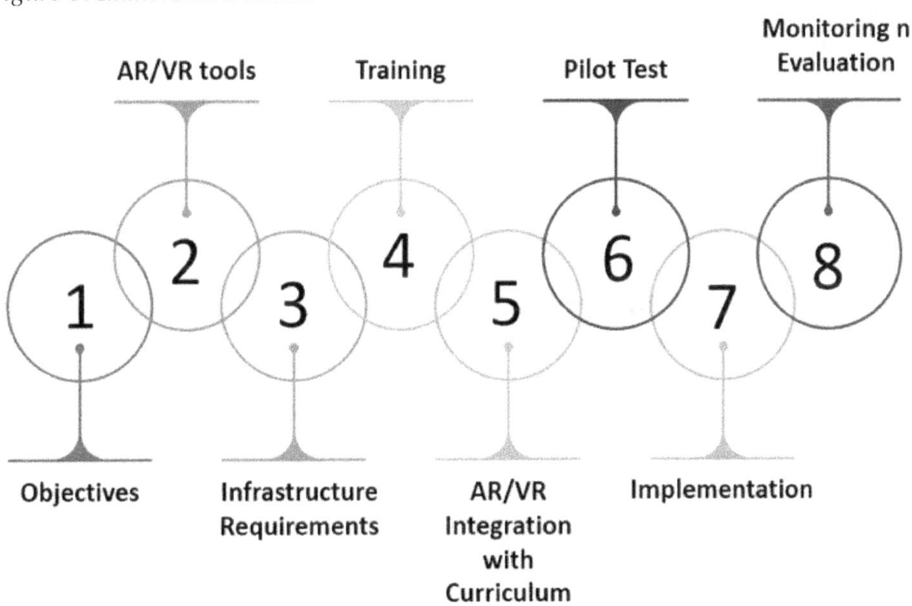

Step 1: Establish Clear Instructional Objectives

It is essential to set clear instructional goals to initiate the development of an immersive educational experience (Bizami et al., 2023; Radianti et al., 2020). The objectives should delineate the specific knowledge or accomplishments students can acquire or accomplish using AR and VR technology. Educators must ascertain precise learning objectives that can be targeted using AR and VR, guaranteeing that these goals align with established curricular requirements. By prioritising how these technologies can improve comprehension, such as by offering visual depictions of intricate ideas or generating interactive simulations, educators can more effectively incorporate these tools into their instructional strategies, resulting in a more significant and efficient learning process.

Step 2: Choosing suitable AR/VR tools

The subsequent phase entails choosing suitable AR and VR instruments for the defined educational objectives (Paszkiewicz et al., 2021). Educators must conduct comprehensive research to investigate AR and VR platforms (Luo et al., 2021). This study should evaluate these platforms based on their instructional content, user-friendliness, and compatibility with the technology already in the classroom. It is crucial to select tools that offer superior quality, precision, and appeal information while being easy for students and teachers. The selected tools should augment the learning experience by rendering intricate subjects more approachable and captivating, bolstering the instructional objectives.

Step 3: Preparing the Required Infrastructure

Once the tools have been chosen, it is crucial to establish the required infrastructure to facilitate the implementation of AR and VR in the classroom (Papanastasiou et al., 2019). This process entails establishing the suitable gear, such as virtual reality headsets, augmented reality-enabled devices, and any essential peripherals, while guaranteeing that all equipment is correctly set up and upheld (Raith et al., 2022). In addition, educators must install and test the necessary software to verify its seamless functionality on the devices. Ensuring a secure physical setting for VR usage is crucial, as students want ample room to navigate without the possibility of harm. By establishing a meticulously prepared and safeguarded setting, educators may effectively enable the smooth incorporation of AR and VR technologies into the educational experience.

Step 4: Providing Training to Educators and Students

To optimise the utilisation of AR and VR technologies, educators and students must receive comprehensive training (Nikimaleki & Rahimi, 2022). The primary objective of educator training sessions should be to acquaint instructors with the technical features of the tools and provide them with pedagogical strategies for effectively incorporating these tools into their classes. Orientation sessions for students should encompass instruction on the fundamental operation of the devices, safety protocols, and the significance of these technologies in their educational endeavours. Highlighting safety precautions is crucial, as it guarantees that educators and students utilise the equipment appropriately and prevent any potential hazards linked to immersive technology (Chittaro & Buttussi, 2015).

Step 5: Incorporating AR/VR into Curriculum Plans

Integrating AR and VR into the curriculum necessitates a purposeful strategy that combines new technologies with conventional teaching approaches (Fu, 2021). Educators should strategically determine the optimal use of AR and VR within the curriculum, creating comprehensive lesson plans that outline the specific ways these AR and VR tools will be used in conjunction with lectures, discussions, and other activities. By integrating immersive experiences with traditional methods, educators can establish a more dynamic and captivating educational setting that accommodates diverse learning preferences and improves students' comprehension of the subject matter.

Step 6: Conducting a Pilot Test

Before fully integrating AR and VR into the entire classroom, it is necessary to carry out a pilot test with a small sample of students. This assessment enables instructors to monitor how students engage with the technology, evaluate their level of involvement, and detect any technical or logistical difficulties. The feedback from the pilot test will be constructive for improving the strategy, as it identifies areas where revisions may be necessary. By conducting a trial run of the integration on a lower level, educators may verify that the AR/VR experience will be successful and advantageous when implemented for the entire class.

Step 7: Refining and Implementing for the Entire Class

According to the pilot test results, appropriate modifications to the lesson plan, technological setup, or instructional approaches are essential. After being improved, lessons enhanced with augmented and virtual reality can be used for the entire class. This step guarantees that all students will benefit from the enhancements implemented during the pilot phase, leading to a uniform and efficient learning experience. Educators can optimise the influence of these technologies on student learning achievements through meticulous refinement and extensive utilisation of AR and VR.

Step 8: Monitoring Progress and Evaluating Learning Outcomes

It is crucial to consistently monitor and assess the development and learning outcomes of integrating augmented and virtual reality. This entails monitoring student performance using tests, quizzes, and classroom observations to ascertain their level of comprehension of the content. Promoting ongoing feedback from both students and educators is essential, as it offers valuable insights into the efficacy of the AR/VR experiences. Educators can evaluate the influence of AR/VR-enhanced classes on student engagement and comprehension by comparing their outcomes with those of traditional teaching methods. This allows them to refine the immersive learning experience continuously.

This systematic and thorough strategy ensures the successful incorporation of AR and VR technologies into the educational setting, offering students immersive and enriching learning possibilities. Moreover, this approach prioritises safety, usability, and educational value.

6. HARNESSING AR/VR TO MITIGATE EDUCATIONAL CHALLENGES

AR and VR technologies offer innovative solutions to the myriad challenges facing the current education system. By integrating these technologies, educational institutions can address a range of issues, from the digital divide to teacher shortages, creating a more equitable and efficient learning environment. Figures 4 and 5 below illustrate the various obstacles in education and the promising potential of AR and VR to alleviate them, inspiring a sense of hope and optimism for the future of education.

Figure 4. Using AR/VR to Mitigate Educational Challenges

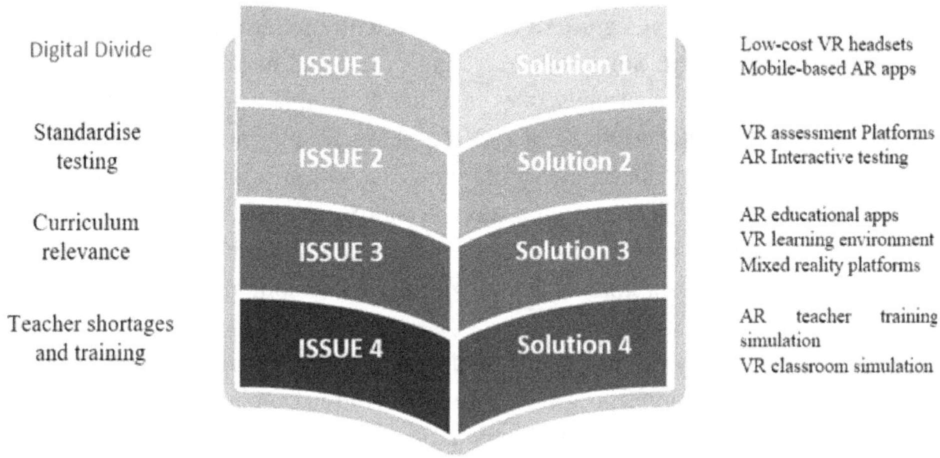

Figure 5. Using AR/VR to Mitigate Educational Challenges

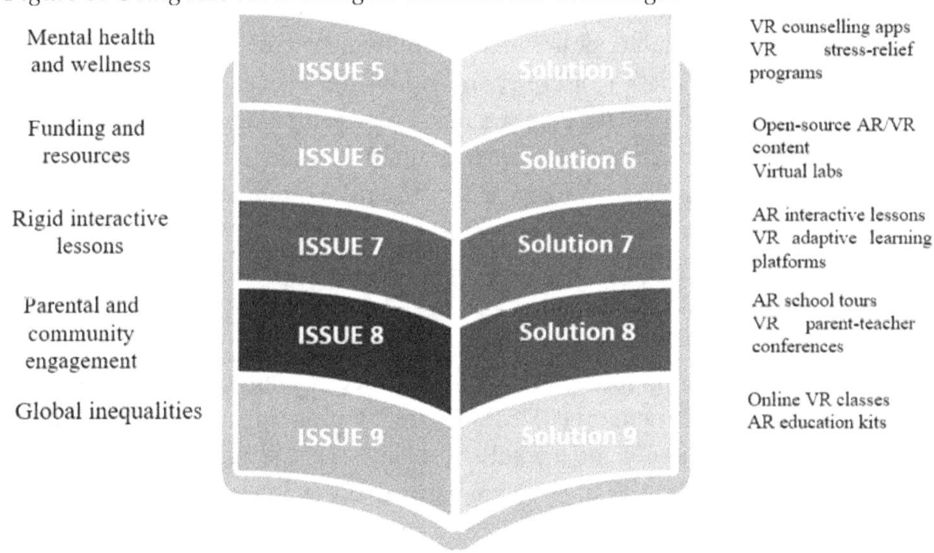

a) Digital divide: The digital divide persists as a significant challenge, as some students lack access to sophisticated technology. Nevertheless, AR and VR have the potential to close this divide. Cost-effective VR headsets and AR applications designed for mobile devices offer economical alternatives for students and educational institutions facing financial constraints. These technologies can enhance the accessibility of immersive learning experiences, ensuring that students from all socio-economic backgrounds can engage with modern educational tools.

b) Standardised testing: Standardised testing's rigid and narrow focus frequently hinders its ability to accurately assess a student's complete potential. AR and VR technology provide options through VR assessment platforms and AR interactive testing. These technologies offer enhanced methods for evaluating student knowledge and skills, enabling a more comprehensive assessment of student abilities. For instance, VR can replicate real-life situations in which students are required to apply their knowledge, providing a more comprehensive evaluation than conventional assessments.

c) Curriculum relevance: The issue of curriculum relevance is a significant concern, as numerous educational programs have challenges in keeping up with the swiftly evolving world. AR educational applications and VR learning environments can be regularly updated to incorporate the most up-to-date knowledge and abilities required in different areas. Mixed reality platforms facilitate students' interaction with knowledge directly pertinent to their interests and future professions, enhancing the significance and practicality of education in their lives.

d) Teacher shortage and training: The scarcity of competent educators and the necessity for continuous professional development are substantial obstacles. AR and VR can assist by offering simulations for teacher training and virtual reality simulations for classroom settings. These technologies enable educators to enhance and perfect their abilities in a supervised and immersed setting, thus enhancing their capacity to manage classrooms and deliver instruction effectively. Using AR for immediate feedback during training sessions, educators can acquire vital insights into their instructional approaches and implement appropriate modifications.

e) Mental health and wellness: Students' emotional and psychological states are crucial factors that might significantly influence academic achievements. VR counselling applications and VR stress-relief programs provide innovative methods to enhance students' mental well-being. These technologies offer secure and immersive settings where students can actively explore coping mechanisms, participate in mindfulness exercises, or receive virtual counselling. AR and VR technologies can enhance the accessibility and engagement of mental health resources, enabling students to better cope with stress and anxiety.

f) Funding and resources: Educational innovation is frequently limited by insufficient money and resources. AR and VR can address this problem by providing accessible educational materials and freely available virtual laboratories. Open-source augmented reality/virtual reality (AR/VR) content offers schools access to superior, cost-free materials that may be tailored to suit their requirements.

Virtual laboratories allow students to perform experiments and investigate intricate ideas without costly tangible apparatus, decreasing expenses while augmenting learning possibilities.

g) Rigid interactive lesson: Conventional interactive classes can occasionally be inflexible and struggle to accommodate each learner's unique demands. AR interactive lessons and VR adaptive learning platforms offer increased flexibility, enabling tailored learning experiences. These technologies can adapt content according to student's performance and learning styles, guaranteeing that each student receives personalised education customised to their needs and learning speed.

h) Parental and community engagement: Student achievement must involve parents and the community in the educational process. AR and VR provide innovative opportunities for engagement, such as AR school tours and VR parent-teacher conferences. AR school tours allow parents to digitally experience the school environment, even if they are unable to visit in person. VR conferences offer a heightened and intimate means of communication with teachers, building better bonds among parents, educators, and the school community.

i) Global inequalities: Differences in access to excellent resources worsen global inequities in education. AR and VR have the potential to create equal opportunities by providing online VR classes and AR education kits that can be accessible globally. These technologies give children in disadvantaged locations equal access to high-quality educational opportunities, fostering global equity and inclusion.

The educational system may efficiently address these notable challenges by utilising AR and VR technology, resulting in a more engaging, fair, and pertinent learning experience for every student. The following sections of this chapter will examine the future possibilities of AR and VR in education and conclude with recommendations for their more comprehensive implementation.

7. CONCLUSION

The incorporation of AR and VR technology in the field of education is transforming the methods by which students interact with learning material. By creating immersive and interactive experiences, these technologies can significantly augment comprehension, retention, and application of knowledge in diverse fields. AR and VR can overcome traditional challenges, such as geographical restrictions and resource limits, by providing students with a more interactive and captivating educational setting. Using AR and VR in education facilitates the cultivation of critical thinking and problem-solving abilities. It accommodates a wide range of learning preferences, enhancing the inclusivity and personalisation of education. Nevertheless, the successful integration of these technologies necessitates a thorough evaluation of aspects such as availability, expense, and the required training of educators to incorporate these tools into their teaching methods. Ultimately, AR and VR have the potential to revolutionise education by offering students enhanced learning opportunities that go beyond the conventional classroom setting. Given the ongoing evolution and increasing accessibility of these technologies, their influence on the future of education is expected to broaden, presenting new opportunities to connect theoretical knowledge with practical implementation.

8. REFERENCES

Al-Ansi, A. M., Jaboob, M., Garad, A., & Al-Ansi, A. (2023). Analyzing augmented reality (AR) and virtual reality (VR) recent developments in education. *Social Sciences & Humanities Open*, 8(1), 100532. DOI: 10.1016/j.ssaho.2023.100532

Al-Emran, M., Malik, S. I., & Al-Kabi, M. N. (2020). *A Survey of Internet of Things (IoT) in Education: Opportunities and Challenges* (pp. 197–209). DOI: 10.1007/978-3-030-24513-9_12

Alexander, C., Wyatt-Smith, C., & Du Plessis, A. (2020). The role of motivations and perceptions on the retention of inservice teachers. *Teaching and Teacher Education*, 96, 103186. DOI: 10.1016/j.tate.2020.103186

AlGerafi, M. A. M., Zhou, Y., Oubibi, M., & Wijaya, T. T. (2023). Unlocking the Potential: A Comprehensive Evaluation of Augmented Reality and Virtual Reality in Education. *Electronics (Basel)*, 12(18), 3953. DOI: 10.3390/electronics12183953

Ali, S., Abdullah, , Armand, T. P. T., Athar, A., Hussain, A., Ali, M., Yaseen, M., Joo, M.-I., & Kim, H.-C. (2023). Metaverse in Healthcare Integrated with Explainable AI and Blockchain: Enabling Immersiveness, Ensuring Trust, and Providing Patient Data Security. *Sensors (Basel)*, 23(2), 565. DOI: 10.3390/s23020565 PMID: 36679361

Almufarreh, A., & Arshad, M. (2023). Promising Emerging Technologies for Teaching and Learning: Recent Developments and Future Challenges. *Sustainability (Basel)*, 15(8), 6917. DOI: 10.3390/su15086917

Alvarez-Marin, A., & Velazquez-Iturbide, J. A. (2021). Augmented Reality and Engineering Education: A Systematic Review. *IEEE Transactions on Learning Technologies*, 14(6), 817–831. DOI: 10.1109/TLT.2022.3144356

Arán Filippetti, V., & Krumm, G. (2020). A hierarchical model of cognitive flexibility in children: Extending the relationship between flexibility, creativity and academic achievement. *Child Neuropsychology*, 26(6), 770–800. DOI: 10.1080/09297049.2019.1711034 PMID: 31913075

Asad, M. M., Naz, A., Churi, P., & Tahanzadeh, M. M. (2021). Virtual Reality as Pedagogical Tool to Enhance Experiential Learning: A Systematic Literature Review. *Education Research International*, 2021, 1–17. DOI: 10.1155/2021/7061623

Barneva, R. P., Kanev, K., Kapralos, B., Jenkin, M., & Brimkov, B. (2017). Integrating Technology-Enhanced Collaborative Surfaces and Gamification for the Next Generation Classroom. *Journal of Educational Technology Systems*, 45(3), 309–325. DOI: 10.1177/0047239516671945

Barteit, S., Lanfermann, L., Bärnighausen, T., Neuhann, F., & Beiersmann, C. (2021). Augmented, Mixed, and Virtual Reality-Based Head-Mounted Devices for Medical Education: Systematic Review. *JMIR Serious Games*, 9(3), e29080. DOI: 10.2196/29080 PMID: 34255668

Beck, D. (2019). Special Issue: Augmented and Virtual Reality in Education: Immersive Learning Research. *Journal of Educational Computing Research*, 57(7), 1619–1625. DOI: 10.1177/0735633119854035

Binytska, K., Bilyakovska, O., Yankovych, O., Buchkivska, G., Binytska, O., Greskova, V., Ocheretna, I., Burov, O., & Lytvynova, S. (2020). Implicit Potential of Immersive Technologies Implementation in the Educational Process at the Universities: World Experience. *Proceedings of the 1st Symposium on Advances in Educational Technology*, 264–274. DOI: 10.5220/0010930700003364

Bizami, N. A., Tasir, Z., & Kew, S. N. (2023). Innovative pedagogical principles and technological tools capabilities for immersive blended learning: A systematic literature review. *Education and Information Technologies*, 28(2), 1373–1425. DOI: 10.1007/s10639-022-11243-w PMID: 35919874

Bujang, S. D. A., Selamat, A., Krejcar, O., Maresova, P., & Nguyen, N. T. (2020). Digital Learning Demand for Future Education 4.0—Case Studies at Malaysia Education Institutions. *Informatics (MDPI)*, 7(2), 13. DOI: 10.3390/informatics7020013

Cevikbas, M., Bulut, N., & Kaiser, G. (2023). Exploring the Benefits and Drawbacks of AR and VR Technologies for Learners of Mathematics: Recent Developments. *Systems*, 11(5), 244. DOI: 10.3390/systems11050244

Cheshmehzangi, A., Zou, T., Su, Z., & Tang, T. (2023). The growing digital divide in education among primary and secondary children during the COVID-19 pandemic: An overview of social exclusion and education equality issues. *Journal of Human Behavior in the Social Environment*, 33(3), 434–449. DOI: 10.1080/10911359.2022.2062515

Chittaro, L., & Buttussi, F. (2015). Assessing Knowledge Retention of an Immersive Serious Game vs. a Traditional Education Method in Aviation Safety. *IEEE Transactions on Visualization and Computer Graphics*, 21(4), 529–538. DOI: 10.1109/TVCG.2015.2391853 PMID: 26357103

Correia, A. P. (2020). Healing the digital divide during the COVID-19 pandemic. *Quarterly Review of Distance Education*.

Dede, C. J., Jacobson, J., & Richards, J. (2017). *Introduction: Virtual*. Augmented, and Mixed Realities in Education., DOI: 10.1007/978-981-10-5490-7_1

Di Natale, A. F., Repetto, C., Riva, G., & Villani, D. (2020). Immersive virtual reality in K-12 and higher education: A 10-year systematic review of empirical research. *British Journal of Educational Technology*, 51(6), 2006–2033. DOI: 10.1111/bjet.13030

Dincelli, E., & Yayla, A. (2022). Immersive virtual reality in the age of the Metaverse: A hybrid-narrative review based on the technology affordance perspective. *The Journal of Strategic Information Systems*, 31(2), 101717. DOI: 10.1016/j.jsis.2022.101717

Dumitru, C. (2023). Distance Learning and Higher Education Hybridization Opportunities and Challenges for Students with Disabilities. In *Advances in Distance Learning in Times of Pandemic* (p. 35).

Farrie, D. (2022). *The Color of Opportunity: How Formula Underfunding Disproportionately Impacts New Jersey's Black and Latino Students*. Education Law Center.

Fu, L. (2021). Research on the Teaching Model of Animation Professional Class Based on AR/VR Technology and 5G Network. *Wireless Communications and Mobile Computing, 2021*(1). DOI: 10.1155/2021/1715909

Fuchs, K. (2022). Bringing Kahoot! Into the Classroom. *International Journal of Information and Education Technology (IJIET)*, 12(7), 3–8. DOI: 10.18178/ijiet.2022.12.7.1662

Ghafar, A. (2020). *CONVERGENCE BETWEEN 21ST CENTURY SKILLS AND ENTREPRENEURSHIP EDUCATION IN HIGHER EDUCATION INSTITUTES: AN EXPLORATORY STUDY FROM UNDERGRADUATES' PERSPECTIVE*. 9582–9582. DOI: 10.21125/inted.2020.0215

Guangul, F. M., Suhail, A. H., Khalit, M. I., & Khidhir, B. A. (2020). Challenges of remote assessment in higher education in the context of COVID-19: A case study of Middle East College. *Educational Assessment, Evaluation and Accountability*, 32(4), 519–535. DOI: 10.1007/s11092-020-09340-w PMID: 33101539

Haleem, A., Javaid, M., Qadri, M. A., & Suman, R. (2022). Understanding the role of digital technologies in education: A review. *Sustainable Operations and Computers*, 3, 275–285. DOI: 10.1016/j.susoc.2022.05.004

Han, D.-I. D., Melissen, F., & Haggis-Burridge, M. (2024). Immersive experience framework: A Delphi approach. *Behaviour & Information Technology*, 43(4), 623–639. DOI: 10.1080/0144929X.2023.2183054

Hooft Graafland, J. (2018). New Technologies and 21st Century Children. *OECD Education Working Papers, 179*(179). DOI: 10.1787/19939019

Jiang, N., Yan-Li, S., Pamanee, K., & Sriyanto, J. (2021). Depression, Anxiety, and Stress During the COVID-19 Pandemic: Comparison Among Higher Education Students in Four Countries in the Asia-Pacific Region. *Journal of Population and Social Studies*, 29, 370–383. DOI: 10.25133/JPSSv292021.023

Kelty, N. E., & Wakabayashi, T. (2020). Family Engagement in Schools: Parent, Educator, and Community Perspectives. *SAGE Open*, 10(4), 215824402097302. DOI: 10.1177/2158244020973024

Ko, A. (2024). VIRTUAL REALITY IN A SYNCHRONOUS CLASSROOM: A STUDY OF LEARNERS'. *IMMERSIVE EXPERIENCE IN THE APPLICATION OF HIVE AND META QUEST*, 3, 10446–10449. DOI: 10.21125/edulearn.2024.2577

Koumpouros, Y. (2024). Revealing the true potential and prospects of augmented reality in education. *Smart Learning Environments*, 11(1), 2. DOI: 10.1186/s40561-023-00288-0

Kubiszyn, T., & Borich, G. D. (2024). *Educational testing and measurement*. John Wiley & Sons.

Kumar, A., Krishnamurthi, R., Bhatia, S., Kaushik, K., Ahuja, N. J., Nayyar, A., & Masud, M. (2021). Blended Learning Tools and Practices: A Comprehensive Analysis. *IEEE Access : Practical Innovations, Open Solutions*, 9, 85151–85197. DOI: 10.1109/ACCESS.2021.3085844

Laine, T. H. (2018). Mobile educational augmented reality games: A systematic literature review and two case studies. *Computers*, 7(1), 19. Advance online publication. DOI: 10.3390/computers7010019

Lampropoulos, G., Fernández-Arias, P., Antón-Sancho, Á., & Vergara, D. (2024). Affective Computing in Augmented Reality, Virtual Reality, and Immersive Learning Environments. *Electronics (Basel)*, 13(15), 2917. DOI: 10.3390/electronics13152917

Lee, J., & Kim, Y. (2023). Sustainable Educational Metaverse Content and System Based on Deep Learning for Enhancing Learner Immersion. *Sustainability (Basel)*, 15(16), 12663. DOI: 10.3390/su151612663

Lee, K. (2012). Augmented Reality in Education and Training. *TechTrends*, 56(2), 13–21. DOI: 10.1007/s11528-012-0559-3

Lembani, R., Gunter, A., Breines, M., & Dalu, M. T. B. (2020). The same course, different access: The digital divide between urban and rural distance education students in South Africa. *Journal of Geography in Higher Education*, 44(1), 70–84. DOI: 10.1080/03098265.2019.1694876

Li, X., Yi, W., Chi, H.-L., Wang, X., & Chan, A. P. C. (2018). A critical review of virtual and augmented reality (VR/AR) applications in construction safety. *Automation in Construction*, 86, 150–162. DOI: 10.1016/j.autcon.2017.11.003

Liberati, N. (2016). Augmented reality and ubiquitous computing: The hidden potentialities of augmented reality. *AI & Society*, 31(1), 17–28. DOI: 10.1007/s00146-014-0543-x

Liu, P., Yang, Z., Huang, J., & Wang, T.-K. (2024). The effect of augmented reality applied to learning process with different learning styles in structural engineering education. *Engineering, Construction, and Architectural Management*. Advance online publication. DOI: 10.1108/ECAM-06-2023-0596

Luo, H., Li, G., Feng, Q., Yang, Y., & Zuo, M. (2021). Virtual reality in K-12 and higher education: A systematic review of the literature from 2000 to 2019. *Journal of Computer Assisted Learning*, 37(3), 887–901. DOI: 10.1111/jcal.12538

Martín-Gutiérrez, J., Fabiani, P., Benesova, W., Meneses, M. D., & Mora, C. E. (2015). Augmented reality to promote collaborative and autonomous learning in higher education. *Computers in Human Behavior*, 51, 752–761. DOI: 10.1016/j.chb.2014.11.093

Mauki, M. M., Kitur, J., Ireri, N. W., & Ngala, F. W. (2020). Competency-Based Curriculum implementation and the role of the universities in Kenya. *Impact: Journal of Transformation*, 45–52.

McGorry, P. D., Mei, C., Chanen, A., Hodges, C., Alvarez-Jimenez, M., & Killackey, E. (2022). Designing and scaling up integrated youth mental health care. *World Psychiatry; Official Journal of the World Psychiatric Association (WPA)*, 21(1), 61–76. DOI: 10.1002/wps.20938 PMID: 35015367

Molina-Carmona, R., Pertegal-Felices, M. L., Jimeno-Morenilla, A., & Mora-Mora, H. (2018). Virtual Reality Learning Activities for Multimedia Students to Enhance Spatial Ability. *Sustainability (Basel)*, 10(4), 1074. DOI: 10.3390/su10041074

Monahan, T., McArdle, G., & Bertolotto, M. (2008). Virtual reality for collaborative e-learning. *Computers & Education*, 50(4), 1339–1353. DOI: 10.1016/j.compedu.2006.12.008

Moreno-Guerrero, A.-J., Alonso García, S., Ramos Navas-Parejo, M., Campos-Soto, M. N., & Gómez García, G. (2020). Augmented Reality as a Resource for Improving Learning in the Physical Education Classroom. *International Journal of Environmental Research and Public Health*, 17(10), 3637. DOI: 10.3390/ijerph17103637 PMID: 32455778

Morze, N., Varchenko-Trotsenko, L., Terletska, T., & Smyrnova-Trybulska, E. (2021). Implementation of adaptive learning at higher education institutions by means of Moodle LMS. *Journal of Physics: Conference Series*, 1840(1), 012062. DOI: 10.1088/1742-6596/1840/1/012062

Mukul, E., & Büyüközkan, G. (2023). Digital transformation in education: A systematic review of education 4.0. *Technological Forecasting and Social Change*, 194, 122664. DOI: 10.1016/j.techfore.2023.122664

Nami, F. (2019). Educational smartphone apps for language learning in higher education: Students' choices and perceptions. *Australasian Journal of Educational Technology*, •••, 82–95. DOI: 10.14742/ajet.5350

Navarro, C., Arias-Calderón, M., Henríquez, C. A., & Riquelme, P. (2024). Assessment of Student and Teacher Perceptions on the Use of Virtual Simulation in Cell Biology Laboratory Education. *Education Sciences*, 14(3), 243. DOI: 10.3390/educsci14030243

Nikimaleki, M., & Rahimi, M. (2022). Effects of a collaborative <scp>AR</scp> -enhanced learning environment on learning gains and technology implementation beliefs: Evidence from a graduate teacher training course. *Journal of Computer Assisted Learning*, 38(3), 758–769. DOI: 10.1111/jcal.12646

Oke, A., & Fernandes, F. A. P. (2020). Innovations in Teaching and Learning: Exploring the Perceptions of the Education Sector on the 4th Industrial Revolution (4IR). *Journal of Open Innovation*, 6(2), 31. DOI: 10.3390/joitmc6020031

Oliver, M., & Trigwell, K. (2005). Can 'Blended Learning' Be Redeemed? *E-Learning and Digital Media*, 2(1), 17–26. DOI: 10.2304/elea.2005.2.1.17

Ortagus, J. C., Kelchen, R., Rosinger, K., & Voorhees, N. (2020). Performance-Based Funding in American Higher Education: A Systematic Synthesis of the Intended and Unintended Consequences. *Educational Evaluation and Policy Analysis*, 42(4), 520–550. DOI: 10.3102/0162373720953128

Pakinee, A., & Puritat, K. (2021). Designing a gamified e-learning environment for teaching undergraduate ERP course based on big five personality traits. *Education and Information Technologies*, 26(4), 4049–4067. DOI: 10.1007/s10639-021-10456-9 PMID: 33613080

Papanastasiou, G., Drigas, A., Skianis, C., Lytras, M., & Papanastasiou, E. (2019). Virtual and augmented reality effects on K-12, higher and tertiary education students' twenty-first century skills. *Virtual Reality (Waltham Cross)*, 23(4), 425–436. DOI: 10.1007/s10055-018-0363-2

Parmaxi, A., Stylianou, K., & Zaphiris, P. (2017). *Leveraging Virtual Trips in Google Expeditions to Elevate Students' Social Exploration* (pp. 368–371). DOI: 10.1007/978-3-319-68059-0_32

Paszkiewicz, A., Salach, M., Dymora, P., Bolanowski, M., Budzik, G., & Kubiak, P. (2021). Methodology of Implementing Virtual Reality in Education for Industry 4.0. *Sustainability (Basel)*, 13(9), 5049. DOI: 10.3390/su13095049

Pellas, N., Mystakidis, S., & Kazanidis, I. (2021). Immersive Virtual Reality in K-12 and Higher Education: A systematic review of the last decade scientific literature. *Virtual Reality (Waltham Cross)*, 25(3), 835–861. DOI: 10.1007/s10055-020-00489-9

Pereira, L. F. B., da Silva, M. M. O., Roberto, R. A., Teichrieb, V., & Cavalcante, P. S. (2020). Virtual Playground: An Augmented Reality Application to Improve English Learning. *2020 22nd Symposium on Virtual and Augmented Reality (SVR)*, 83–91. DOI: 10.1109/SVR51698.2020.00027

Phakamach, P., Senarith, P., & Wachirawongpaisarn, S. (2022). The Metaverse in Education: The Future of Immersive Teaching & Learning. *RICE Journal of Creative Entrepreneurship and Management, 3*(2), 75–88. https://doi.org/DOI: 10.14456/rjcm.2022

Potkonjak, V., Gardner, M., Callaghan, V., Mattila, P., Guetl, C., Petrović, V. M., & Jovanović, K. (2016). Virtual laboratories for education in science, technology, and engineering: A review. *Computers & Education*, 95, 309–327. DOI: 10.1016/j.compedu.2016.02.002

Puncreobutr, V. (2016). Education 4.0: New challenge of learning. *St. Theresa. Journal of the Humanities and Social Sciences*, 2(2), 92–97.

Radianti, J., Majchrzak, T. A., Fromm, J., & Wohlgenannt, I. (2020). A systematic review of immersive virtual reality applications for higher education: Design elements, lessons learned, and research agenda. *Computers & Education*, 147, 103778. DOI: 10.1016/j.compedu.2019.103778

Raith, A., Kamp, C., Stoiber, C., Jakl, A., & Wagner, M. (2022). Augmented Reality in Radiology for Education and Training—A Design Study. *Health Care*, 10(4), 672. DOI: 10.3390/healthcare10040672 PMID: 35455849

Rana, B., Singh, Y., & Singh, P. K. (2021). A systematic survey on internet of things: Energy efficiency and interoperability perspective. *Transactions on Emerging Telecommunications Technologies*, 32(8), e4166. Advance online publication. DOI: 10.1002/ett.4166

Richards, S. (2023). Student Engagement Using HoloLens Mixed-Reality Technology in Human Anatomy Laboratories for Osteopathic Medical Students: An Instructional Model. *Medical Science Educator*, 33(1), 223–231. DOI: 10.1007/s40670-023-01728-9 PMID: 36691419

Sadler, R., & Thrasher, T. (2023). XR: Crossing Reality to Enhance Language Learning. *CALICO Journal*, 40(1), i–xi. DOI: 10.1558/cj.25517

Sancar, R., Atal, D., & Deryakulu, D. (2021). A new framework for teachers' professional development. *Teaching and Teacher Education*, 101, 103305. DOI: 10.1016/j.tate.2021.103305

Šašinka, Č., Stachoň, Z., Sedlák, M., Chmelík, J., Herman, L., Kubíček, P., Šašinková, A., Doležal, M., Tejkl, H., Urbánek, T., Svatoňová, H., Ugwitz, P., & Juřík, V. (2018). Collaborative Immersive Virtual Environments for Education in Geography. *ISPRS International Journal of Geo-Information*, 8(1), 3. DOI: 10.3390/ijgi8010003

Selvan, C., Vidhya, K., Senthil Kumar, R., & Veningston, K. (2024). *Comparative Overview of Augmented Reality*. Virtual Reality, and Mixed Reality., DOI: 10.4018/979-8-3693-1123-3.ch001

Tan, Y., Xu, W., Li, S., & Chen, K. (2022). Augmented and Virtual Reality (AR/VR) for Education and Training in the AEC Industry: A Systematic Review of Research and Applications. *Buildings*, 12(10), 1529. DOI: 10.3390/buildings12101529

Tang, Y. M., Chau, K. Y., Kwok, A. P. K., Zhu, T., & Ma, X. (2022). A systematic review of immersive technology applications for medical practice and education - Trends, application areas, recipients, teaching contents, evaluation methods, and performance. *Educational Research Review*, 35, 100429. DOI: 10.1016/j.edurev.2021.100429

Tritthart, M. (2021). Animated Urban Surfaces: Spatial Augmented Reality in Public Discourse. *International Journal of Film and Media Arts*, 6(2), 75–94. DOI: 10.24140/ijfma.v6.n2.05

Tsirulnikov, D., Suart, C., Abdullah, R., Vulcu, F., & Mullarkey, C. E. (2023). Game on: Immersive virtual laboratory simulation improves student learning outcomes & motivation. *FEBS Open Bio*, 13(3), 396–407. DOI: 10.1002/2211-5463.13567 PMID: 36723273

Tyner, A. (2023). *Think Again: Is Education Funding in America Still Unequal? Thomas B*. Fordham Institute.

Vallée, A., Blacher, J., Cariou, A., & Sorbets, E. (2020). Blended Learning Compared to Traditional Learning in Medical Education: Systematic Review and Meta-Analysis. *Journal of Medical Internet Research*, 22(8), e16504. DOI: 10.2196/16504 PMID: 32773378

Vreuls, J., Koeslag-Kreunen, M., van der Klink, M., Nieuwenhuis, L., & Boshuizen, H. (2022). Responsive curriculum development for professional education: Different teams, different tales. *Curriculum Journal*, 33(4), 636–659. DOI: 10.1002/curj.155

Whalley, B., France, D., Park, J., Mauchline, A., & Welsh, K. (2021). Towards flexible personalized learning and the future educational system in the fourth industrial revolution in the wake of Covid-19. *Higher Education Pedagogies*, 6(1), 79–99. DOI: 10.1080/23752696.2021.1883458

Yung, R., & Khoo-Lattimore, C. (2019). New realities: A systematic literature review on virtual reality and augmented reality in tourism research. *Current Issues in Tourism*, 22(17), 2056–2081. DOI: 10.1080/13683500.2017.1417359

Chapter 7
Digital Tools to Support Executive Functioning for Persons With Attention-Deficit Hyperactivity Disorder

Terri Jackson
https://orcid.org/0009-0007-9247-8499
University of Prince Edward Island, Canada

ABSTRACT

Lagging executive functioning skills are common amongst individuals diagnosed with ADHD (Adler et al., 2017; Ustun et al., 2017) and research indicates, these skill deficits can make the activities of daily living challenging (Loe et al., 2007). Many assistive technology and artificial intelligence tools can support executive functioning. This chapter explores how emotional regulation, organizational skills, sustained attention, and working memory can be enhanced through the use of assistive technology and artificial intelligence. When used effectively, Summit, Headspace, Goblin Tools, Motion, Positive Feedback Bot, Speechify, TickTick, and Bear can raise, reinforce, and consolidate executive functioning skills in persons diagnosed with ADHD thereby fostering independence, confidence, and self-efficacy.

DOI: 10.4018/979-8-3693-9811-1.ch007

INTRODUCTION

Attention-Deficit Hyperactivity Disorder (ADHD) is a common neurobehavioral diagnosis in children (Vadala et al., 2011), particularly in North America, where diagnosis rates are higher than in other regions (Abdelnour et al., 2022). For individuals with ADHD, core symptoms of inattentiveness, hyperactivity, and/or impulsivity can make navigating the workplace, school system, and/or daily living particularly challenging (Loe et al., 2007). Lagging executive functioning skills, common among persons with ADHD diagnoses, make the management of ADHD symptoms more difficult (Rodden, 2021; Silverstein et al., 2018). Research indicates that lagging executive functioning skills are a central characteristic of adults with ADHD (Adler et al., 2017; Ustun et al., 2017). In fact, a diagnosis of ADHD was identified as the most negative of the 252 effects studied in Hattie's 2009 meta-analysis, making ADHD's impact on learning significant (Hattie, 2009). While the context of Hattie's research was the K-12 school system, for many people with ADHD diagnoses, inattentiveness, hyperactivity, and/or impulsivity persist into adulthood (Vadala et al., 2007).

This chapter will explore numerous assistive technologies and/or artificial intelligence tools through the lens of supporting lagging executive functioning skills. Each evidence-backed assistive technology and/or artificial intelligence tool shared effectively accommodates one or more executive functioning skills. Specifically, tools that support emotional regulation, organizational skills, sustained attention, and working memory will be explored. First, *Summit* and *Headspace* are shared as tools to support emotional regulation. Organization, planning, and prioritizing tools, *Goblin Tools,* and *Motion* are then shared. Next, the *Positive Feedback Bot* and *Speechify* are discussed in the context of supporting sustained attention. And finally, *TickTick* and *Bear,* tools to support working memory, are shared. This chapter aims to support the independence, confidence, self-efficacy and executive functioning skills of persons diagnosed with ADHD by integrating assistive technology and/or artificial intelligence.

LITERATURE REVIEW

ADHD & Executive Functioning: Definition

Attention-deficit hyperactivity disorder (ADHD) is among the most common neurobehavioral diagnoses in childhood (Vadala et al., 2007). While the core symptoms of inattentiveness, hyperactivity, and/or impulsivity are not exclusive to a diagnosis of ADHD, persons with an ADHD diagnosis meet clinical markers as their daily functioning is impacted by the symptoms (Barkley, 1988; Murphy & Gordon, 1998). Symptoms of ADHD can impact individuals in the workplace, educational, and/or daily living settings. For many, the symptoms of ADHD persist into adulthood (Vadala et al., 2007); in general, adults with ADHD have a significantly greater history of academic underachievement, substance use, poor occupational adjustment, antisocial behavior, relationship difficulties, mood or affective disorders, and/or personality disorders (Young et al., 2003). In recent years, a growing movement has called ADHD a disorder of executive functioning, with up to ninety percent of persons diagnosed with ADHD struggling with executive functioning skills (Rodden, 2021).

Executive function skills are the cognitive or mental abilities one needs to pursue goals actively (Rodden, 2021). Executive functioning supports our immediate actions and helps us monitor our behavior and motivate us to act upon our future goals (Rodden, 2021). There is a strong correlation between executive functioning skills and self-regulation (Silverstein et al., 2018). Executive functioning skills and self-regulation collaboratively enable us to "plan, focus attention, remember instructions, and juggle multiple tasks successfully" (Together for Children – Sunderland, 2024, p.1). In general, the term executive functioning refers to seven essential skills: (i) self-awareness; (ii) inhibition or self-restraint; (iii) non-verbal working memory; (iv) verbal working memory; (v) emotional self-regulation; (vi) self-motivation; and (vii) planning and problem-solving (Barkley, 2019). Our brains must filter distractions, prioritize tasks, set and achieve goals, and control impulses to move through life successfully (Rodden, 2021). Executive functioning skills are thought to develop in the order listed above. Persons with ADHD develop these skills about thirty or forty percent more slowly, making age-appropriate executive functioning skills delayed in their acquisition (Barkley, 2019).

Barkley (2019) described four circuits in the neurobiology of the ADHD brain and executive functioning. Each circuit is highly connected to the prefrontal cortex and supports executive functioning. The *what* circuit connects to working memory and supports tasks to occur in the future, namely planning and goal setting. The *when* circuit connects in the cerebellum and connects to the area of the brain that supports time orientation and time management. The *why* circuit of the brain connects our limbic system through the amygdala and controls emotional responses and motivation. Finally, the *who* circuit connects to self-awareness, our understanding of our work, and what's happening. Barkley purports that understanding the neurocircuitry helps us not only with the variation in symptoms but also with the accommodations that could offer support to individuals.

Studies have indicated that, in particular, persons with ADHD struggle with executive functioning skills that impact their educational experience (Rosello et al., 2018). This makes success in an academic setting more difficult. Moreover, the transition to post-secondary is particularly challenging for persons diagnosed with ADHD as fewer external structures in addition to their lagging internal structures make managing the schedule, the enhanced autonomy, and decreased support difficult (Turnock et al., 1998; Quinn et al., 2000; Farrell, 2003; Knouse & Fleming, 2016). For some post-secondary students with ADHD, hyper-focus can present a challenge as well, as it results in fixation and can lead to neglecting other essential tasks (Garcia Pimenta et al., 2024).

Assistive Technology & Artificial Intelligence as Accommodations

Assistive technology (AT) and artificial intelligence (AI) are of great assistance to persons with disabilities, including those with ADHD (Orcam, 2023). For persons with ADHD, taking advantage of AI and AT tools to help manage lagging executive functioning skills and support independence by accommodating these skills in a specific and meaningful way.

Assistive Technology includes "products ranging from physical products such as wheelchairs, glasses, prosthetic limbs, white canes, and hearing aids to digital solutions such as speech recognition or time management software and captioning" so long as the primary purpose of the item is to help improve or maintain an individual's functioning related to "cognition, communication, hearing, mobility, self-care and/or vision" (World Health Organization, 2024, paragraph 2). For persons with ADHD, assistive technology supports have often been connected to environmental compensatory supports like noise-canceling headphones, wobbly seats, and fidgets. While these physical tools are helpful, opportunities to harness the power of AI further opportunities for independence and self-efficacy are growing more abundant.

International Business Machines (IBM) defines AI as "technology that enables computers and machines to simulate human intelligence and problem-solving capabilities" (IBM, 2024, paragraph 8). For example, artificial intelligence is embedded into speech-to-text software, learning with the user. It is also embedded into word prediction software, which predicts words based on user frequency of use and level of vocabulary. Huge leaps in AI have occurred in mainstream communities in recent years with the public awareness of *ChatGTP* and similar OpenAI tools.

For persons with ADHD, artificial intelligence and assistive technology intersect as these innovations "offer substantial potential for enhancing communication, interaction, and social engagement" (Iannone & Giansanti, 2023, p. 41). While this body of research is in its infancy, Iannone & Giansanti (2024) purport that massive potential exists when AT tools are integrated with AI; specifically, those tools support communication, social interaction, and engagement in persons diagnosed with autism spectrum disorder. Iannone & Giansanti argue that AI will play a "pivotal role in bridging various domains, and integrated ATs with AI are positioned to act as crucial connectors" (p.1). The ADHD Centre (UK) suggests that persons with ADHD can use AI in many ways, "it can help you stay organized, remember important tasks, and even improve your focus" (The ADHD Center, 2023). AT and AI will transform the lives of persons with ADHD, allowing them to lead increasingly independent, fulfilling lives (Orcam, 2024).

In recent years, significant increases have been seen in the lay public's access to artificial intelligence. Researchers, educators, and the general public are seeking accommodations that leverage the power of AI and AT to support their lagging executive functioning skills due to a diagnosis of ADHD. In subsequent sections of this chapter, we will discuss specific AT and/or AI tools that can help persons with ADHD manage the challenges their lagging executive functioning skills present. Connected to a particular functioning of executive skill, each tool will be overviewed functionally and by sharing anecdotal uses. It should be noted that this is not an exhaustive list of useful tools but simply an opportunity to consider how AI and AT tools can support executive functioning skills in persons with ADHD diagnoses.

SUPPORTING EXECUTIVE FUNCTIONING THROUGH THE USE OF AI & AT

Improve Emotional Control, Patience, & Stress Tolerance

Emotional control, patience, and stress tolerance are executive functions that connect to emotion regulation, which is the executive function that supports one's ability to manage and regulate emotions (Center for the Developing Child – Harvard University, 2024). Seven major types of self-regulatory behaviors have been connected to executive functioning and include (i) self-awareness, (ii) self-restraint, (iii) non-verbal working memory, (iv) verbal working memory, (v) emotional, (vi) self-motivation, and (vii) planning and problem-solving. Emotional control, patience, and stress tolerance challenge the quality of life and daily functioning, making supporting the development of this skill essential (Hammud et al., 2023). Research indicates a strong correlation between emotional regulation and emotional and behavioral challenges (Fernandez et al., 2023). For persons with ADHD, studies suggest that lagging self-regulatory skills are common and may be correlated with psychiatric comorbidities and criminal activity (Soler-Gutierrez et al., 2023). As such, seeking ways to foster self-regulation skills is necessary.

Supporting emotional control, patience, and stress tolerance is challenging for persons with ADHD, particularly when inhibition control is a primary symptom; slowing down, learning to recognize emotional shifts, and labeling emotions can help foster self-regulation (Lovering, 2022). Promising research indicates that mindfulness, in particular, can support individuals in building emotional regulation skills (Guendelman et al., 2017); by controlling their attention, persons with ADHD become more emotionally aware and are less likely to act impulsively (Levine, 2024). Regular mindfulness practice has been linked to a reduction in ADHD symptoms, specifically those connected to impulsivity (Levine, 2024). Mindful practices include meditation, journaling, walking, yoga, or other repetitive tasks like swimming or jogging (Levine, 2024).

In this section, we will look at how *Summit* ADHD coaching and *Headspace*, a meditation and mindfulness tool, can support the executive functioning skills of emotional control, patience, and stress tolerance in persons with ADHD. You will see how these tools can act as an accommodation for persons with ADHD, allowing individuals to receive support in a dignified and individualized way.

ADHD Coaching: Summit

With its origins in the 1990s, ADHD coaching has been used increasingly to help individuals diagnosed with ADHD to "identify and employ strategies and skills to help both minimize the effective of ADHD symptoms on their daily lives and more easily achieve their personal goals" (Ahmann et al., 2018, p. 17). In other words, ADHD coaching is often used to help build executive functioning skills in persons with ADHD. Using a life coaching model, individuals build their understanding of their diagnosis and the structures that can support their success (Wright, 2014). ADHD coaching can be seen as an overall tool that supports individuals in developing the skills needed for all aspects of their executive functioning, including self-regulation. ADHD is increasingly seen as an implementation issue, so "by necessity, ADHD coaching addresses the need for development of skills and strategies to manage the often significant pragmatic challenges faced by achieving one's aspirations while living with ADHD" (Ahmann et al., 2018, p. 19). ADHD coaching is intended to build overall skill-building in terms of both executive function and self-regulation.

Because executive functioning skills are a significant area of concern for persons with ADHD, supporting this is an essential aspect of ADHD coaching (Wright, 2014). Executive functioning skills like procrastination, lack of concentration, ineffective self-regulation, poor planning, anxiety, social incompetence, and time management present significant challenges (Ahmann et al., 2018). "The ultimate goal of the coaching process is for the client to develop internal and external structures independently" (Ahmann et al., 2018, p. 19). This research concluded that ADHD coaching improved ADHD symptoms and supported executive functioning skill development and is mirrored in an earlier study, where outcomes in a case study reported improvements in academic outcomes and self-report measures of study skills and learning strategies as a result of ADHD coaching (Schwartz et al., 2005).

Although the gold standard in ADHD coaching continues to be in-person support, there are a growing number of virtual coaches or AI assistants available to coach persons with ADHD. For instance, the app *Summit* promises to help persons with ADHD "better manage their days by helping organize goals tasks and giving an always available coach to talk to you any time" (Summit, 2024). These virtual assistants help to support the general development of executive function and self-regulation skills in persons with ADHD. Dubbed "the AI life coach," *Summit* boasts an advisory board of accredited coaches who support the creation and integration of AI. *Summit* is considered a single space where AI helps users break down big goals and track progress via text, phone, or email exchange; AI mirrors the experience the user responds best to – whether it be a kind ear or a drill sergeant, as it has been developed using "world-class real-life coach methodologies" and the AI coach integrates learning from previous conversations to help move the user towards their goals. For persons with ADHD, the opportunity for differentiated coaching promises potential (Summit, 2024).

Summit is an AI tool that can assist individuals with ADHD by coaching the development of many executive functioning skills. However, supporting the development of stress tolerance and patience through this AI allows individuals to build a bank of strategies to use independently. Furthermore, *Summit* differentiates its support to meet the needs of individuals.

Mindfulness: Headspace

Self-regulation is notoriously difficult for persons with ADHD (Center for the Developing Child – Harvard University, 2024). Recognizing and regulating one's emotions can be challenging, especially when impulsivity is a primary trait connected to the diagnosis (Center for the Developing Child – Harvard University, 2024). Studies indicate that there is a correlation between self-regulation skills and academic achievement (Sahranavard et al., 2018) and overall socio-emotional and behavioral development (Eisenberg et al., 2010).

Science has demonstrated that mindfulness and meditation positively benefit self-regulation (Mayo Clinic, 2023). In fact, research indicates that meditation can improve your emotional and physical well-being by giving you a new way to look at things that cause stress, building skills to manage your stress, making you more self-aware, focusing on the present, reducing negative feelings, helping you be more creative, helping you be more patient, and lowering your resting heart rate (Mayo Clinic, 2024). Further, research indicates that the benefits of meditation don't end when your session does; meditation can help one to move more calmly through their day (Behan, 2020). Although this research occurred during the pandemic, Behan (2020) found that "Meditation and mindfulness are useful skills that can help us to sit with our fears and our circumstances and to observe that like our thoughts, this period in our lives too shall pass" (p.2).

There are many ways to meditate effectively, and individuals should try various methods to find the one best suited to their lives and lifestyles (Mayo Clinic, 2024). Meditation aims to teach an individual to self-monitor (Behan, 2020). Studies have indicated that meditation can reduce ADHD-like symptoms; meditation can foster and improve on-task behavior, sustained attention, reduce mind wandering, and regulate emotions (Rahl et al., 2018). Another study indicated that meditation improved adults' emotional regulation and executive functioning skills (Mitchell et al., 2017). In this study, it was explicitly indicated through self-reports of dysregulation, ADHD and executive functioning-related symptoms were reduced.

Apps like *Headspace* leverage AI to create guided meditation sessions, sleep aids, and relaxation exercises for users (Headspace, 2024). All recommendations within the app are personalized based on user preferences. Specifically, for persons diagnosed with ADHD, mindfulness can help with awareness of attention and emotional regulation, choosing your actions, modulating your stress levels, and communicating with others (Zylowska et al., 2008). In a 2022 study, a digital mindfulness program using *Headspace* indicated that after four weeks, participants had significantly reduced anxiety and sleep problems (Fried et al., 2022). Their findings indicated that when used in collaboration with other treatments, *Headspace* can benefit individuals with ADHD, particularly those with comorbid anxiety and/or sleep problems. On their site, *Headspace* encourages users with ADHD to "make the experience their own" by determining how to best build mindfulness into their life; for example, for those who benefit from routine, a class may be the best choice (Headspace, 2024). No matter what individuals choose, practicing self-compassion is vital to supporting the successful integration of mindfulness practice; "Mindfulness is like a muscle, and the more we use it, the easier it gets" (Headspace, 2024, paragraph 9).

Headspace is an AI tool that can assist individuals with ADHD in building self-regulation skills, stress management, and patience. Supporting the development of these skills may help persons with ADHD deal with challenges in daily living more successfully.

Improve Organization, Planning, and Prioritizing

Organizational skills, planning, and prioritizing include how our mental and physical environments unite (Prada, 2024). Developing organizational skills is complex, making this task particularly daunting for people with lagging executive functioning skills (Langberg et al., 2013). Individuals with lagging organizational skills may need help to find placed items, arrive in situations unprepared, and struggle with effective communication. For persons with ADHD, these lagging skills can lead to academic struggles as task and material management can present challenges. In daily living, forgetfulness, tardiness, incomplete tasks, failure to follow instructions, and an inability to adapt to changing environments can present further challenges.

Developing effective organizational skills requires a functional approach and a clear definition of success with accommodations to simplify the process (Durand et al., 2020). In a 2008 study, children who received an 8-week organizational skills intervention showed decreased homework problems and corresponding increases in academic performance (Langberg & Epstein, 2008), showing that with support, reinforcement, and the suitable accommodations, persons diagnosed with ADHD can improve their organization, planning and prioritizing executive functioning skills. To help simplify the process, individuals can work on setting limits on decision-making, ensuring individuals do not spend days agonizing over a choice (Kolberg, 2023). One blogger suggested a "place for the misplaced" where families can place stray items, making it easier for forgotten items to be reclaimed (Prada, 2024). Another suggested a "launchpad" by the front door to store items like keys, wallets, and jackets (Kolberg, 2023). Graphic organizers, timers, color coding, lists, and routines are helpful in school and the workplace (Prada, 2024; Kolhberg, 2023).

In this section, we will look at how *Goblin Tools*, a task organizational tool, and *Motion,* a time management system, can support the executive functioning skills of organization, planning, and prioritizing in persons with ADHD. You will see how these tools can act as supportive accommodations for persons with ADHD, allowing individuals to receive support in a dignified and individualized way.

Task Analysis & Initiation: Goblin Tools

Task initiation requires a solid initial effort to get started; for many persons with ADHD, a task is difficult because their internal motivation system does not provide sufficient energy to begin (Hitomi, 2021). Perfectionistic tendencies may make it challenging to get started, as one expects ideal conditions to present before they gets started (Durand et al., 2020). Further, time blindness, struggles prioritizing, and challenges managing the sequence required to complete tasks make it even more difficult for a person with ADHD to complete tasks (ADDitude Editors, 2022).

Task overwhelm is relatively common for persons with ADHD (Durand et al., 2020). For some, it can lead to meltdowns or a flight, flight, or freeze response, and for others, it can lead to flooding or something known as ADHD overwhelm paralysis (ADHD Centre, 2023b). One anonymous individual indicates on the ADDitude site, a community for persons with ADHD, that "I have a hard time starting a task because I get so overwhelmed when I see what has to be done" (ADDitude Editors, 2022).

Symptoms of overwhelm can be caused by managing a fast-paced workload (i.e., managing a household, cooking a meal, organizing a wedding, checking a heaping email inbox). Structures that help the individual organize the task at hand and to see what needs to be done in sequence can be followed to support their success.

Goblin.tools is a free web-based assistant that allows users to put in any task and have it broken down into parts using their "magic to do" technology (Goblin. tools, 2024, title). It streamlines workflow and simplifies tasks by breaking them into manageable units (LMS AI, 2024). Users can customize the level of detail in the task lists, adjust the level of formality of the task, provide time allocation estimates, and support the organization of thoughts into a streamlined workflow. The user then can check off items as they are completed. Simple in its interface and interactivity, *Goblin.tools* is an easy entry point for persons looking to build their organizational skills, specifically as a support for individuals needing task analysis and task initiation support. Tools like Goblin tools help ADHDers identify the task they need to complete and break it down into manageable chunks.

Goblin.tools is an AI tool that can assist individuals with ADHD in breaking complex tasks into manageable chunks, building a sense of self-efficacy, and leading to greater organizational, planning, and self-management skills. Supporting the development of these skills may help persons with ADHD to deal with challenges in daily living more successfully.

Time Management & Organization: Motion

Time blindness is a common trait in persons diagnosed with ADHD (Mate, 1999; Toplak & Tannock, 2005). This refers to executive functioning struggles that individuals may have in timekeeping, perceiving, and managing time. Persons with time blindness may struggle to estimate how long an activity will take, stick to a schedule, and know when to start or finish a task. Procrastination, chronic lateness, missed deadlines, overcommitting, difficulties with transitions, and impulsivity are common signs of time blindness (Pedersen, 2023). Cognitive overload and the feeling that time may be slipping away can lead to distress in persons with ADHD (Mate, 1999; Nazari et al., 2018) and make attempting to navigate professional, school, and daily living obstacles challenging (Weigard & Huang-Pollock, 2017). Recent studies indicate that structures in the brain may be connected to time blindness. However, this same study suggests that accommodations in the school setting could help mitigate many of these challenges (Ptacek et al., 2019).

Managing time blindness requires thoughtful accommodation, including timers, alarms, visual schedules, thoughtful planning and chunking of tasks, routines, mindfulness, self-awareness, and time management software (Pedersen, 2023). In a study with teenagers, challenges in continuous performance tasks were found, particularly in their ability to manage time (Yin et al., 2005), indicating that methods like the Pomodoro Technique may be supportive (Ditzell, 2021). In the Pomodoro Technique, individuals set a timer and work continuously for twenty-five minutes. At this point, individuals take a short break before continuing for another twenty-five-minute block. For persons with ADHD, the Pomodoro Technique "may be helpful because it structures tasks into short bursts of focus time. It also sets a time limit for work, which can help prevent hyper-focus on a specific task for too long" (Ditzell, 2021, paragraph 11).

Motion is a "comprehensive, AI-powered task management and calendar tool that allows users to optimize their day" by analyzing an individual's workload and organizing tasks into the most effective schedule (Suarez-Angelino, 2024). Users enter a prioritized task list and time availability into the calendar, and then Motion's AI automatically schedules tasks; this automation is said to improve productivity by twenty-five percent (Posey, 2024). The AI generates personalized recommendations that synch automatically with your Google or Microsoft Outlook calendar. There is a feature that allows you to book meetings; this results in the AI automatically rearranging existing tasks within the calendar. For persons with ADHD, the issues of time blindness are mitigated by AI, as schedules are created automatically.

Motion has two spaces within it – schedules and workspaces. Workspaces are the categories of the projects or tasks to be completed (i.e., marketing, readings), and schedules are the hours you want to be scheduled. Cashion (2023), a *Motion* user diagnosed with ADHD, describes how she uses *Motion* as a graduate student,

> *"I create schedules in motion for work and school... Next, I create a separate Workspace for each class [I am enrolled in that semester... In the course workspace, I create a project for each week. In that project, I added tasks for each lecture video to watch, such as reading assignments, quizzes, other assignments, etc. I set hard deadlines for the most important tasks like papers or big projects, and soft deadlines for readings, etc. I use the Start Date and Blockers feature to make sure everything is scheduled in a reasonable order. Otherwise, I'd have quizzes and reflections scheduled before watching the lectures or reading the assigned chapters... I create a Project template in the workspace for each course with these tasks in this order. A bonus to using the Template is that I can set a default Schedule for each task... Now that the courses are set up in Motion step back and look at your calendar. Your assignments for the next two weeks will be planned out for you! ... As I work through my course assignments for the term, I can feel confident that Motion will find time to complete them all and help me adapt to disruptions in my schedule."*

The ability to organize and prioritize information using AI supports *Motion* users in minimizing time blindness by supporting these executive functioning skills through smart scheduling. Further *Motion* supports persons who experience time blindness by blocking time, ensuring all essential tasks are addressed adequately. Overall, this assistive technology tool provides the accommodation necessary to minimize many of the negative impacts of delayed organizational and time management skills on an individual diagnosed with ADHD.

Improve Sustained Attention

Attention is "the act of tuning out some information to focus on other information deemed more important at the moment" (Prada, 2024, p. 9). Sustained attention predicts positive outcomes in real-world scenarios, including academics (Gallen et al., 2023). Difficulties sustaining attention can result in incomplete work, missing instructions, challenges with task initiation, forgetfulness, 'careless' errors, not learning from mistakes, and difficulty engaging due to lethargy or overstimulation (Prada, 2024). For many persons with ADHD, sustained attention is problematic as individuals struggle to "resist distractions so you can focus, concentrate, and pay attention, and to sustain that focus even when the material is boring" (Diamond, 2024, paragraph 6). As my colleague, an academic diagnosed with ADHD has said, "I have lots of attention… it's knowing where to put it that presents a problem!" (Nicholls, personal communication, 2024). For individuals with ADHD, difficulty sustaining attention can result in significant challenges, including incomplete work, chores, sporting activities, and even conversations (Bubnik et al., 2015).

The first step in improving one's attention is understanding one's attention profile (Prada, 2024). To understand an individual's attention profile, it is critical to consider three aspects: (i) How long can the individual sustain attention? (ii) What kinds of stimuli (internal, external) are the most distracting to the individual? and (iii) What types of activities present the most significant challenges? Once the attention profile has been determined, strategies like mindfulness, movement breaks, neurofeedback, limiting distractions, positive reinforcement, listening rather than reading, and dedicated workspaces can help accommodate lagging sustained attention skills (Prada, 2024; Slattery et al., 2022).

In this section, we will look at how *Positive Feedback Bot*, a positive reinforcement AI tool, and the text-to-speech software *Speechify,* can support the executive functioning skill of sustained attention in persons with ADHD. You will see how these tools can act as supportive accommodations for persons with ADHD, allowing individuals to receive support in a dignified and individualized way.

Positive Reinforcement: Positive Feedback Bot

Research indicates a correlation between sustained attention and positive reinforcement, particularly in persons diagnosed with ADHD, particularly when they are engaging in longer tasks (Bubnik et al., 2015). Studies with young children have found positive reinforcement essential for building appropriate behaviors and skills (Hardy & McLeod, 2020). Further studies on school-aged children indicate positive reinforcement can improve student motivation, increase academic behaviors, and reduce negative behaviors (Khattak & Ahmad, 2018). There is the suggestion that individuals diagnosed with ADHD may respond to reinforcement and punishment differently than their neurotypical peers; in studies with children, persons with ADHD benefitted considerably from positive reinforcement but did not have the same response to punishment that their neurotypical peers did (Tripp, 2024). Some researchers believe that persons with ADHD have lower levels of dopamine in the brain, which may result in dopamine-seeking behaviors, whether they be pro-social (with positive reinforcement serving as the dopamine trigger) or anti-social (with punishment serving as the dopamine trigger) (Duggal, 2023). Dopamine allows us to regulate our emotional response and to take specific actions to receive rewards. Dopamine is responsible for our feelings of pleasure and reward (Tripp, 2024). Dopamine seeking may look like seeking instant gratification, looking for praise, and/or a higher affinity towards addiction (Jaksa, 2022).

Attaboy.ai was created just over a year ago when two Podcast hosts shared a story about how receiving an "attaboy" for completing simple chores and tasks around their home increased their likelihood of completing the tasks (attaboy.ai, 2024). The hosts suggested that an app that used AI to reward their behaviors might increase their overall helpfulness around the house. Together, they built *attaboy.ai*, an AI tool that you text a message to when you have something you could finish – the app then sends you a custom response to encourage you to complete the job well done (attaboy.ai, 2024). These virtual "attaboys" trigger dopamine and encourage the user to continue with a task (Jaska, 2022). Simple to use, this AT tool assists people who struggle with the executive functioning skill of self-motivation and sustained attention by giving them a hit of dopamine each time they complete a task. Theoretically, this should help sustain motivation and encourage the individual to complete further tasks.

Attaboy.ai is an AI tool that can assist individuals with ADHD in sustaining attention by providing timely positive reinforcement. Supporting the development of these skills may help persons with ADHD deal with challenges in daily living more successfully.

Text-to-Speech Software: Speechify

Sustained attention is positively correlated to academic performance; in a 2023 study, sustained attention was linked both to math fluency and reading comprehension (Gallen et al., 2023). Although it is typical for attention to wane over the course of an activity, for many persons who struggle to sustain attention, the impact of distraction impedes their ability to engage fully in their environment and can impact learning. Research indicates that the reading comprehension is significantly impaired by lagging sustained attention skills (Segal, 2023) making accommodations for this lagging executive functioning skill a requirement for children diagnosed with ADHD.

Diamond (2024) found that listening to stories improved sustained attention, especially when done in a storytelling format. Without visuals, listeners are required to focus more closely on content. For many persons with learning disabilities, speech-to-text software has been the gold standard accommodation, allowing users to access cognitively appropriate content without the burden of fluency or decoding as a barrier (Raffoul & Jaber, 2023). Recent studies have extended to include persons diagnosed with ADHD; in a study on the role of text-to-speech software in improving reading in e-learning environments, persons with impulsive control and hyperactive behavior demonstrated improved reading skills and enhanced spelling skills when using text-to-speech software (Herawati et al., 2022). Therefore, when sustained attention is a barrier an individual with ADHD faces, text-to-speech software could be an effective accommodation.

Thanks to AI, significant advances have arisen in text-to-speech software (Bonaifacci et al., 2021). Specifically, the cold and robotic voices found in the original manifestations of text-to-speech software have been made more expressive and human-like. In addition, significant breakthroughs in real-time translation are reducing barriers for language learners (Chandra, 2024). In a study focused on evaluating the effectiveness of text-to-speech software on the sustained attention of mind wandering and reading comprehension in young students with dyslexia, it was found that text-to-speech software significantly reduced rates of mind wandering (Bonaifacci et al., 2021). Another study mirrored these findings, indicating that text-to-speech software positively improved reading comprehension (Keelor et al., 2020).

Speechify is one example of a text-to-speech software program that allows users to listen to articles and interact with computer applications like Microsoft Word (Speechify, 2023). Using advanced AI, *Speechify* converts text into speech; users indicate the text they would like to listen to, and after selecting a voice option (including Snoop Dogg!), users choose the narration rate and begin listening (Chandra, 2024). Speechify can be used across platforms, with extensions for Google Chrome, Microsoft Edge, iPhones, iPads, and Android devices. The mobility of *Speechify* makes it supportive both in academic, professional, and personal settings. For persons with ADHD, using these tools can improve reading comprehension fluency rates, help individuals adapt to new learning methods, and improve mental health as they gain confidence in the utility and functionality of *Speechify*. One *Speechify* user with an ADHD diagnosis indicates,

> *"I just read 48 pages in an hour & I remember everything, thanks to this sweet little app that allowed me to screenshot each page of the book, then press play and let it read to me while I highlighted my book & took notes in the margins. I honestly haven't been able to read past chapter 1 in an educational or self-help book in maybe 20 years because of my ADHD & in one hour, I got through 4 chapters & retained it." (Speechify, 2023, User Candice81).*

The opportunity to support sustained attention skills through text-to-speech software will help individuals with ADHD diagnoses effectively and efficiently navigate written text. The utility of this integrated AT/AI tool is game-changing for many persons diagnosed with ADHD.

Improve Working Memory

Working memory is the ability to hold information in a "readily accessible form" (Cowan, 2014, p. 198). Working memory allows individuals to understand the world, solve problems, rationalize and reason through events, and plan forward. Using top-down manipulation of the world, working memory allows us to hold important information in short-term memory, readily accessible (Kasper et al., 2012). Working memory plays a vital role in learning. For persons with ADHD, it is challenging to "consistently anticipate, plan, enact, and maintain goal-oriented actions" (Kolfer et al., 2018, p. 59) as ADHD impedes one's ability to regulate attention. Kolfer et al. (2020) found that persons with ADHD diagnoses had higher magnitude impairments in working memory, resulting in difficulties managing the cognitive load associated with daily tasks. As a result, many individuals diagnosed with ADHD may find themselves with memory gone missing and lost keys, wallets, and ideas.

Accommodating lagging working memory skills can include: (i) reducing the workload, (ii) encouragement, (iii) monitoring closely, (iv) overt instructional strategies, (v) testing alternatives, and (vi) extra time (Houston, 2016). Each of these accommodations supports individuals with lagging working memory skills by compensating in an area of difficulty. For example, chunking a workload into smaller sections and presenting each individually reduces the cognitive load on an individual and the working memory required to complete the task. Frequent check-ins and reminders encourage the continuation of tasks. Assistive technology and artificial intelligence can accommodate individuals with working memory challenges by organizing information, providing visual aids, and simplifying information into manageable chunks.

In this section, we will look at how *TickTick*, a task reminder tool, and *Bear*, an organizational management app, can support the executive functioning skill of working memory in persons with ADHD. You will see how these tools can act as supportive accommodations for persons with ADHD, allowing individuals to receive support in a dignified and individualized way.

Task Reminders & To-Do Lists: TickTick

For many persons with ADHD, forgetfulness is a massive barrier to success in academic, professional, and social domains (Kolfer et al., 2020). Whether it be an inability to remember important events, one's keys, or information a moment after it was shared, challenges with working memory impact daily living.

Using task reminders or to-do list software can help to prompt memory, making it easier for persons to be successful. Further, whenever something is crossed off the list, dopamine rewards the user by sending a rush of positive neurotransmitters to the brain, increasing the likelihood that the user will continue.

TickTick is a web-based application that synchronizes across ten platforms, making integration between devices simple (TickTick, 2024). It purports to allow one to "organize all aspects of your life… get to-dos out of your mind and get them done in less time" and features options to quickly add tasks, deadline management support, focus supports, and collaboration features built into the interface. To manage task lists, *TickTick* offers voice input features and the opportunity to turn emails into list items. The opportunity to enter information efficiently supports those with weak working memory and/or those who are highly distractible. More importantly, for individuals who struggle with email management, the rapid ability to turn emails into list items promotes efficiency and reduces the likelihood that something will be forgotten.

Deadline management is supported by intelligent date parsing, task duration, multiple reminders, and the opportunity to set recurring tasks. By entering a due date and time information when creating a new task, *TickTick* will automatically parse your task into a reminder. For essential tasks, *TickTick* has an option for an "annoying alert." This alert creates multiple reminders for one task to help users avoid missing important events. The recurring tasks feature allows users to create repeats without inputting the event multiple times for regularly scheduled events. These features make it possible for users to manage deadlines more effectively.

Organizational features allow users to prioritize to-dos. Folders, lists, and check items organize information at four different levels. The Eisenhower Matrix is embedded into the software, allowing users to prioritize tasks. Progress features and incentivization provide dopamine surges, increasing the likelihood that a user will continue. Irish (2023), a Zapier blogger, indicated, "*TickTick* not only helps you focus – it *rewards* focus with a fun little achievements system." This supports users by helping detect inefficiencies and encouraging focus on the most critical tasks. Daily summaries further encourage productivity and make great progress monitoring tools.

TickTick is just one example of an AI-optimized support system that allows users to analyze tasks and monitor for productivity. Many other tools exist; users should consider this self-assessment when choosing one: (i) is the app simple to use? (ii) is it flexible and adaptable to the way I work? (iii) will it help me manage my time and my tasks?; (iv) Will it celebrate my accomplishments? (Irish, 2023).

For persons with weak working memory, tools like *TickTick* harness the power of an AI assistant to support task completion and increase productivity. Supporting the development of these skills may help persons with ADHD to more successfully deal with challenges in daily living.

Keeping Track of Information: Bear

Effectively managing files and paperwork can be difficult for people with working memory difficulties. As described in her 2020 blog post, Woodruff calls labels persons with these challenges "paper pilers." She indicates,

> *"It starts with a small pile of mail on the kitchen counter—items you need to reconcile or pay. You don't file these items because they need action, so you leave them in plain sight. Only things don't get done, and they gather paper friends. Papers to be filled out, rebates to be sent in, coupons to use, and correspondence to read. The last paper to pile on is the to-do lists, reminders to yourself, and notes scribbled on envelopes so you "don't forget." We all have these stacks. We let them go until we can't stand it anymore. Then we sit down and process as much as we can until things are more manageable again" (paragraph 5).*

Lost or forgotten paperwork can have a significant impact on the daily living of persons with ADHD, making it challenging for them to function in their professional, educational, and work settings.

Moving to a virtual filing system is a better solution for paper pilers, as it reduces the quantity of paper available to gather and grow. The reduced quantity of piles reduces the burden on working memory and leads to a less cluttered learning and working space.

The *Bear* app for iOS allows users to insert content and hold it in a virtual filing cabinet. On their website, Bear purports, "Bear is a beautiful, powerfully simple Markdown note-taking app to capture, write, and organize your life" (Bear.app, 2024, paragraph 5). This information can be sorted, stored, and found using hashtags as search parameters. Further, one can search for words within the app, allowing users to find things by labeling them. This makes it easier for people to find what they are looking for.

The seamless markdown process lets users capture multiple information sources in one note – including text, photos, and tables. The flexible tags make it easy to find information. One user reported, "My life is organized by tags – I use it for creating notes from meetings, one-on-one sessions, and I'm able to keep the same under various hashtags" (Bear.app, 2024). The platform's simplicity reduces the load on working memory by providing users with multiple options for search – one could search based on the assigned hashtag, a word used in the document, or the date the file was created. Creating multiple modes for retrieval is supportive.

Because it integrates with other apps, Bear helps users manage their workflow. For example, *TickTick,* the to-do app mentioned previously, integrates with Bear through automation making the integration between various platforms powerful, thereby managing the paperwork associated with daily living more effectively. The opportunity to move between devices and platforms makes integration with the Bear app valuable and seamless. Bear's organization's simplicity helps eliminate paper waste, makes it easier for users to contain information, and makes it a readily available storage center for important receipts.

Bear is an AI tool that can assist individuals with ADHD in managing information, reducing the burden on working memory, and organizing their environment. Supporting the development of these skills may help persons with ADHD to deal with challenges in daily living more successfully.

CONCLUSION

Assistive technology and artificial intelligence are accommodations that can scaffold the lagging executive functioning skills in persons with ADHD diagnoses. These tools support improved working memory, help foster sustained attention, improve organization, prioritizing, and planning, and can facilitate the development of self-regulation and executive functioning skills. When used effectively, *Summit, Headspace*, *Goblin Tools*, *Motion, Positive Feedback Bot, Speechify, TickTick,* and *Bear* can accommodate the lagging executive functioning skills of persons diagnosed with ADHD. As discussed in this chapter, lagging executive functioning skills can result in academic underachievement, substance use, poor occupational adjustment, antisocial behavior, relationship difficulties, mood and affective disorders, and personality disorders (Young et al., 2022), rendering these accommodations critical to enhancing opportunities for persons with ADHD diagnoses to succeed in the workplace, school system and beyond.

AT and AI add a layer of support that can result in increasingly positive outcomes for persons with ADHD. Employers, educators, student support personnel, and the lay public should consider how these accommodations, or similar, could benefit in their particular context. Although a relatively new field that should be explored in greater depth, when used effectively, AT and AI help to level the playing field for persons with ADHD by supporting their executive functioning skills.

REFERENCES

Abdelnour, E., Jansen, M. O., & Gold, J. A. (2022). ADHD diagnostic trends: Increased recognition or overdiagnosis? *Missouri Medicine*, 119(5), 467–473. PMID: 36337990

ADDitude Editors. (2022 May 13). "I'm a serial procrastinator:" Why getting started is endlessly hard. ADDitude. https://www.additudemag.com/getting-started-adhd-challenges/

Adler, L. A., Faraone, S. V., Spencer, T. J., Berglund, P., Alperin, S., & Kessler, R. C. (2017). The structure of adult ADHD. *International Journal of Methods in Psychiatric Research*, 26(1), e1555. DOI: 10.1002/mpr.1555 PMID: 28211596

Ahmann, E., Tuttle, L. J., Saviet, M., & Wright, S. D. (2018). A descriptive review of ADHD coaching research: Implications for college students. *Journal of Postsecondary Education and Disability*, 31(1), 17–39.

Attaboy.ai. [Computer software]. (2024). Available from https://www.attaboy.ai/

Barkley, R. A. (1988). Attention deficit disorder with hyperactivity. In Mash, E. J., & Terdal, L. G. (Eds.), *Behavioral assessment of childhood disorders* (2nd ed., pp. 69–104). Guilford Press.

Barkley, R. A. (1997). Behavioral inhibition, sustained attention, and executive functions: Constructing a unifying theory of ADHD. *Psychological Bulletin*, 121(1), 65–94. DOI: 10.1037/0033-2909.121.1.65 PMID: 9000892

Bear. [Computer software]. (2024). Available from https://bear.app/

Behan, C. (2020). The benefits of meditation and mindfulness practices during times of crisis such as COVID-19. *Irish Journal of Psychological Medicine*, 37(4), 256–258. DOI: 10.1017/ipm.2020.38 PMID: 32406348

Bonifacci, P., Colombini, E., Marzocchi, M., Tobia, V., & Desideri, L. (2021). Text-to-speech applications to reduce mind wandering in students with dyslexia. *Journal of Computer Assisted Learning*, 38(2), 440–454. Advance online publication. DOI: 10.1111/jcal.12624

Bubnik, M. G., Hawk, L. W.Jr, Pelham, W. E.Jr, Waxmonsky, J. G., & Rosch, K. S. (2015). Reinforcement enhances vigilance among children with ADHD: Comparisons to typically developing children and to the effects of methylphenidate. *Journal of Abnormal Child Psychology*, 43(1), 149–161. DOI: 10.1007/s10802-014-9891-8 PMID: 24931776

Cashion, N. (2023 September 15). This AI productivity tool helps me manage my time as a graduate student. [Blog Post] Retrieved from: https://www.nathancashion.com/blog/2023/9/this-ai-productivity-tool-helps-me-manage-my-time-as-a-graduate-student

Center for the Developing Child – Harvard University. (2024). Executive function and self-regulation. Center for the Developing Child – Harvard University. https://developingchild.harvard.edu/science/key-concepts/executive-function/

Chandra, D. (2024 April 21). Speechify review 2024: Best AI Text-to-speech app? [Review] https://theprocesshacker.com/blog/speechify-review/

Cowan, N. (2014). Working memory underpins cognitive development, learning, and education. *Educational Psychology Review*, 26(2), 197–223. DOI: 10.1007/s10648-013-9246-y PMID: 25346585

Diamond, A. (2024 May 15). How to sharpen executive functions: Activities to hone brain skills. ADDitude. https://www.additudemag.com/how-to-improve-executive-function-adhd/#:~:text=Inhibitory%20Control%20at%20the%20Level%20of%20Attention,is%20boring%20(sustained%20attention)

Duggal, N. (2023 December 15). Attention deficit hyperactivity disorder (ADHD): The role of dopamine. Healthline. https://www.healthline.com/health/adhd/adhd-dopamine

Durand, G., Arbone, I. S., & Wharton, M. (2020). Reduced organizational skills in adults with ADHD are due to deficits in persistence, not in strategies. *PeerJ*, 8, e9844. DOI: 10.7717/peerj.9844 PMID: 33194353

Eisenberg, N., Spinrad, T. L., & Eggum, N. D. (2010). Emotion-related self-regulation and its relation to children's maladjustment. *Annual Review of Clinical Psychology*, 6(1), 495–525. DOI: 10.1146/annurev.clinpsy.121208.131208 PMID: 20192797

Farrell, E. F. (2003). Paying attention to students who can't. *The Chronicle of Higher Education*, 50(5), A50–A51.

Fernandes, B., Wright, M., & Essau, C. (2023). The role of emotion regulation and executive functioning in the intervention outcome of children with emotional and behavioural problems. *Children (Basel, Switzerland)*, 10(1), 139. DOI: 10.3390/children10010139 PMID: 36670689

Fried, R., DiSalvo, M., Farrell, A., Biederman, (2022). Using a digital meditation application to mitigate anxiety and sleep problems in children with ADHD. *J Atten Disord. 26(*7), 1033–1039.

Gallen, C. L., Schaerlaeken, S., Younger, J. W., Younger, J. W., O'Laughlin, K. D., Anguera, J. A., Bunge, S. A., Ferrer, E. E., Hoeft, F., McCandliss, B. D., Mishra, J., Rosenberg-Lee, M., Gazzaley, A., Uncapher, M. R., Anguera, J. A., & Gazzaley, A. (2023). Contribution of sustained attention abilities to real-world academic skills in children. *Scientific Reports*, 13(1), 2673. DOI: 10.1038/s41598-023-29427-w PMID: 36792755

Garcia Pimenta, M., Gruhnert, R. K., Fuermaier, A. B. M., & Groen, Y. (2024). The role of executive functions in mediating the relationship between adult ADHD symptoms and hyperfocus in university students. *Research in Developmental Disabilities*, 144, 104639. DOI: 10.1016/j.ridd.2023.104639 PMID: 38039699

Goblin.tools. [Computer Software]. (2024) Available from https://goblin.tools

Guendelman, S., Medeiros, S., & Rampes, H. (2017). Mindfulness and emotion regulation: Insights from neurobiological, psychological, and clinical studies. *Frontiers in Psychology*, 8, 220. DOI: 10.3389/fpsyg.2017.00220 PMID: 28321194

Hammud, G., Avital-Magen, A., Schusheim, G., Barzuza, I., & Engel-Yeger, B. (2023). How self-regulation and executive functions deficits affect quality of life of children/adolescents with emotional regulation disorders. *Children (Basel, Switzerland)*, 10(10), 1622. DOI: 10.3390/children10101622 PMID: 37892283

Hardy, J., & McLeod, R. (2020). Using positive reinforcement with young children. *Beyond Behavior*, 29(2), 107429562091572. DOI: 10.1177/1074295620915724

Hattie, J. (2012). *Visible learning for teachers: Maximizing impact on learning.* Routledge/Taylor & Francis Group. https://doi.org/DOI: 10.4324/9780203181522

Headspace. (2024). Meditation and ADHD: An unexpected (but worthwhile) match. Headspace. https://www.headspace.com/articles/meditation-and-adhd

Herawati, D., Widajati, W., & Sartinah, E. (2022). The role of text to speech assistive technology to improve reading ability in E-Learning for ADHD students. *Journal of ICSAR*, 6(2), 169. DOI: 10.17977/um005v6i22022p169

Hitomi, M. (2021 June 21). ADHD and task initiation: Getting the ball rolling. Timeo. https://www.tiimoapp.com/blog/adhd-task-initiation

Houston, L. (2016 March 2). Workin memory difficulties: Strategies for the classroom. LD@School. https://www.ldatschool.ca/working-memory-difficulties/

Iannone, A., & Giansanti, D. (2023). Breaking barriers - the intersection of AI and assistive Technology in Autism care: A narrative review. *Journal of Personalized Medicine*, 14(1), 41. DOI: 10.3390/jpm14010041 PMID: 38248742

International Business Machines. (2024) What is automation? IBM. https://www.ibm.com/topics/automation#:~:text=Artificial%20intelligence%2C%20or%20AI%2C%20is,are%20fields%20of%20artificial%20intelligence

Irish, J. (2024 March 20). 3 to-do list apps that actually work with ADHD. [Blog Post] Retrieved from: https://zapier.com/blog/adhd-to-do-list/

Jaska, P. (2022 March 31). The ADHD & addiction link: Addictive behaviours in adults explained. [Blog Post] Retrieved from: https://www.additudemag.com/addictive-behaviors-adhd/

Kasper, L. J., Alderson, R. M., & Hudec, K. L. (2012). Moderators of working memory deficits in children with attention-deficit/hyperactivity disorder (ADHD): A meta-analytic review. *Clinical Psychology Review*, 32(7), 605–617. DOI: 10.1016/j.cpr.2012.07.001 PMID: 22917740

Keelor, J. L., Creadhead, N., Silbert, N., & Horowitz-Kraus, T. (2020). Text-to-speech technology: Enhancing reading comprehension for students with reading difficulties. *Assistive Technology Outcomes and Benefits*, 14, 19–35.

Khattak, A. & Ahmad, A. (2018). Effects of positive reinforcement on student's academic performance. 01. 220-225.

Knouse, L. E., & Fleming, A. P. (2016). Applying cognitive-behavioral therapy for ADHD to emerging adults. *Cognitive and Behavioral Practice*, 23(3), 300–315. DOI: 10.1016/j.cbpra.2016.03.008

Kofler, M. J., Sarver, D. E., Harmon, S. L., Moltisanti, A., Aduen, P. A., Soto, E. F., & Ferretti, N. (2018). Working memory and organizational skills problems in ADHD. *Journal of Child Psychology and Psychiatry, and Allied Disciplines*, 59(1), 57–67. DOI: 10.1111/jcpp.12773 PMID: 28714075

Kofler, M. J., Singh, L. J., Soto, E. F., Chan, E. S. M., Miller, C. E., Harmon, S. L., & Spiegel, J. A. (2020). Working memory and short-term memory deficits in ADHD: A bifactor modeling approach. *Neuropsychology*, 34(6), 686–698. DOI: 10.1037/neu0000641 PMID: 32437194

Kolberg, J. (2023 November 29). 33 ADHD-friendly ways to get organized. ADDitude. https://www.additudemag.com/how-to-get-organized-with-adhd/

Langberg, J., Epstein, J., Urbanowicz, C., Simon, J., & Graham, A. (2008). Efficacy of an organization skills intervention to improve the academic functioning of students with Attention-Deficit/Hyperactivity Disorder. *School Psychology Quarterly*, 23(3), 407–417. DOI: 10.1037/1045-3830.23.3.407

Levine, H. (2024 March 13). Meditation and yoga for ADHD. WebMD. https://www.webmd.com/add-adhd/adhd-mindfulness-meditation-yoga

LMS AI. (2024 February 20). Make your research plan easier using Goblin Tools. [Blog Post]. Retrieved from: https://www.rs.lmssolution.net/post/make-your-research-plan-easier-using-goblin-tools

Loe, I. M., & Feldman, H. M. (2007). Academic and educational outcomes of children with ADHD. *Journal of Pediatric Psychology*, 32(6), 643–654. DOI: 10.1093/jpepsy/jsl054 PMID: 17569716

Lovering, N. (2022 May 18). The Connection Between Emotional Regulation and ADHD. Healthline. https://www.healthline.com/health/adhd/emotional-regulation

Maté, G. (1999). *Scattered: How attention deficit disorder originates and what you can do about it*. Penguin.

Mayo Clinic Staff. (2023 December 13). Meditation: A simple, fast way to reduce stress. Mayo Clinic. https://www.mayoclinic.org/tests-procedures/meditation/in-depth/meditation/art-20045858

Mitchell, J. T., McIntyre, E. M., English, J. S., Dennis, M. F., Beckham, J. C., & Kollins, S. H. (2017). A pilot trial of mindfulness meditation training for ADHD in adulthood: Impact on core symptoms, executive functioning, and emotion dysregulation. *Journal of Attention Disorders*, 21(13), 1105–1120. DOI: 10.1177/1087054713513328 PMID: 24305060

Murphy, K. R., & Gordon, M. (1998). Assessment of adults with ADHD. In Barkley, R. A. (Ed.), *Attention-deficit hyperactivity disorder: A handbook for diagnosis and treatment* (2nd ed., pp. 345–369). The Guilford Press.

Nazari, M. A., Mirloo, M. M., Rezaei, M., & Soltanlou, M. (2018). Emotional stimuli facilitate time perception in children with attention deficit/hyperactivity disorder. *Journal of Neuropsychology*, 12(2), 165–175. DOI: 10.1111/jnp.12111 PMID: 27566601

Orcam. (2023). Empowering students with ADHD and dyslexia. Orcam https://www.orcam.com/en-ca/orcam-learn

Pederson, T. (2023 September 29). What is time blindness? Healthline. https://www.healthline.com/health/time-blindness

Posey, J. (2024 May 9). 10 best AI tools for project managers in 2024. [Blog Post] Available at: https://www.usemotion.com/blog/ai-tools-for-product-managers

Prada, J. (2024). EF skills: Organization. Smart Kids with ADHD._https://www.smartkidswithld.org/getting-help/executive-function-disorder/ef-skills-organization/

Ptacek, R., Weissenberger, S., Braaten, E., Klicperova-Baker, M., Goetz, M., Raboch, J., Vnukova, M., & Stefano, G. B. (2019). Clinical implications of the perception of time in attention deficit hyperactivity disorder (ADHD): A review. *Medical Science Monitor*, 25, 3918–3924. DOI: 10.12659/MSM.914225 PMID: 31129679

Quinn, P. O., Ratey, N. A., & Maitland, T. L. (2000). Working with an ADD coach. In Quinn, P. O., Ratey, N. A., & Maitland, T. L. (Eds.), *ADD and the college student: A guide for high school and college students with attention deficit disorder* (pp. 99–109). Magination Press.

Raffoul, S., & Jaber, L. (2023). Text-to-Speech software and reading comprehension: The impact for students with learning disabilities. *Canadian Journal of Learning and Technology*, 49(2), 1–18. DOI: 10.21432/cjlt28296

Rahl, H. A., Lindsay, E. K., Pacilio, L. E., Brown, K. W., & Creswell, J. D. (2017). Brief mindfulness meditation training reduces mind wandering: The critical role of acceptance. *Emotion (Washington, D.C.)*, 17(2), 224–230. DOI: 10.1037/emo0000250 PMID: 27819445

Rodden, J. (2021 January 14). What is executive dysfunction? Signs and symptoms of EFD. ADDitude._https://www.additudemag.com/what-is-executive-function-disorder/

Roselló, B., Berenguer, C., Baixauli, I., Mira, Á., Martinez-Raga, J., & Miranda, A. (2020). Empirical examination of executive functioning, ADHD associated behaviors, and functional impairments in adults with persistent ADHD, remittent ADHD, and without ADHD. *BMC Psychiatry*, 20(1), 134. DOI: 10.1186/s12888-020-02542-y PMID: 32204708

Sahranavard, S., Miri, M. R., & Salehiniya, H. (2018). The relationship between self-regulation and educational performance in students. *Journal of Education and Health Promotion*, 7(1), 154. DOI: 10.4103/jehp.jehp_93_18 PMID: 30693291

Segal, D. (2023). Sustained attention plays a critical role in adults with and without ADHD. *Learning and Individual Differences*, 105. Advance online publication. DOI: 10.1016/j.lindif.2023.102300

Silverstein, M. J., Faraone, S. V., Leon, T. L., Biederman, J., Spencer, T. J., & Adler, L. A. (2020). The relationship between executive function deficits and DSM-5-defined ADHD symptoms. *Journal of Attention Disorders*, 24(1), 41–51. DOI: 10.1177/1087054718804347 PMID: 30296883

Slattery, E. J., O'Callaghan, E., Ryan, P., Fortune, D. G., & McAvinue, L. P. (2022). Popular interventions to enhance sustained attention in children and adolescents: A critical systematic review. *Neuroscience and Biobehavioral Reviews*, 137, 104633. DOI: 10.1016/j.neubiorev.2022.104633 PMID: 35337900

Soler-Gutiérrez, A. M., Pérez-González, J. C., & Mayas, J. (2023). Evidence of emotion dysregulation as a core symptom of adult ADHD: A systematic review. *PLoS One*, 18(1), e0280131. DOI: 10.1371/journal.pone.0280131 PMID: 36608036

Speechify. [Computer Software]. (2024). Available at: https://speechify.com/blog/how-to-overcome-adhd-through-immersion-reading/

Suarez-Angelino, L. (2024). Motion review 2024: Pros & cons, cost and who it's right for. Choosing Therapy. https://www.choosingtherapy.com/motion-app-review/

Summit. [Computer Software] (2024). Your ADHD AI coach. https://www.summit.im/features/ai-adhd-coach#:~:text=Summit%20helps%20those%20with%20ADHD,to%20chronic%20illness%20and%20ADHD

Swartz, S. L., Prevatt, F., & Proctor, B. E. (2005). A coaching intervention for college students with attention deficit/hyperactivity disorder. *Psychology in the Schools*, 42(6), 647–656. DOI: 10.1002/pits.20101

The ADHD Centre. (2023 September 27). How AI is changing the game for ADHD management. The ADHD Centre. https://www.adhdcentre.co.uk/how-ai-is-changing-the-game-for-adhd-management/

The ADHD Centre. (2023b July 9). Overcoming ADHD procrastination with these productivity-boosting strategies. [Website] Available at: https://www.adhdcentre.co.uk/overcome-adhd-procrastination-with-these-productivity-boosting-strategies/

TickTick. [Computer Software] (2024) Available at: https://ticktick.com/?language=en_US

Together for Children – Sunderland. (2024) [PDF document] Retrieved from: https://www.togetherforchildren.org.uk/media/27515/One-minute-guide-Self-Regulation-and-Executive-Function/pdf/SELF_REGULATION_AND_EXECUTIVE_FUNCTION.pdf?m=637970361370630000

Toplak, M. E., & Tannock, R. (2005). Time perception: Modality and duration effects in attention-deficit/hyperactivity disorder (ADHD). *Journal of Abnormal Child Psychology*, 33(5), 639–654. DOI: 10.1007/s10802-005-6743-6 PMID: 16195956

Tripp, G. (2024 June 11). Carrots vs. sticks: The science of rewards and punishment for children with ADHD. ADDitude. https://www.additudemag.com/positive-reinforcement-reward-and-punishment-adhd/

Turnlock, P., Rosen, L., & Kaminski, P. L. (1998). Differences in academic coping strategies of college students who self-report high and low symptoms of Attention Deficit Hyperactivity Disorder. *Journal of College Student Development*, 39(5), 484–493.

Ustun, B., Adler, L. A., Rudin, C., Faraone, S. V., Spencer, T. J., Berglund, P., Gruber, M. J., & Kessler, R. C. (2017). The World Health Organization adult Attention-Deficit/Hyperactivity Disorder self-report screening scale for DSM-5. *JAMA Psychiatry*, 74(5), 520–527. DOI: 10.1001/jamapsychiatry.2017.0298 PMID: 28384801

Vadalà, R., Giugni, E., Pichiecchio, A., Balottin, U., & Bastianello, S. (2011). Attention deficit hyperactivity disorder (ADHD): From a childhood neuropsychiatric disorder to an adult condition. *Functional Neurology*, 26(3), 117–119. PMID: 22152431

Weigard, A., & Huang-Pollock, C. (2017). The role of speed in ADHD-related working memory deficits: A time-based resource-sharing and diffusion model account. *Clinical Psychological Science*, 5(2), 195–211. DOI: 10.1177/2167702616668320 PMID: 28533945

Woodruff, L. (2020, October 19). Filing cabinets don't work for ADHD minds: Help for paper pilers. ADDitude. https://www.additudemag.com/how-to-organize-paperwork-adhd-clutter/

World Health Organization. (2024 January 2). Assistive technology. World Health Organization. https://www.who.int/news-room/fact-sheets/detail/assistive-technology

Wright, S. (2014). *ADHD coaching matters: The definitive guide*. ACO Books.

Yin, H., Sun, Z., Li, D., & Shi, J. (2005). Time perception characteristic in children with attention-deficit/hyperactivity disorder. *East Asian Archives of Psychiatry*, 29(1), 22–27.

Young, S., & Thome, J. (2011). ADHD and offenders. *The World Journal of Biological Psychiatry*, 12(1), 124–128. DOI: 10.3109/15622975.2011.600319 PMID: 21906010

Zylowska, L., Ackerman, D. L., Yang, M. H., Futrell, J. L., Horton, N. L., Hale, T. S., & Smalley, S. L. (2008). Mindfulness meditation training in adults and adolescents with ADHD: A feasibility study. *Journal of Attention Disorders*, 11(6), 737–746. DOI: 10.1177/1087054707308502 PMID: 18025249

Chapter 8
Exploring the Application of Immersive Virtual Reality in English Speaking Instruction for Chinese College Students

Meng Huan
Universiti Sains Malaysia, China

Mageswaran Sanmugam
https://orcid.org/0000-0003-3313-4462
Universiti Sains Malaysia, Malaysia

ABSTRACT

This study explores the application of immersive virtual reality (VR) in English speaking instruction for Chinese university students. With the continuous development and popularity of immersive VR, education based on this technology has emerged as a new field. As an internationally recognized language, English is widely used and promoted globally, making the learning of English speaking crucial for students' development. Therefore, utilizing immersive VR holds significant practical value for enhancing the quality and effectiveness of English speaking instruction.

DOI: 10.4018/979-8-3693-9811-1.ch008

INTRODUCTION

Immersive Virtual Reality (IVR) is an advanced, ideal, and complex virtual reality system (Fan, Y. L., & Ding, W., 2020). It utilizes enclosed scenes and audio systems to isolate the user's auditory and visual senses from the outside world, immersing them completely in a computer-generated environment. Users can input relevant data and commands through input devices such as spatial position trackers, data gloves, and 3D mice. The computer measures the user's movements and postures based on the data collected, feeding this information back into the generated visuals, creating a sense of presence and complete engagement in the experience. In the implementation of immersive virtual reality, Head-Mounted Displays (HMDs) are among the most common and widely used tools. HMDs integrate high-resolution displays and sensors into a device worn on the head, providing users with a comprehensive immersive experience. This device can directly present the virtual environment in the user's field of vision and accurately track head movements, enabling real-time interaction with the virtual world. The convenience and immersion offered by such devices have led to their widespread use in various applications, including gaming, education, and training. Continuous advancements in HMD technology, particularly in high-resolution displays and precise motion tracking, provide a solid foundation for immersive learning and the application of virtual reality.This study aims to explore the application of immersive virtual reality (VR), specifically focusing on HMD tools in the English speaking instruction of Chinese university students, with a particular emphasis on analyzing their effectiveness in teaching. By constructing a relevant application framework, we hope to provide students with authentic language communication scenarios and assess the impact of immersive VR on their speaking skills. The research will leverage immersive VR to enhance students' speaking practice experiences and explore its potential in improving English teaching outcomes.

RESEARCH BACKGROUND

In the context of rapid development in virtual reality (VR) technology, immersive virtual reality (IVR) has shown extensive potential in education by creating highly realistic interactive learning environments that help students understand knowledge more intuitively (Xiao, J., & Xie, G. M. 2019). Among emerging technologies, IVR seems particularly well-received and applied in language education (Wu, J. G., Miller, L., Huang, Q., & Wang, M. 2023). IVR is a technology that creates artificial virtual environments to replace users' real-world environments (Lee, S.-M., Yang,

Z., & Wu, J. G. 2023). Key features of IVR include complete immersion, physical presence, utilization of virtual avatars, and interaction.

Previous research has identified numerous educational benefits of virtual reality and IVR in English speaking education, such as vocabulary development (Alfadil, M. 2020) and cultural competence (Tai, T.-Y. 2022). Compared to traditional classrooms, the contextualized and immersive IVR worlds are more likely to facilitate effective English learning, where knowledge is constructed and internalized, intrinsic motivation is enhanced and sustained, and interaction is encouraged and customized (Taguchi, N. 2021). Additionally, the diverse forms of IVR offer various resources for users, allowing English teachers and learners to choose suitable VR tools according to their budget and needs, which further promotes the popularity of IVR in English speaking education (Chen, C. Y. 2022).

In IVR related research, IVR developed specifically for language learning provides various scenarios where students can learn and use English in specific locations and situations. Users in IVR can create their own virtual avatars and communicate through synchronized voice and text. This technology also offers multiple teaching functions, such as video demonstrations, assembly (gathering students), grouping (organizing students), focusing (directing students' attention to the teacher), and object placement. According to Kaplan-Rakowski and Gruber (referenced in Kaplan-Rakowski & Gruber 2019), high-immersive IVR with visual, auditory, and kinesthetic modes allows for interaction between people and with objects (Sadler & Thrasher, referenced in Sadler & Thrasher 2023).

Currently, there are several issues in college English speaking instruction. Firstly, traditional teacher-centered and exam-oriented teaching methods neglect the development of students' speaking skills and learning interests, leading to a monotonous classroom model that fails to meet the demand for high-level English talents (Fu, X. 2013). Secondly, large class sizes limit teacher-student interaction and personalized teaching, reducing students' sense of participation and opportunities for practical speaking practice (He, D. C. 2018). Moreover, the overemphasis on exams like CET-4 and CET-6 makes students view exams as the primary goal, leading to insufficient learning motivation and poor teaching effectiveness. Although some universities have attempted tiered teaching and small class reforms, there is still significant room for improvement in college English speaking instruction.

However, this study focuses on the application of immersive VR technology due to its integration of advanced technologies such as graphics, multimedia, sensing, and artificial intelligence, which can highly simulate and construct authentic language learning scenarios, meeting the need for "contextualized" interactive language learning (He, J. H., Huang, X. L., Han, G. X., et al. 2019). Zheng Chunping and others have identified four major advantages of using immersive VR in language teaching: 1) 3D immersive virtual simulation user interface; 2) anonymity of virtual avatars;

3) real-time interactive communication technology; and 4) task-oriented language teaching design. For college English, applying immersive VR to course instruction can not only promote reform in college English, facilitate the transition of teacher roles, diversify teaching methods, but also enhance students' intrinsic motivation, optimize classroom experiences, and improve overall teaching effectiveness, potentially reducing teaching costs in the long run (Zheng, C. 2019).

In this study, we will focus on the application of head-mounted virtual reality (HMD) within immersive VR for English speaking instruction. HMD VR, by providing an all-encompassing immersive experience, allows students to practice in simulated language environments, effectively enhancing their speaking abilities.

DEVELOPMENT OF IMMERSIVE VIRTUAL REALITY

Immersive virtual reality is considered one of the most popular emerging technologies today.

- 1956: Filmmaker Morton Heilig created a virtual reality system called Sensorama, which provided synchronized visual, auditory, olfactory, and tactile stimulation to viewers (Heilig, 1962).
- 1961: Cumeau and Bryan developed the first head-mounted display with motion tracking capabilities. This device allowed users to control a remote camera through head movements, enabling a natural observation of the camera's surroundings (Comeau, 1961).
- 1968: Ivan Sutherland developed the first head-mounted display with tracking and positioning capabilities that could interact with a computer. Due to its cumbersome design, the device needed to be suspended from the ceiling and was thus named the "Sword of Damocles" (Sutherland, 1965).
- 1985: The company VPL (short for "Visual Programming Language") was established in the United States. It was the first commercial company to sell virtual reality headsets. Its founder, Jaron Lanier, popularized the term "Virtual Reality," which became a standard term in the field of research (Zhao Q, 2009). Despite its development, immersive virtual reality did not achieve widespread acceptance due to high hardware costs and suboptimal user experiences. Its applications were mainly limited to specialized fields such as aerospace, military, and industrial needs.
- Post-2012: The release of consumer-oriented virtual reality devices such as Oculus Rift, HTC Vive, and Sony PSVR allowed individual consumers to experience high-quality virtual reality products. This led to increased interest from commercial companies in the virtual reality field, accelerating the

development of immersive virtual reality across various sectors including industry, medicine, education, culture, and entertainment. This expansion has greatly advanced related research.

OBJECTIVES

The primary objectives of this chapter are as follows:

i. Summarize the concept of immersive virtual reality (IVR) and its development history.
ii. Analyze the characteristics and system types of immersive virtual reality.
iii. Evaluate the types of immersive virtual reality devices and their application advantages.
iv. Explore specific application models of immersive virtual reality in college English speaking teaching.
v. Review the current research status and trends of immersive virtual reality both domestically and internationally.
vi. Establish a theoretical framework for immersive virtual reality and define operational terminology.
vii. Discuss the significance, limitations, and ethical considerations of immersive virtual reality research.

CHARACTERISTICS OF IMMERSIVE VR

i. **Real-Time Performance:** To achieve the same sensation as the real world in an immersive virtual reality system, high real-time performance is essential. For example, when the user's head moves and changes the view, the system's tracking devices must detect this movement promptly, and the computer must calculate and display the corresponding scene with minimal delay, ensuring that changes are continuous and smooth.
ii. **High Immersion:** Immersive VR systems use various input and output devices to create a virtual world that looks, sounds, and feels real. This requires a high degree of immersion, ensuring that the user is completely isolated from the real world and unaffected by external stimuli.

iii. **Good System Integration:** To provide a comprehensive sense of immersion, the system must integrate multiple devices and related software technologies, ensuring that these components interact seamlessly without interference. Therefore, the system must have good system integration.
iv. **Good Openness:** Immersive VR systems should leverage the latest hardware and software technologies. This requires the immersive VR system to be easily adaptable for hardware and software improvements, necessitating a more flexible approach to constructing the system's hardware and software architecture.
v. **Support for Parallel Operation of Multiple Input and Output Devices:** To achieve a full sense of immersion, multiple devices may need to be used in combination and work synchronously. The immersive VR system should support the parallel operation of various input and output devices.

In this study, the immersive VR system, by providing high immersion and interactive experiences, enables students to simulate real-life scenarios in a virtual English learning environment, enhancing their language application skills. This system not only adapts to students with different learning styles but also boosts their intrinsic motivation through real-time feedback and interaction.

TYPES OF IMMERSIVE VIRTUAL REALITY SYSTEMS

Common Types of Immersive Virtual Reality include Head-Mounted VR, Cave Automatic Virtual Environment (CAVE), Cockpit VR, Projection VR, and Telepresence Systems (Shen, W., & Xia, L. 2020).

i. **Head-Mounted Virtual Reality System (HMD):** This system uses a head-mounted display to provide a stereoscopic visual and auditory output for a single user, allowing them to be fully immersed in the virtual environment.
ii. **Cave Automatic Virtual Environment (CAVE):** This system is a projection-based, collaborative environment that uses multi-channel visual synchronization technology and stereoscopic display technology. It allows multiple participants to be immersed in a virtual environment surrounded by stereoscopic projections. Using appropriate virtual reality interactive devices, participants experience an immersive environment with six degrees of freedom interaction.
iii. **Cockpit Virtual Reality System (COCKPIT):** This system is an airplane simulation cockpit installed on a motion platform. Users sit in the cockpit and perform operations such as flying or driving through instruments and displays. They can observe changes in the outside environment through "windows" and feel the rotation and tilting of the cockpit, creating the sensation of being in a

realistic virtual world. This system is mainly used for flight and vehicle driving simulations.

iv. **Projection-Based Virtual Reality System (PROJECTION):** This system uses one or more large-screen projections to achieve stereoscopic visual and auditory effects on a large scale, allowing multiple users to feel fully immersed simultaneously.

v. **Remote System (REMOTE):** Through computers and networks, users receive realistic sensory feedback and interaction, as if they are physically present in the location. They can remotely control operations at the site.

COMPARISONS OF IMMERSIVE VR

This study provides a detailed comparative analysis of the five common types of immersive virtual reality (VR) systems, including head-mounted VR systems (HMD), Cave Automatic Virtual Environment (CAVE) systems, cockpit-style VR systems, projection-based VR systems, and telepresence systems. Each of these systems has unique characteristics and is widely applied across various fields, showcasing distinct advantages in different areas.

Head-mounted VR systems (HMD) are currently the most widely used immersive VR devices. In his study, Zhang Maojun (2001) detailed the technical features and application scenarios of this system. He emphasized that HMD systems provide a strong sense of immersion by using a display worn on the user's head, allowing users to become fully immersed in the virtual environment. Additionally, HMD systems offer high convenience, allowing users to operate and move easily, making them widely used in fields such as education, entertainment, and healthcare.

CAVE systems excel in immersion and visual display. According to research by Zhang Muhua, Liu Ziyi, and Li Yan (2024), CAVE systems use multiple projectors to display images on four or six walls. The user stands in the center of the system and experiences immersion through a panoramic visual effect. Their meta-analysis of 44 empirical studies published domestically and internationally between 2011 and 2022 indicates that CAVE systems not only perform well in terms of immersion but also have a significant positive impact on enhancing student learning outcomes, particularly in science and engineering education.

Cockpit-style VR systems are primarily applied in high-complexity simulation scenarios. Jia Yufei et al. (2018) explored the application of cockpit-style VR systems in the simulation of special vehicles, emphasizing the value of these systems in complex simulation tasks. Cockpit systems typically simulate specific environments, such as an airplane cockpit or a car driving simulator, offering highly realistic interactive experiences. Therefore, they are widely used in aerospace, military, and specialized equipment training.

Projection-based VR systems are more commonly applied in education and training, especially in medical education. Huang et al. (2016) examined the technical features of projection-based VR systems and their acceptance in medical teaching. This system projects virtual images into a physical space, allowing multiple users to participate and interact simultaneously. Studies show that projection-based VR systems can effectively enhance learners' motivation and engagement, making them particularly suitable for collaborative learning and training.

Remote systems represent a unique form of immersive VR, primarily used in remote operations and laboratory simulations. Tzafestas et al. (2006) provided a detailed analysis of its application in remote operations, especially in laboratory environments. These systems offer users an immersive operational experience by remotely controlling equipment with real-time feedback. Telepresence systems have significant application value in fields such as telemedicine, remote education, and industrial control.

Through the analysis of five major immersive virtual reality (VR) systems, we find that head-mounted (HMD) systems offer significant advantages in terms of immersion, interactivity, and portability. Although Cave Automatic Virtual Environment (CAVE) systems excel in immersion and interactivity, their high cost and complex equipment limit their widespread application. Cockpit VR systems are suitable for flight and driving simulation training, projection-based VR systems are commonly used for exhibitions and multi-user experiences, and Remote systems primarily serve remote meetings and medical fields. While these systems perform well in specific applications, their high cost and complex equipment restrict their general applicability. Due to its high immersion, good portability, and moderate cost, head-mounted VR has become a mainstream technology in the market, particularly suitable for education, training, and entertainment.

In the context of English speaking skills, head-mounted VR demonstrates unique advantages. Its high level of immersion and interactivity effectively simulates real language environments, enhancing students' speaking practice and language abilities. Compared to other immersive VR systems, head-mounted VR provides a more personalized and flexible learning experience, allowing students to achieve higher engagement and motivation in simulated conversations and language training.

TYPES OF IMMERSIVE VR DEVICES

i. **PC VR:** A VR display device that relies on an external computer or console for operation. It uses multiple sensing devices, such as gyroscopes, to track user movements and enhance the sense of immersion. Often referred to as an "upgraded display with sensing and interactive capabilities," its data processing and image transmission are handled by the PC. It features high technical content, offering the strongest immersion and interactivity, but with poor portability.
ii. **Mobile VR:** A display device that uses a smartphone as the content running system. It mainly consists of a phone case with two lenses, with the phone serving as the platform for computing and display. It has a simple structure and is low-cost. Its additional equipment, such as Bluetooth controllers or touchpads, is typically used to control the smartphone. This VR device has low technical content, offering a less immersive experience and poor interactivity.
iii. **All-in-one VR:** An independent device that integrates both the content platform and the display headset into one unit. It is equivalent to embedding the configuration of a high-end smartphone into the headset, with the advantage of being highly portable. It doesn't require a phone, an external computer, or a gaming console. Equipped with sensors like gyroscopes for positional tracking, it can perform independent processing and input/output functions.

With the continuous development of immersive virtual reality, selecting suitable VR devices is crucial for enhancing the learning experience. Each type of device has unique advantages and limitations that affect users' sense of immersion and interaction. Console VR devices, with their superior performance and high interactivity, are suitable for learning scenarios that require deep interaction and experience. In contrast, mobile VR offers greater accessibility due to its convenience and low cost, although it may compromise on technical capabilities. All-in-one VR headsets perfectly combine the advantages of both, providing a convenient user experience and good performance, making them an increasingly popular choice in the field of education.

In the practice of English speaking instruction for Chinese university students, the rational selection of immersive VR devices can create a richer language learning environment, stimulating students' interest and engagement. Therefore, future research and practice can explore how these devices can leverage their respective strengths in different teaching scenarios, particularly regarding interactivity and immersion in language learning. Additionally, studies should focus on how to align specific teaching objectives and student needs to design more effective teaching strategies, promoting innovation and development in English speaking instruction and ultimately enhancing students' language proficiency and confidence. Through such exploration, immersive VR is expected to play a larger role in education, supporting students' comprehensive development.

ADVANTAGES OF USING IMMERSIVE VR IN COLLEGE ENGLISH SPEAKING INSTRUCTION

With the rapid development of virtual reality (VR) technology, immersive virtual reality (IVR) has demonstrated broad application potential in the educational field, particularly in creating diverse simulated learning environments. IVR offers students an immersive learning experience, simulating scenarios and tasks that are difficult to obtain or replicate in the real world, making it especially suitable for language learning that requires contextual communication and task interaction (Natale et al., 2020). In college English teaching, IVR exhibits four significant advantages:

i. Cross-temporal and Spatial Immersive Interaction:

Immersive VR provides a cross-temporal and spatial interactive experience, allowing students to interact with native speakers in a virtual environment. This dual stimulation of visual and auditory senses enhances language learning effectiveness (Feng, Q. 2016). For example, VR can transform abstract concepts from textbooks into visual content, helping students retain language knowledge more effectively.

ii. Task-based Language Learning:

IVR supports task-based language learning, enabling students to engage in role-playing and scenario simulations guided by specific task objectives. This approach improves the relevance and fluency of language use (Fang, Y. 2015).

iii. Low-pressure Virtual Learning Space:

IVR provides a low-pressure virtual learning space where students can interact anonymously or through avatars. This setup enhances self-confidence and increases motivation for autonomous learning (Zhang, N. 2017).

iv. Enhanced Interaction and Equal Teacher-Student Relationship:

IVR fosters more interactive and equitable teacher-student relationships, encouraging students to actively participate in the classroom and stimulating their interest in learning. It also offers teachers a richer set of teaching tools (Chen, Y. 2003).

CHALLENGES OF USING IMMERSIVE VR IN COLLEGE ENGLISH SPEAKING INSTRUCTION

Despite the numerous advantages of immersive virtual reality (IVR) in education, its implementation faces several challenges.

i. Financial Constraints:

Schools may encounter financial difficulties when adopting IVR, as the high cost of equipment and maintenance makes it challenging to scale up (Fu, X. Q. 2013).

ii. Technological Limitations:

Technologically, IVR is not yet fully mature. The virtual task libraries available for English learning are relatively limited, and there remains a gap between promotional claims and actual usage effectiveness. Additionally, integrating IVR seamlessly into daily teaching has yet to be achieved (Liu, D. J., & Huang, R. H. 2016).

iii. Lack of Familiarity:

There is a varying degree of familiarity with modern information technology among teachers and students. Many educators and learners lack sufficient understanding of IVR, and some have doubts about its role in education (Gao, Y., Liu, D. J., et al. 2016).

iv. Experimental Stage:

The application of IVR in higher education English teaching is still in the experimental phase, lacking systematic and sustained application models (Chung, L. Y. 2011).

Given these advantages and challenges, this study aims to explore the impact of immersive virtual reality on English speaking skills among Chinese university students, particularly its role in enhancing students' intrinsic motivation. Specifically, the research will focus on the application of head-mounted VR as a significant form of immersive VR, which, through highly immersive experiences, facilitates students' entry into learning contexts and improves their language learning outcomes.

RESEARCH ON THE APPLICATION OF IMMERSIVE VR IN COLLEGE ENGLISH SPEAKING INSTRUCTION

The effective use of immersive VR in the teaching process of college English speaking can create a conducive English-speaking environment for students, significantly enhancing the teaching effectiveness. Currently, during the period of reform in college English education, the introduction of immersive VR can establish an excellent teaching environment, thus stimulating students' initiative and ultimately improving their overall competencies. Effective use of immersive VR can help students connect with global standards, adapting to the rapid development of globalization.

i. Creating a New Teaching Environment

The core technologies of immersive VR in English speaking include eye-tracking systems, VR virtual scene construction, AI dialogue, and speech recognition systems. These technologies create virtual speaking training scenarios where students can interact with AI in unexpected situations, enhancing their speaking skills. At the same time, the system records students' subconscious actions and psychological traits during these interactions, improving their speaking and social skills from multiple angles. Using VR eye-tracking system, the system can track and record students' eye movement and fixation duration during the virtual speaking simulation, aggregating the data into heat maps. These heat maps can be analyzed to understand students' psychological and behavioral patterns during the speaking process. Through AI technology and speech recognition, the system selects situational dialogues based on keywords from students' speech, converting their spoken words into text for an evaluation of their actual speaking abilities. The virtual simulation system recreates various linguistic environments and usage scenarios, simulating emergencies that help students improve their speaking skills. Immersive VR integrates virtual 3D scenes into actual classroom settings, enhancing the scientific and innovative aspects of the teaching environment. With this technology, teachers can shift away from traditional lecture-based methods, ensuring timely and innovative learning experiences for the new generation of students, helping them access the most cutting-edge knowledge. A well-matched teaching environment can lead to greater teaching effectiveness. Immersive VR can create a 360° fully immersive learning scene, improving students' overall competencies while enhancing the quality of instruction.

ii. Creating a Positive Learning Atmosphere

When using immersive VR to create a speaking environment for students, it is important to integrate English lessons into real-world applications. Teachers can effectively combine textbook knowledge with students' real-life experiences, allowing students to integrate what they've learned in specific English-speaking environments. Immersive VR can vividly recreate scenes necessary for speaking practice in a highly immersive audiovisual format. For example, when students enter the workforce and need to use English for foreign trade, teachers can use immersive VR to simulate a business negotiation environment. Students can role-play as business managers and engage with virtual foreign traders. These traders are rendered via immersive VR with pre-recorded voices tailored to different scenarios. In this environment, students can experience a sense of immersion, both visually and aurally, and get a feel for real-world job tasks. This experience can stimulate students' interest in learning and help them gain valuable experience before entering the workforce. Using immersive VR in English speaking instruction can not only improve teaching efficiency but also provide students with access to the most advanced knowledge. In the era of fast-developing online videos, traditional paper textbooks and multimedia classrooms struggle to meet the learning needs of the new generation of college students. By using computer technology to create immersive VR scenes, embedding vivid images and videos into these scenes, students' auditory and visual senses can be fully engaged, providing them with a learning environment that differs from traditional methods. This creates a more up-to-date learning atmosphere, bringing a fresh experience and helping to capture students' attention. VR environments can generate a sense of novelty, motivating students to actively engage in learning.

iii. Stimulating Learning Interest

According to real data, individual differences exist among students. Some students have strong adaptability and can quickly adjust to changes in the environment, mastering and applying textbook knowledge proficiently. Under the influence of new educational reform policies, the introduction of immersive VR can effectively support immersive English-speaking learning. With the rapid development of science and technology, many colleges and universities are already equipped with advanced electronic devices. Students' learning spaces are no longer limited to physical platforms but can access excellent virtual platforms through immersive VR. In these virtual environments, students can obtain high-quality learning resources, broaden their horizons, and extend their knowledge. Immersive VR provides tailored English-speaking learning environments for students, shifting the focus of the classroom from teachers to students, who are given more opportunities to lead. Students can choose tasks and roles in the pre-set scenarios that match their interests and learning styles for speaking practice. These highly realistic scenarios can stimulate students' enthusiasm for learning, unleash their creativity, and offer them a different learning experience. This aligns with the concept of personalized education.

iv. Improving the Teaching Evaluation System

During the reform of college English-speaking instruction, the importance of the teaching evaluation system should also be emphasized. Teachers can use immersive VR to make the teaching evaluation system more scientific and reasonable, promoting a more modernized model for English-speaking instruction. Through big data technology, teachers can track students' performance and learning status in virtual reality scenarios. Based on system-generated data, using text analysis and index calculation, teachers can objectively analyze students' subjective performances. This analysis data can then be input into the information system, allowing for an objective evaluation of current learning levels. As a result, teachers gain a precise understanding of students' fundamental learning conditions, providing them with reasonable suggestions and guiding assistance for further in-depth English learning. Finally, using immersive VR, teachers can design appropriate learning tasks based on students' individual levels, guiding them with positive feedback to practice repeatedly, thereby improving their English-speaking skills.

ANALYSIS OF IMMERSIVE VR CLASSROOM TEACHING APPLICATION MODELS

Based on the characteristics of different VR device types, this study proposes four different VR teaching application models, with detailed analysis as follows:

i. Mobile VR and All-in-One VR Application Model

The principle of Mobile VR and All-in-One VR is to use binocular parallax and gyroscopes to achieve stereoscopic visual display and image rotation tracking. Interaction is accomplished by continuously gazing at a hot spot or button to open or close embedded images, videos, and voice overs within panoramic scenes, or to activate embedded 360-degree sequence images that present 3D objects. In the classroom, students need to prepare their own smartphones that meet the necessary specifications, install the appropriate app, insert the phone into a VR headset (or directly wear the All-in-One VR headset) to experience the VR content. This model is the most cost-effective solution, often used for viewing panoramic images and videos. It is suitable for VR experiences based on image synthesis in panoramic galleries and allows for simultaneous use by multiple students in a regular classroom or laboratory.

ii. PC VR Application Model

The PC VR system, which uses head-mounted displays (HMD) and motion controllers, offers the best immersion and interactivity, providing students with the optimal individual experience. However, the headset and controllers need to be connected to a high-performance PC, which makes the overall setup quite expensive. Additionally, each system can only be used by one student at a time, and operation requires a certain amount of space.

iii. PC VR + Large Screen Application Model

This model uses one VR system to output a large video display through a projector. The projector maps the images seen by the teacher in the headset onto a large screen, allowing students to follow the teacher's route and viewpoint while listening to the teacher's narration. The teacher acts as a visual guide, providing explanations and organizing activities, with the students seeing and hearing the pre-planned content. This model highlights both teacher-virtual scene interaction and teacher-student interaction. Additionally, students can also act as guides, leading others through the virtual environment. The advantage of this model is that it allows for large group teaching, solving the issue of many students experiencing the content simultaneously. The downside is that students do not personally enter the VR scene, resulting in weaker immersion and experience. However, compared to watching videos, students see real-time interactions between the teacher and the virtual environment, providing a stronger sense of presence. When classroom activities are well-designed and thoroughly prepared, this model can still achieve good teaching results.

iv. VR Stereoscopic Projection Application Model

This model builds upon the previous one by upgrading the display system to a multi-channel stereoscopic projection system. Students can wear 3D glasses to see stereoscopic images with depth, enhancing the visual experience from two-dimensional to three-dimensional, greatly improving the sense of presence and immersion. However, due to the requirements of viewing distance and position for stereoscopic displays, students who are too far away or off-center will not experience good stereoscopic effects. Thus, the number of participants at one time is limited.

Depending on the size of the panoramic screen, the ideal number of students is between 30 and 50 per session. This model also requires higher investment in equipment and content development compared to the third model. First, multi-channel stereoscopic projection systems are expensive, and second, the software needs to output dual images suitable for stereoscopic display in real-time, posing higher technical demands on developers.

This study provides an in-depth analysis of the application models of VR in classroom instruction, revealing the characteristics of different types of VR devices and their potential in English speaking education. Whether it's the economical mobile VR and all-in-one VR headsets or the highly immersive console VR systems, various models demonstrate the possibility of enhancing student engagement and learning outcomes. The combination of console VR with large screens and VR stereoscopic projection further enriches teaching methods, offering new opportunities for interaction between teachers and students. Future research could focus on how to integrate these application models to optimize classroom teaching design and enhance students' language learning experiences. As technology continues to advance and educational needs evolve, the prospects for immersive virtual reality in the education sector are broad, warranting ongoing attention and research. Through continuous innovation and practice, immersive VR is expected to bring about profound changes in language education, promoting students' holistic development.

CURRENT RESEARCH STATUS AND TRENDS OF IMMERSIVE VR IN ABROAD AND CHINA

- Current Status of Research Abroad

Immersive Virtual Reality (VR) has been widely applied across various disciplines in education. To specifically understand the research status of immersive VR in college English speaking classes, this study conducted a search in the Web of Science Core Database using the keywords "head-mounted virtual reality" and "English speaking learning," covering the time span from January 1, 2001, to September 6, 2024. A total of 177 articles were retrieved, and after refining the papers to exclude irrelevant ones and reviews, 116 articles were identified. Based on the relevance of the topics, 46 pertinent documents were selected. Through the organization and analysis of the aforementioned literature, insights were gained regarding the research status, outcomes, and trends of immersive VR in English speaking education.

i. Analysis of Literature Distribution

After filtering the literature based on relevance, the number of publications and their publication dates were statistically analyzed. Between 2001 and 2007, there were no relevant studies on the application of immersive VR in English education. In the following years, only a few articles were published sporadically, but starting in 2019, a steady stream of articles on this research topic emerged. The number of publications surged in 2021, and by September 2024, a total of 31 papers had been published. From the publication trend, it can be inferred that future articles related to immersive VR in English speaking education will likely increase in foreign contexts, indicating that this area is becoming a research hot spot as new teaching models are continually being explored in English education.

ii. Keyword Analysis

To understand the research status, outcomes, and trends of immersive VR in English speaking education, this study employed Cite Space, a visualization analysis software, to analyze the 46 relevant articles. The core keywords identified in this research include "immersive virtual reality," "English speaking," "education," "language," "applications in subject area," and "interactive learning environment." The research on the application of immersive VR in English speaking education abroad is relatively advanced. Many studies incorporate additional learning strategies or artificial intelligence technologies based on the introduction of virtual reality technology to explore their impact on the effectiveness of English speaking teaching.

For example, Chen and Hsu designed an English speaking learning program based on a virtual reality game and employed self-regulation strategies to investigate students' English speaking learning outcomes from cognitive and psychological perspectives. Their experimental results indicated that VR enhanced students' learning potential, and the immersion provided by VR increased students' self-efficacy, thereby reducing their anxiety in English speaking learning (Chen, Y. L., & Hsu, C. C., 2020). Similarly, Wang allowed students in the experimental group to use an artificial intelligence-based college English tutoring system and expert system within a virtual reality environment to aid their English speaking learning, while the control group utilized multimedia devices and traditional teaching methods. The results showed that the experimental group outperformed the control group in English speaking proficiency (Wang, X., 2022).

The keywords "higher education" and "cognitive style" are both relevant to the research subjects. Approximately one-third of the literature focuses on university students. For instance, Bacca-Acosta et al. explored scaffolded English learning in immersive virtual reality technology using eye-tracking methods, selecting 44 university students for their study. They found that natural scaffolding in immersive virtual reality environments was most beneficial for students' English learning (Bacca-Acosta, J., Tejada, J., Fabregat, R., et al., 2022).

Additionally, several studies abroad have examined the differing learning outcomes of students with varying cognitive styles in immersive virtual reality environments when learning English speaking skills. For example, Chen and Hwang investigated whether interactive spherical video-based virtual reality technology would affect the English speaking performance and learning anxiety of students with different cognitive styles. The research indicated that the interaction between learning methods and cognitive styles significantly influenced students' speaking performance and learning anxiety (Chen, M. R. A., & Hwang, G. J., 2022).

- **Current State of Domestic (China) Research**

Research on immersive virtual reality (IVR) in China began relatively late. This study conducted a search in the CNKI database using the keywords "immersive virtual reality" and "English speaking," covering the period from January 1, 2001, to September 6, 2024. After filtering for research relevance, a total of 100 papers were identified. A review of these papers provides a clear understanding of the current state, hot topics, and trends of research on the application of IVR in English speaking.

i. Literature Distribution

Statistical analysis of the selected literature reveals trends in the application of IVR to English speaking in China. Since 2004, the use of immersive virtual reality technology in English speaking has gradually emerged. From 2009 to 2022, the number of published papers fluctuated, peaking in 2020 with 21 publications. Subsequently, the number declined until 2024, when it reached a peak of 31 papers.

ii. Keyword Analysis

Using the visualization software Citespace, the 100 selected papers were analyzed to better understand the research status and hot topics in this field. The keywords "college English speaking" and "English teaching in higher education" indicate that the primary focus of domestic research is on university students. Some studies apply to general college English, while others focus on English in specialized courses.

The application of immersive virtual reality in English speaking teaching is showing a gradual upward trend. Both domestically and internationally, as research deepens and technology advances, immersive virtual reality is becoming a popular research direction in the field of education. Although international research began to grow significantly in 2019, domestic research has also seen notable growth in recent years, despite its later start. This indicates a sustained global interest and research focus on this technology.

The future application of immersive VR technology in English teaching holds great potential and value. By creating highly immersive learning environments, students can engage more actively in the learning process, enhancing their motivation, participation, and learning outcomes. This technology is particularly suitable for language disciplines as it offers learners a more interactive and authentic language practice environment. Continued exploration of immersive VR applications in English speaking teaching will not only promote innovative teaching methods but also provide optimized solutions for personalized learning, playing a significant role in future educational reforms.

THEORETICAL FRAMEWORK

The purpose of the theoretical framework is to help identify concepts related to the research problem and focus on determining solutions. This study's theoretical framework is based on three main theories:

- Cognitive Load Theory(CLT) (Sweller, 1994)
- Cognitive Theory of Multimedia Learning (CTML) (Mayer, 2001)
- Cognitive Affective Theory of Learning with Multimedia (CATLM) (Moreno & Mayer, 2007)

These theories constitute the theoretical framework of this study and are depicted in the research framework.

Cognitive Load Theory(CLT)

Cognitive architecture consists of limited working memory and partially independent visual and auditory information processing units, which interact with unlimited long-term memory. Cognitive Load Theory involves methods to reduce working memory load, aiming to alleviate the changes in long-term memory associated with schema acquisition (Sweller, 1994).

Cognitive Load Theory (CLT) was first proposed by educational psychologist John Sweller in the 1980s. Cognitive load refers to the total amount of information that can be processed during human information processing, primarily encompassing the total volume of information that working memory can store and process (Sweller, J. 2008). CLT is based on the theories of limited resources and schemas, examining learning and problem-solving from the perspective of resource allocation. The theory of limited resources posits that human resources are finite; engaging in multiple activities simultaneously raises issues of resource distribution, governed by the basic principle that "the total amount remains constant, so if one increases, another decreases." During the learning or problem-solving process, each cognitive activity consumes cognitive resources. If the total amount of resources required by these activities exceeds the individual's available resources, issues of insufficient resource allocation arise, thereby impacting the effectiveness of learning or problem-solving.

In this study, Cognitive Load Theory is selected as the theoretical framework, focusing primarily on its impact on intrinsic motivation. This theory posits that the cognitive load experienced during learning affects learners' attention and information processing abilities. In the context of using immersive virtual reality (VR) for English learning, students may encounter rich information and complex tasks that require them to manage this information effectively, potentially influencing their intrinsic motivation. By analyzing the cognitive load levels of Chinese university students of different genders when using immersive VR to learn English, and how this cognitive load impacts their intrinsic motivation, this study aims to explore the relationship between intrinsic motivation and cognitive load. This will provide a theoretical basis for designing more effective teaching strategies and virtual learning environments, enhancing students' learning experiences and motivation.

Mayer's Cognitive Theory of Multimedia Learning (CTML)

Mayer's Cognitive Theory of Multimedia Learning (CTML) (2001) describes how people learn from words and pictures and how they process information through two basic channels: verbal and visual. Mayer's theory is based on meaningful learning theory, drawing on work related to generation theory by Wittrock et al. and dual coding theory by Paivio et al. (Mayer, 1997). Generation theory helps understand that meaningful learning occurs when learners select relevant information from the presented content, organize it into coherent mental representations, and integrate these newly constructed representations with existing ones (Mayer, 1997). It explains that cognitive processes occur in two distinct information processing systems: the visual system for processing visual information and the verbal system for processing verbal information (Mayer, 1997). Mayer (2001) proposed a cognitive model of multimedia learning that illustrates the human information processing system, divided into three stages: sensory memory, working memory, and long-term memory.

Using Mayer's Cognitive Theory of Multimedia Learning as the theoretical framework for this study is justified because the theory emphasizes the impact of multimedia learning environments on cognitive processing, particularly in terms of information presentation. In the context of using immersive VR to learn English speaking, learners interact with information through visual, auditory, and interactive learning content, aligning with the multi-sensory information processing principles emphasized by Mayer's theory. Therefore, Mayer's theory provides strong cognitive support to explain how immersive VR affects the English speaking learning of Chinese university students.

Cognitive Affective Theory of Learning with Media (CATLM)

The Cognitive Affective Theory of Learning with Media (CATLM) (Moreno, R. 2005) expands upon the Cognitive Theory of Multimedia Learning (Mayer, R. E. 2005a) to include media such as VR, case-based learning environments, and agent-based learning, where learners are presented with instructional materials beyond just words and pictures. Figure 1.6 shows the Cognitive Affective Theory of Learning with Media (CATLM) (Moreno, 2005).

This study chooses CATLM because the theory encompasses both cognitive and affective processes that learners experience when using media for learning. In the context of using immersive VR for English speaking, learners receive information through media and process and understand this information cognitively, while also being influenced by emotional factors such as interest, motivation, and mood. The CATLM theory emphasizes the importance of learners' cognitive and emotional responses to media on learning outcomes, as well as the impact of media design on emotions and cognition. In this study, CATLM helps to gain deeper insights into the cognitive and emotional processes of Chinese university students of different genders when using immersive VR for English speaking, further exploring the impact of media learning on students. Thus, the choice of CATLM provides a comprehensive theoretical framework for explaining and analyzing the role of immersive VR in English speaking learning and offers valuable theoretical support for the research.

DEFINITION OF OPERATIONAL TERMS

- **Immersive Virtual Reality (IVR)-** refers to the use of computer-generated simulations of a 3D virtual world, engaging multiple sensory modalities such as visual, auditory, and tactile sensations, to create an immersive experience that makes the user feel as if they are physically present in that virtual environment (Luo, T., & Leng, W. 2021).

In this study, immersive VR provides a novel, highly interactive, and immersive environment for learning English speaking. This technology can simulate real language environments, allowing students to practice language in scenarios that closely resemble real-life situations, thus enhancing their language proficiency and learning outcomes. Additionally, immersive VR helps boost students' intrinsic motivation by stimulating their curiosity and desire for exploration, leading to increased engagement in the learning process.

- **Head-Mounted VR -** is a type of immersive virtual reality technology that provides an immersive visual and auditory experience through a device worn on the head, known as a head-mounted display. This device typically includes one or more screens that project virtual environment images directly in front of the user's eyes, combined with head-tracking technology to allow users to move and interact freely within a three-dimensional space, thereby achieving an immersive experience (Anglin, J. M., Sugiyama, T., & Liew, S. L. 2017).

Head-mounted VR offers a highly immersive learning experience, which holds significant potential for English speaking education. By fully placing users in a virtual environment, head-mounted VR allows them to learn and practice language through natural visual and auditory interactions, which enhances the realism and engagement of the learning experience. Compared to other VR technologies, head-mounted VR better simulates real-life contexts, thus improving students' language application skills and communication abilities. Furthermore, the motion-tracking capabilities of head-mounted VR support multidimensional interactions, providing a personalized learning experience that is especially important for students with varying learning styles. Therefore, researching the application of head-mounted VR in English speaking education not only provides insights into its impact on students' intrinsic motivation but also offers valuable perspectives for future developments in educational technology.

- **Intrinsic Motivation-** Motivation is the driving force behind human behavior. It can stem from physiological or psychological needs, thoughts, or emotions (Baumeister, 2016). Our physiological needs help maintain optimal bodily function, while psychological needs typically support our growth. Motivation can be internal (intrinsic) or external (extrinsic). Ryan and Deci define intrinsic motivation as engaging in an activity for the inherent satisfaction it brings, rather than for some separable outcome (Ryan, R. M., & Deci, E. L. 2000). When driven by intrinsic motivation, an individual acts for the inherent pleasure or challenge of the activity itself, rather than for external rewards, pressure, or outcomes. Thus, when we do something simply because we want to, and because the act itself brings us joy based on our natural interests, values, and passions, it is intrinsic motivation at play.

In this study, intrinsic motivation is a key factor driving students' sustained engagement in learning. In the context of immersive VR for English speaking, an engaging and interactive learning environment can stimulate students' interest and curiosity, thereby enhancing their intrinsic motivation. Consequently, investigating changes in intrinsic motivation can help us understand how immersive VR impacts students' learning attitudes and behaviors.

- **College Students-** are those who have completed their secondary education and are pursuing higher education in a university setting (Yao, X. 2003). The age range for university students can vary by country, but typically, they are between 18 and 22 years old. They are engaged in undergraduate studies or other higher education qualifications and usually receive specialized training, beginning to develop their knowledge and skills in specific fields.

In this study, focusing on college students is crucial because the university stage is a critical period of personal learning and growth, marked by the transition from high school to higher education and significant changes in learning tasks and environments. By exploring intrinsic motivation among university students, we can better understand their learning needs and challenges during this pivotal phase, providing important references and theoretical foundations for the research.

RESEARCH SIGNIFICANCE

This study's significance can be understood from multiple perspectives regarding its theme. It focuses on the application of immersive virtual reality (VR) in English speaking instruction for Chinese university students, exploring how this technology can enhance language learning and classroom interaction. This research not only enriches the theoretical framework of educational technology and foreign language teaching but also provides valuable references and insights for future teaching practices.

Firstly, the research contributes to the development of personalized learning models. In traditional teaching models, it is often challenging for teachers to address each student's learning differences and needs fully, leading to uneven teaching outcomes. By introducing the concept of learning styles, this study aims to construct an immersive VR English learning framework that can deliver precise instruction based on individual differences. This learning model can provide customized learning content and interactive methods according to students' learning characteristics and styles, thus better meeting their needs and enhancing their learning achievements.

Secondly, the study holds significant practical value. With the continuous development and popularization of immersive VR, its applications in education are becoming more widespread. This research empirically verifies the effectiveness and feasibility of immersive VR in English speaking learning, providing valuable experiences and insights for future teaching practices. Additionally, the results can serve as references for immersive VR teaching in other subjects, promoting its broad application and further development in the field of education.

Lastly, the research promotes interdisciplinary collaboration and integration. In the era of information explosion, interdisciplinary research has become a crucial trend. This study combines theories and methods from educational psychology and educational technology, offering new perspectives and directions for interdisciplinary research through comprehensive analysis and in-depth exploration.

RESEARCH SCOPE AND LIMITATIONS

The limitations of a study refer to issues and events that occur beyond the control of the researchers (Simon & Goes, 2013). The limitations of this study are outlined as follows:

i. The limited time span of the study may not adequately capture the changes in students' speaking abilities and learning experiences during long-term use of immersive virtual reality (VR) technology.
ii. The limitation of the sample size. The study may focus only on students from specific universities, resulting in a relatively small sample size that may not comprehensively reflect the actual situation of Chinese university students in English speaking learning nationwide.
iii. The implementation and application of immersive VR also have limitations. Although immersive VR provides an immersive and interactive experience for English speaking learning, its practical application in teaching is still subject to certain constraints. For instance, factors such as the availability of devices, technological stability, cost, and students' acceptance may all affect the widespread use of immersive VR.
iv. The limitations of research methods and the challenges of data collection cannot be overlooked. Both quantitative and qualitative research have their advantages and disadvantages. Choosing the appropriate research method and effectively collecting and analyzing data are crucial to the success of the study. Additionally, ensuring the reliability and validity of the research, while avoiding sample bias and subjective interference, are important issues that need to be addressed during the research process.

RECOMMENDATIONS FOR FUTURE RESEARCH

After exploring the application of immersive virtual reality (IVR) in English speaking instruction for Chinese university students, future research can further deepen and expand in the following areas:

i. Investigating the impact of Different Types of Immersive VR Devices:

This study primarily focuses on one type of immersive VR device, namely head-mounted VR. Future research could compare the different effects of various VR devices (such as desktop VR) on students' English speaking learning outcomes, providing a more comprehensive understanding of the educational potential and limitations of each type of device.

ii. Expanding the Diversity of Research Samples:

This study targets Chinese university students; future research could broaden the sample range to include students from different regions, age groups, or disciplines, exploring the potential impact of cultural background, language proficiency, and learning stages on immersive VR learning experiences. Additionally, examining the performance of students with varying levels of English speaking proficiency in immersive VR environments would help reveal the applicability of this technology among different learners.

iii. Assessing the Adaptability of Different Student Groups to Immersive VR

Research could compare the adaptability and learning outcomes of students of different genders, learning styles, and foundational English levels in immersive virtual reality environments, exploring how to adjust teaching strategies based on student characteristics.

iv. Exploring the Integration of immersive VR with Other Teaching Methods:

Future research could investigate the combined effects of immersive VR with other teaching methods (such as flipped classrooms, blended learning, and online learning platforms), assessing its potential applications in comprehensive teaching strategies. This would reveal the synergistic effects of immersive VR with other methods, aiding in the optimization of instructional design.

v. Studying the impact of Interaction Design in immersive VR on Speaking Learning:

The interaction design within virtual reality (such as interaction with virtual objects or virtual tutors) may have varying degrees of influence on learners' speaking skills. Future studies could further explore how interaction methods, feedback mechanisms, and social interactions in virtual environments enhance students' learning experiences.

vi. Researching the Optimization of Immersive VR Content Design:

The design of immersive VR content is crucial to students' learning experiences. Future research could explore how to optimize task design, language context settings, and feedback mechanisms within immersive VR to improve engagement and effectiveness. Additionally, studies could examine the impact of different types of content on learning outcomes, such as game-based VR tasks or deeply simulated immersive language contexts.

ETHICAL CONSIDERATIONS AND CHALLENGES

In applying immersive virtual reality (IVR) to research on English speaking learning among Chinese university students, ethical considerations and challenges are crucial. The rights and interests of university students as research subjects must be thoroughly protected. This involves respecting their autonomy, as well as ensuring privacy and data security in virtual environments. The following are key ethical considerations and challenges:

i. Autonomy and Informed Consent:

University students, as adults, should have their autonomy respected in the research process. Prior to participation, it is essential to ensure that all participants are fully informed about the study's purpose, procedures, and potential risks. Informed consent must be obtained from all participants. Additionally, participants should have the right to withdraw from the study at any time without facing any penalties or negative consequences.

ii. Privacy and Data Security:

Protecting the personal privacy and data security of university students is paramount. Personal data collected during the study (including behavioral data and language expression records) must be processed with strict encryption and anonymization measures to prevent misuse or unauthorized disclosure of participants' information.

iii. Psychological Well-being and Immersion:

The use of immersive VR technology, which creates highly realistic virtual environments, may have profound effects on students' psychological states. The study must ensure that the design and interaction of VR content are appropriate for students' psychological tolerance levels to avoid inducing anxiety, stress, or other psychological discomfort. Additionally, the potential impact of VR use on students' academic performance and social interactions should be monitored to ensure that it does not adversely affect their long-term mental health.

iv. Technological Accessibility and Equity:

Ensuring that all participants have equal access to and can adapt to the technology is important to avoid excluding some students due to a lack of equipment or technical barriers. Given that immersive VR devices can be expensive, it is crucial to provide all participants with the opportunity to use and become familiar with the technology to ensure fairness in the study's results.

v. Professional Development and Ethical Oversight:

Educators and researchers need to continually enhance their professional skills to adhere to the latest standards and ethical guidelines in educational technology. An effective oversight mechanism should be established to regularly assess the long-term impact of the technology on university students, ensuring that its use is ethically sound and supports students' overall development.

REFERENCES

Anglin, J. M., Sugiyama, T., & Liew, S. L. (2017). Visuomotor adaptation in head-mounted virtual reality versus conventional training. *Scientific Reports*, 7(1), 45469. DOI: 10.1038/srep45469 PMID: 28374808

Anglin, J. M., Sugiyama, T., & Liew, S. L. (2017). Visuomotor adaptation in head-mounted virtual reality versus conventional training. *Scientific Reports*, 7(1), 45469. DOI: 10.1038/srep45469 PMID: 28374808

Bacca-Acosta, J., Tejada, J., Fabregat, R., Kinshuk, , & Guevara, J. (2022). Scaffolding in immersive virtual reality environments for learning English: An eye tracking study. *Educational Technology Research and Development*, 70(1), 339–362. DOI: 10.1007/s11423-021-10068-7

Baumeister, R. F. (2016). Toward a general theory of motivation: Problems, challenges, opportunities, and the big picture. *Motivation and Emotion*, 40(1), 1–10. DOI: 10.1007/s11031-015-9521-y

Chen, C. Y. (2022). Using immersive virtual reality for training pre-service teachers in managing challenging student behaviors: A preliminary study. *British Journal of Educational Technology*, 53(4), 998–1024. DOI: 10.1111/bjet.13181

Chen, M. R. A., & Hwang, G. J. (2022). Effects of experiencing authentic contexts on English speaking performances, anxiety, and motivation of EFL students with different cognitive styles. *Interactive Learning Environments, 30*(9), 1619–1639.

Chen, Y. (2003). On the Advantages of Extracurricular Language Activities in Promoting Language Acquisition. [Social Science Edition]. *Journal of Liaocheng University*, 2003(6), 126–127.

Chen, Y. L., & Hsu, C. C. (2020). Self-regulated mobile game-based English learning in a virtual reality environment. *Computers & Education*, 154, 103910. DOI: 10.1016/j.compedu.2020.103910

Chung, L. Y. (2011). Using avatars to enhance active learning: Integration of virtual reality tools into college English curriculum. In The 16th North-East Asia Symposium on Nano, Information Technology, and Reliability (pp. 29-33). IEEE.

Comeau, C. (1961). Headsight television system provides remote surveillance. *Electronics (Basel)*, 34(2), 86–90.

Dooly, M., Thrasher, T., & Sadler, R. (2023). "Whoa! Incredible!:" Language Learning Experiences in Virtual Reality. *RELC Journal*, 54(2), 321–339. DOI: 10.1177/00336882231167610

Fan, Y. L., & Ding, W. (2020). Review of the application of immersive virtual reality technology in science teaching. *Chemical Educator (Boise, Idaho)*, 41(5), 84–90.

Fang, Y. (2015). Application of virtual reality technology in college English oral teaching. *Shanhai Classic*, 2015(16), 43–44.

Feng, Q. (2016). Exploring the application of virtual reality technology in foreign language teaching. *Theoretical Observation*, 2016(5), 170–175.

Fu, X. Q. (2013). Observations on the current situation and existing problems in college English teaching. *University Education*, 2013(7), 58–60.

Fu, X. Q. (2013). Observations on the current situation and existing problems in college English teaching. *University Education*, 2013(7), 58–60.

Gao, Y., & Liu, D. J.. (2016). Core Elements and Challenges of Virtual Reality Technology in Promoting Learning. *Research in Dance Education*, 2016(10), 77–87.

He, D. C. (2018). Analysis of the current status and prospects of the VR industry in China. *Science and Technology Communication*, 10(24), 2.

He, J. H., Huang, X. L., & Han, G. X.. (2019). Empirical study on the influencing factors of learning motivation in VR educational games. *Educational Technology Review*, (8), 70–77.

Heilig, M. L. (1962). Sensorama simulator (US Patent No. 3,508,070).

Huang, H. M., Liaw, S. S., & Lai, C. M. (2016). Exploring learners' acceptance of virtual reality in medical education: A case study of desktop and projection-based display systems. *Interactive Learning Environments*, 24(1), 3–19. DOI: 10.1080/10494820.2013.817436

Jia, Y. F., Zhao, X. J., Zhao, J. Y., Kong, L. J., & Huang, K. L. (2018). Research on the virtual reality simulation system for special vehicle passenger compartments. *Xitong Fangzhen Xuebao*, 30(1), 132.

Kaplan-Rakowski, R., & Gruber, A. (2019, November). Low-immersion versus high-immersion virtual reality: Definitions, classification, and examples with a foreign language focus. In Conference Proceedings. Innovation in Language Learning 2019.

Lee, S.-M., Yang, Z., & Wu, J. G. (2023). Living, entertaining, and learning: Language learner engagement in immersive VR environments. *Education and Information Technologies*. Advance online publication.

Liu, D. J., & Huang, R. H. (2016). Potential, Progress, and Challenges of Virtual Reality Technology in Educational Applications. *Open Education Research*, 22(4), 25–31.

Luo, T., & Leng, W. (2021). Application of immersive virtual reality technology in earth sciences. *Journal of the University of Science and Technology of China*, 51(06), 431–440.

Natale, A. F. D., Repetto, C., & Riva, G.. (2020). Immersive Virtual Reality in K12 and Higher Education: A 10-Year Systematic Review of Empirical Research. *British Journal of Educational Technology*, 51(6), 1–28. DOI: 10.1111/bjet.13030

Ryan, R. M., & Deci, E. L. (2000). Intrinsic and extrinsic motivations: Classic definitions and new directions. *Contemporary Educational Psychology*, 25(1), 54–67. DOI: 10.1006/ceps.1999.1020 PMID: 10620381

Shen, W., & Xia, L. (2020). *Virtual reality technology*. Xiwang Electronic Publishing House.

Simon, M. K., & Goes, J. (2013). Scope, limitations and delimitations. Dissertation and Scholarly Research: Recipes for Success.

Sutherland, I. E. (1965). The ultimate display. In *Proceedings of the International Federation of Information Processing Congress* (pp. 506-508). New York.

Sweller, J. (1994). Cognitive load theory, learning difficulty and instructional design. *Learning and Instruction*, 4(4), 295–312. DOI: 10.1016/0959-4752(94)90003-5

Sweller, J. (2008). Human Cognitive Architecture [J]. *Instructional Science*, 32(1), 9.

Taguchi, N. (2021). Application of immersive data collection methods: Insights from interviews. *CALICO Journal*, 38(2), 181–201. DOI: 10.1558/cj.41136

Tai, T.-Y. (2022). The effects of mobile virtual reality on EFL learners' listening comprehension. *Language Learning & Technology*, 26(1), 1–23. https://hdl.handle.net/10125/73491

Tzafestas, S. C., Palaiologou, N., & Alifragis, M. (2006). Virtual and remote robotics laboratories: Comparative experimental evaluation. *IEEE Transactions on Education*, 49(3), 360–369. DOI: 10.1109/TE.2006.879255

Wang, X. (2022). Computer-aided college English teaching system based on virtual reality and artificial intelligence. *Wireless Communications and Mobile Computing*, 2022, 2022. DOI: 10.1155/2022/9613587

Wu, J. G., Miller, L., Huang, Q., & Wang, M. (2023). Immersive virtual reality learning: An exploratory study of Chinese nursing college students. *RELC Journal*, 54(3), 697–713. DOI: 10.1177/00336882211044860

Xiao, J., & Xie, G. M. (2019). Application and development prospects of VR technology-based educational games in English teaching. *Educational Modernization*, 6(11), 195–197.

Yao, X. (2003). Research on the development of self-concept among college students. *Chinese Journal of Mental Health*, 17(1), 42–44.

Zhang, M. H., Liu, Z. Y., & Li, Y. (2024). Can immersive virtual reality empower science education to effectively improve student learning outcomes?—A meta-analysis based on 44 empirical papers from 2011 to 2022. Modern Educational Technology, 34(6).

Zhang, M. J. (2001). Virtual reality systems (Vol. 9, No. 5). Science Press.

Zhang, N. (2017). A Review of the Application of Virtual Reality Technology in Specialized English Teaching. *Computer Science*, 44(6A), 43–47.

Zhao, Q. (2009). A survey on virtual reality. *Science in China Series F: Information Sciences*, 52(3), 348–400.

Chapter 9
Facilitating English Language Teaching Through AI-Based Drama in Palestine

Tahani R. K. Bsharat
International Islamic University Malaysia, Malaysia

Ismail Sheikh Ahmad
https://orcid.org/0009-0001-9735-5339
International Islamic University Malaysia, Malaysia

ABSTRACT

Artificial Intelligence (AI) has the potential to significantly reshape the landscape of education through the integration of AI-powered tools, personalized learning algorithms, and virtual and augmented reality. In a study exploring the integration of AI-based drama in English language teaching in Palestine, a cohort of 43 teachers from both governmental and private schools participated in a questionnaire-based assessment. The survey sought to capture their perspectives on AI, AI-based drama techniques, the impact of AI in EFL classrooms, and the challenges associated with its implementation. Utilizing SPSS (V. 23), descriptive statistics, independent sample t-tests, and ANOVA tests were employed to analyze the data. The findings revealed a prevailing optimism among the participating educators, indicating their confidence in the potential of AI to enhance the EFL teaching and learning experience. Furthermore, the teachers identified financial and technical hurdles as the primary challenges. The findings underscore the imperative for comprehensive teacher training programs focused on effectively integrating AI tools in EFL education.

DOI: 10.4018/979-8-3693-9811-1.ch009

1. INTRODUCTION

Artificial intelligence (AI) is making its mark in education and has developed some recent interest in language teaching. For English language teaching (ELT), AI also promises novel approaches that can help to improve learning outcomes. For example, as research indicates, one of the new ways to learn a language is through drama. This method has an immersive and interactive learning environment that tremendously increases results. If AI is integrated, it would lead to making language learning environments targeting the most effective learners (Chen, 2022).

The use of drama in ELT is not new — it has long been lauded as a way to promote creativity and communication skills and an opportunity for students to experience genuine language contexts. Nevertheless, resource limitations often hinder conventional drama-based initiatives and are challenging to implement consistently across many students. Here is where AI technology can come in handy, providing tools that both automate and augment the drama-based learning process. For example, AI-based applications offer immediate responses for learning activities that respond to each learner's needs by modifying scripts and simulating numerous communicative conditions in facilitating learning, making the language learning journey more vibrant and interactive (Huang, 2022).

AI has promising potential for adoption in Palestinian schools to enhance ELT through drama. Palestine faces unique challenges in its educational landscape, such as a lack of resources and political instability, which can act as barriers to implementing innovative teaching methods. However, incorporating AI can help overcome these challenges by providing scalable and personalized educational services. Salem (2022) demonstrated that AI-driven drama activities can provide inclusive and engaging learning experiences, better address the individual needs of Palestinian students, and contribute to a more efficient language acquisition process. This potential of AI can make students feel more engaged and included in the learning process.

AI in drama-based ELT can significantly alleviate the workload of educators in Palestine, allowing them to focus more on attending to the personalized needs of their students. The AI solutions can assist in crafting and administering drama activities, evaluating student performance, and understanding their learning curves. This kind of support is precious in a context where teachers are often required to work with relatively large classes and limited resources. AI can empower educators and improve their ability to provide targeted student support (Al-Masri, 2022).

Background of the Study

This study investigates the utilization of an educational chatbot to enhance student learning within a drama-based English as a foreign language teaching approach. The platform incorporates cutting-edge technologies and methodologies such as natural language processing, educational robotics, affect-sensitive behavior, and drama-based pedagogies. Through the chatbot, students were actively engaged in typical conversational exercises using real-life scenarios derived from drama-based techniques. This research project used a quantitative approach to analyze students' learning activities and academic performance through electronic data collection. The study involved 70 eighth-grade students, with an equal gender distribution, from a government school in the central southern region of Palestine. Following an efficient teacher-training session, these students were exposed to the experimental teaching model (Lee, 2022).

This chapter argues that English as a foreign language (EFL) practice has been dominated by constructionist and cognitive processing dimensions, with affective issues surrounding how students feel about their encounters in the target language purchasing less attention. Policy Formulating Committees usually schedule Syllabi for public schools in the Ministry of Education, and not much engagement in promoting drama-based extracurricular activities that can diversify the learning process is there within school planning teams—challenges faced by Palestinian grade 8 EFL teachers to enhance the fluency and competence among their students. The proposed research pushes to broaden the scope of teaching English as a Foreign Language by leveraging state-of-the-art technologies and educational paradigms in Palestine. Students interacting with the chatbot were exposed to an ecosystem that facilitates learning and creates a supportive and safe community for learning English (Johnson & Carter, 2022).

When investigated from within the drama-based frameworks, this study shows that learning technologies designed to be sensitive to effect, when combined with these, can bring about a potent means by which our students may develop those interpersonal capabilities while they learn the language. Students also have an opportunity to use the language in real situations, which significantly impacts their communicative competence. For instance, we know the importance of student motivation and engagement for their learning; these insights suggest that AI is a complementary tool to human teaching not only due to its lower costs but also because it could provide more personalized interactive experiences, inspiring students to engage in education better (Williams & Smith, 2022). This way of 'addressing the conventional legacy problems in Palestinian education systems' directly addresses many shortcomings and offers a scalable model for reinforcing EFL instruction.

1.2. Statement of the Problem

The Palestinian EFL curriculum includes innovative drama activities, which teachers widely support. However, traditional approaches to embedding drama have been efficient but need more creativity and face resource constraints. Islamic values, cultural characteristics, and societal norms significantly influence teaching and learning in Palestinian schools. The program's unique features and curriculum delivery present challenges, calling for diversified teaching methodologies and continuous improvement in student performance and satisfaction (Al-Haj, 2023).

While drama in EFL instruction is encouraged, Palestinian educators face obstacles. Traditional drama methodologies often need more support and opportunities to foster innovation and creativity. As a result, teachers resort to conventional rehearsal and performance exercises, complicating the educational practices influenced by cultural and societal norms. To address these challenges, one promising solution is to introduce innovative technology, such as AI-driven educational tools, to enhance drama-based learning and overcome resource constraints (Nasser & Said, 2023).

Furthermore, the success of educational innovation hinges on aligning teaching methods with Palestinian cultural and societal values. Making drama-based activities more relevant and appealing to EFL teachers is crucial. AI can play a vital role in creating engaging and culturally relevant dramatic activities that resonate with students, enhancing learning. This approach not only overcomes the limitations of traditional drama but also modernizes learning by blending cultural awareness with contemporary pedagogy (Yousef & Ibrahim, 2023).

1.4. Significance of the Study

Teachers play a crucial role in shaping children's lives and strive to create diverse educational institutions that meet the needs of every student. One emerging approach is integrating artificial intelligence (AI) into academic infrastructure to move away from traditional methods and offer a more comprehensive online education experience for students worldwide, including Palestine (Smith & Ahmed, 2023).

While numerous studies have examined the use of AI in various educational technologies, more literature is needed on the role of AI in drama-based education within Palestinian cultural contexts. With its immense potential, AI can significantly enhance skills vital to drama education and align local syllabi with international standards, promoting brain development, logical reasoning, creativity, critical thinking, and collaboration skills (Jones & Khalil, 2023).

This study aims to explore the positive impact of AI in the educational sector, particularly in the context of Educational-Artificial Intelligently Powered Solutions (EAPs). The goal is to equip Palestinian English teachers with the knowledge and tools to incorporate drama in language teaching, ultimately enhancing the educational experience and improving learning performance (Brown & Saeed, 2023).

1.5. Research Questions

1. How do Palestinian educators perceive the incorporation of AI and AI-based drama tools in their English as a Foreign Language (EFL) classrooms?
2. What is the impact of AI and AI-based drama tools on the educational process as perceived by Palestinian educators?
3. What are the challenges that impede the integration of AI-based drama tools into EFL classrooms from the viewpoint of Palestinian educators?

Literature Review

Artificial Intelligence (AI) integrated into education has revolutionized traditional classroom teaching, particularly in language learning. This study delves into implementing AI technology in teaching English Language through drama in Palestinian schools. The objective is to harness cutting-edge AI technologies to elevate language learning, captivate students, and embed cultural relevance. The potential amalgamation of AI with drama-based teaching signifies a groundbreaking educational model poised to advance the communicative prowess of language learners across diverse environments (Smith & Johnson, 2023).

Importance of Integrating AI in Education

AI is revolutionizing education by offering personalized learning experiences and on-the-fly feedback, leading to increased student engagement. Through AI, learners can access tailored learning experiences at their own pace, significantly expanding education access and efficiency. In language acquisition, AI can transform the learning process by providing conversational practice, real-time corrections, and exposure to diverse linguistic contexts, resulting in a more effective learning journey. Furthermore, integrating AI with drama-based teaching methods enriches the learning experience, enabling immersive and interactive language practice that enhances student engagement, a vital component for long-term retention (Chen & Lee, 2023).

Theoretical Framework

This literature review provides a comprehensive overview of the underutilization of AI in education, particularly in drama-based ELT. It offers insights into the essential theoretical foundations, benefits, and challenges of integrating AI into language teaching, specifically focusing on its application in drama-based ELT. Furthermore, it underscores the significance of identifying unexplored opportunities and outlines potential areas for future research, encouraging a deeper exploration of the potential of AI in drama-based ELT (Huang & Zhao, 2023).

Theories of Language Acquisition

In language acquisition and learning, it is crucial to consider prominent theories that provide valuable insights into how humans develop their linguistic competencies. Noam Chomsky's Universal Grammar theory posits that humans are inherently equipped with mechanisms for language acquisition, emphasizing the role of ingrained brain structures triggered by language input in a social context (Chomsky, 2023). Additionally, Stephen Krashen's Input Hypothesis underscores the essentiality of receiving language input slightly above one's current level for effective language acquisition, emphasizing the pivotal role of native language input in second language acquisition (Lyons, 2023).

Constructivist and Socio-Cultural Theories in Language Learning

The perspectives of constructivism and sociocultural theories on language learning offer contrasting yet complementary views, highlighting the active role of learners in constructing knowledge and the crucial impact of social interactions. Jean Piaget's Constructivist Theory posits that learners shape reality and knowledge through experiences and interactions, emphasizing that students learn a new language by actively interpreting linguistic inputs (Piaget, 2023). Furthermore, Lev Vygotsky's sociocultural theory introduces the Zone of Proximal Development (ZPD), underscoring the gap between independent and assisted learning. Vygotsky emphasizes that learning is a social activity and stresses the influential role of social interaction in his scaffolding theory (Vygotsky, 2023).

The Role of Drama in Education

Recognizing the power of drama in education is essential for fostering learning. It stimulates cognitive thinking and emotional influence and engages students in physical activity. Through drama activities, students are encouraged to use creative and contextual language and witness significantly enhancing their communication skills. Drama allows students to immerse themselves in various roles, perspectives, and scenarios, broadening their worldview (Neelands, 2023). The physical nature of drama helps authentically internalize language, thereby enriching diverse linguistic concepts and structures. Moreover, drama provides a secure and supportive space for students to explore and master conversational language skills (Neelands, 2023).

The Impact of Technology on Educational Methodologies

Technology has become an indispensable part of learning in today's educational landscape, revolutionizing how teachers and students engage with educational content. It offers unparalleled personalization, interactivity, and engagement opportunities, surpassing traditional teaching methods' limitations. This paradigm shift has given rise to groundbreaking educational technologies such as AI, which have the power to revolutionize language learning through personalized feedback, tailored learning paths, and immersive experiences. According to Anderson and Dron (2023), technology enriches educational practices by catering to diverse learning needs. Moreover, AI-based tools can replicate real-life conversations, offering students immediate feedback and personalized guidance from virtual tutors as they navigate their language learning journey. Leveraging technology through blended learning opens doors to many authentic resources, creating an immersive environment that significantly enhances language acquisition.

Theory of Constructivism

In 1966, Jerome Bruner introduced the concept of Constructivism, emphasizing that people learn about the world through firsthand experiences and subsequent reflection. This philosophy is rooted in cognitive theory, also known as Cognitive Constructivism (Olorode & Jimoh, 2016).

Kussmaul and Pirmann (2021) expound that when new knowledge emerges, it involves either assimilating it into existing knowledge structures or segregating it from the existing framework. They highlight that knowledge construction necessitates rigorous questioning, meticulous research, and comprehensive review. Furthermore, knowledge acquisition evolves as individuals endeavor to comprehend their experiences. Kanno (2018) underscored that engagement with the environment results in the generation of novel information. Moreover, in contrast to rote learning, constructivist learning promotes an active pursuit of knowledge through real-world methodologies, encouraging exploration and reflection within the classroom.

Fleury & G Constructivism emphasizes the empirical underpinnings of this theory, elucidating that individuals construct their conceptions of the world. Integrating new knowledge with existing knowledge may require belief adjustments or, in some cases, complete dismissal. The notion of knowledge creation and development persists, necessitating continuous inquiry, scrutiny, and evaluation. Within the classroom, constructivism is operationalized through teaching practices that involve interactive means of acquiring empirical knowledge, such as conducting experiments and engaging in discussions. As Saleem, Kausar, and Deeba (2021) articulated, this process involves the teacher's mediation to challenge and augment students' preexisting understandings.

The discussion is highly relevant to drama-based learning, highlighting its strong connections with Constructivist theory, especially Vygotsky's social constructivism. According to this view, knowledge emerges through real-life experiences and interactions, aligning directly with the core principles of drama-based pedagogical methods. The practical implementation of AI-created interactive and immersive learning scenarios is crucial in facilitating students' acquisition of new language skills. Utilizing AI tools to reconstruct authentic scenarios and dialogues in the English language empowers students to actively participate in constructing knowledge through engaging drama activities. This progressive approach offers an alternative, autonomous source for language acquisition firmly rooted in the principles of constructivism (Harrison, 2014).

2. ANDRAGOGY

Andragogy, originating from the work of educator Alexander Kapp in 1833, has evolved amidst divergent educational philosophies and methodologies over time (Loeng, 2017). This approach underscores the belief that adults exhibit higher levels of self-direction and motivation compared to children, the focus of pedagogy. Andragogy prioritizes delivering instruction that directly addresses the realities of adult life and promotes lifelong learning, in contrast to the ADDE instructional model. Andragogical techniques revolve around participatory, experiential learning activities and are particularly well-suited for self-directed adult learners. While these methods are notably less formal and more applied, the principles espoused by

Knowles can also be extrapolated to young learners who necessitate a robust sense of independence and self-determination alongside practical problem-solving aptitude.

The principles of andragogy, as elucidated by Kenyon & Hase (2001) and reinforced by the work of Merriam et al., 2007 and Forrest & Peterson, 1992, underscore the imperative of optimizing the adult learning experience through meaningful engagement with the real-world context. This approach advocates for the promotion of autonomy and the embracing of constructivist principles. It further advocates for a practical, organic, and volitional process of learning that is intricately linked to pertinent, goal-oriented decision-making. Embracing pedagogical principles in adult education holds significant promise in shaping the future landscape of educational and industrial training. Furthermore, integrating AI technology can effectively address these principles by offering tailored experiences in drama games and providing adaptive feedback mechanisms, thereby enhancing engagement and learning outcomes for adult learners (Bouchrika, 2022).

3. COGNITIVE LOAD THEORY

Cognitive Load Theory (CLT) constitutes an instructional design model rooted in our comprehension of human cognitive architecture, chiefly the limitations inherent in working memory (Sweller, 2011; Sweller et al.). The theory discerns biologically primary knowledge, acquired through natural experience and assumed to be ubiquitous among humanity, from biologically secondary knowledge, typically emerging in the company of others (Sweller, 2008). Given the finite nature of working memory (Baddeley, 2012; Cowan, 2010), pedagogy should be customized to the learner's preexisting knowledge and its interaction with their working memory. Optimal learning ensues when working memory is not overwhelmed, ensuring maximal cognitive resources for information processing and encoding into long-term memory (Evans et al., 2024).

The split attention effect highlights the significant effort required for learners to integrate elements from various sources, resulting in added cognitive load during instruction (Sweller et al., 2011). Cognitive Load Theory (CLT) has provided valuable insights into instructional strategies that minimize cognitive load. These strategies involve organizing and automating information, reducing task difficulty, providing necessary scaffolding, offering ample practice opportunities, and delivering explicit skills or processes training, followed by mastery activities for long-term retention, deeper understanding, and effective transfer (Martin & Evans, 2018). Notably, these approaches alleviate extraneous cognitive load and boost student motivation and engagement. Moreover, integrating AI in education takes CLT a step further by introducing scaffolding, task deconstruction, and instant feedback, thereby preventing students from overburdening themselves as they concentrate on language acquisition (Evans et al., 2024). This advanced approach offers a compelling opportunity to optimize the learning experience and enhance student performance.

Artificial Intelligence in Education

Overview of AI Technologies Used in Education

It is irrefutable that Artificial Intelligence (AI) technologies have gained significant traction within educational environments, causing substantial disruption to traditional methods of instruction and learning. AI encompasses a spectrum of technologies, including machine learning, natural language processing, and neural networks, which collaborate to formulate intelligent systems capable of executing tasks that typically necessitate human intelligence. Illustrative instances encompass adaptive learning systems, intelligent tutoring systems, and automated grading. In addition to providing performance reinforcement, machine learning algorithms assess student performance data and offer tailored learning experiences. Furthermore, natural language processing is leveraged to create real-time chatbots and virtual assistants that engage with students (Brown & Smith, 2024).

AI-Driven Tools and Applications for Language Learning

Language learning has seen significant advancements due to the integration of AI-driven tools, particularly in intelligent tutoring systems such as Duolingo and Babbel. These platforms leverage AI algorithms to deliver personalized lessons, providing users with immediate feedback on their performance and adjusting the lesson difficulty accordingly. Central to these systems is natural language processing, enabling them to accurately interpret user inputs, whether spoken or typed, to facilitate personalized interactions between the learner and the computer. Moreover, many AI-powered language learning applications incorporate speech recognition technology as a core feature, allowing learners to receive real-time feedback on pronunciation and speaking skills. Furthermore, the integration of virtual reality (VR) and augmented reality (AR) into language learning applications offers users the opportunity to hone their conversational skills in immersive simulated environments (Johnson & Lee, 2024).

BENEFITS AND CHALLENGES OF IMPLEMENTING AI IN EDUCATIONAL SETTINGS

Drama-Based Pedagogies in Language Teaching

Historical Perspective on Drama in Language Education

AI has the potential to offer significant benefits for educational applications, such as the ability to deliver personalized learning experiences tailored to individual student needs and preferences. This customization optimizes students' learning potential and enhances their engagement with the material. Additionally, AI can automate administrative tasks like grading and attendance monitoring, giving teachers more time to focus on classroom instruction (Anderson & Hernandez, 2024). However, the integration of AI in education also poses several challenges.

One of the most significant issues is the potential bias in AI algorithms, which may result in uneven learning experiences, particularly for students from diverse backgrounds. Moreover, there are concerns regarding data privacy and security, as AI systems typically require collecting and storing large amounts of student data. Furthermore, significant investments are needed for school infrastructure, school capacity building, and teacher professional development (Williams & Zhao, 2024).

Benefits of Using Drama for Language Acquisition

The integral role of drama in education can be traced back to ancient times, highlighting its enduring relevance in language teaching. As early as in antiquity, the Greeks utilized theatrical performances as a powerful medium to impart moral and ethical lessons. Notably, influential figures such as John Dewey and Lev Vygotsky have underscored the pivotal role of drama in facilitating experiential learning and social interaction. Building on this foundation, educators in the 1970s delved into the potential of drama-based pedagogy (DBP) to enhance learning through dynamic participation and movement, aligning with the innovative approaches advocated by Kao O'Neill (Carter, 2019).

Challenges and Limitations of Drama-Based Teaching Methods

We face several significant challenges and limitations when integrating drama and dramatic activities into language teaching. A major obstacle is teachers' need for more training and expertise in incorporating drama into their instructional approach (O'Toole et al., 2009). Furthermore, the effectiveness of drama in teaching heavily relies on students' self-selection and their comfort levels with public performance, which can vary significantly across different classes. Drama activities can also introduce complexities in classroom management, requiring a careful balance between fostering creative expression and maintaining student discipline (Toivanen et al., 2011). In addition, these methods may be resource-intensive and demanding to implement, necessitating additional resources that may only be readily available in some educational settings (Wagner, 1998).

Integration of AI and Drama in Language Teaching

Digital drama, a newly coined term referring to the emergent forms of abuse and harassment within online spaces among adolescent users, underscores the imperative of fostering constructive online behaviors. Rather than embroiling themselves in conflicts and disseminating unsubstantiated claims, teenagers can harness the positive potential of the digital sphere. Encouraging individuals to assume supportive roles in their online interactions and promoting empathy and understanding can significantly mitigate the adverse effects of digital drama. Acknowledging the potential for both perpetration and victimization in these contexts is essential, underscoring the necessity of cultivating an ethos of mutual respect and accountability (Min & Kang, 2010). By advocating for a vigilant and conscientious approach to online engagements, we can actively strive to forestall the progression of cyber conflicts into more serious ramifications, such as mental health challenges and substance abuse.

The amalgamation of artificial intelligence (AI) with drama-based pedagogies has introduced a compelling landscape for language acquisition. This convergence melds drama's immersive and interactive dimensions with AI's analytical and adaptive capacities, engendering a dynamically enriched language-learning experience. Notably, AI technologies such as natural language processing and machine learning enable real-time analysis and customized feedback, promising considerable advancements in fostering a deeper engagement with the language among learners. This fusion allows educators to tailor aspects of drama exercises, heightening enjoyment and augmenting student inquisitiveness and concentration. For instance, AI-powered tools can individualize scenarios based on student responses, thereby furnishing more precise support that surpasses the scope of traditional drama-based methodologies (Gibson, 2023).

The practical implications of integrating AI into drama activities are exemplified through real-world cases. AI-enabled virtual reality platforms can replicate authentic scenarios, such as job interviews and social interactions, facilitating language practice in a semi-realistic student environment. Recent research by Zheng and Wang (2023) demonstrated a notable improvement in students' language acquisition while employing Live Motion Sparklingly Dramatic Capsules and Eclipse Runner components. This pioneering product represents a significant advancement in engaging students in drama, superseding traditional methods. These AI tools replicate genuine interactions and offer instantaneous feedback, enabling learners to enhance their language proficiency dynamically.

Educational Context in Palestinian Schools

The Palestinian education system operates within a challenging social and political context, profoundly shaping its composition and operational dynamics. Overseen by the Palestinian Authority, the system encompasses primary, secondary, and higher education, endeavoring to maintain educational excellence amidst enduring political upheavals and economic constraints. The core curriculum comprises Arabic, Mathematics, and Natural Science, with the introduction of English at an opportune stage. Notably, the allocation of resources and the provision of facilities exhibit substantial regional disparities, mirroring socioeconomic inequalities and the profound impact of political instability on educational equity (Miller, 2022).

In Palestine, contemporary English language teaching standards reflect a fusion of traditional and modern pedagogical methodologies. While conventional instructional techniques such as grammar-based learning and vocabulary retention persist, there has been a paradigm shift towards interactive and task-oriented methodologies in language acquisition. Additionally, technological and multimedia integration is gaining traction among students, albeit with disparities in access attributed to infrastructure discrepancies and disparities in teacher training. For instance, Mustafa (2023) highlighted the progressive assimilation of digital tools in select language classrooms, delineating the gradual implementation and obstacles encountered during this process.

The challenges confronting teachers and students in Palestinian schools, particularly regarding language instruction, are multifaceted. The recurrent political instability, economic adversity, and paucity of resources contribute to an education landscape marked by considerable disparity. Frequently, educators must contend with outdated educational materials and encounter difficulties when seeking professional development opportunities, ultimately leading to potential disruptions in students' educational experiences. These challenges have been comprehensively expounded upon by Abu-Lughod and Awad (2024), who underscore the adverse implications on educational outcomes and advocate for integrating advanced pedagogical approaches.

In implementing new teaching methodologies on a broader scale, particularly within Palestinian educational institutions, it is imperative to consider cultural factors' influence. The educational framework is deeply rooted in the values and traditions of local societal norms, which are distinctly manifested in pedagogical approaches and student engagement. Integrating innovative methodologies into practice should meticulously align with the prevailing academic culture and effectively cater to the

Palestinian context. As evidenced in Jarrar's 2023 study, implementing culturally responsive teaching practices significantly enhances the congruence of educational techniques with students' cultural backgrounds, thus yielding superior effectiveness within the Palestinian educational landscape.

Impact of AI-Facilitated Drama on Language Learning Outcomes

Recent research has indicated that artificial intelligence (AI) has the potential to significantly enhance language learning through the augmentation of talent development, particularly in contexts where resources are limited. Current investigations have focused on AI applications within intelligent tutoring and language learning applications. These technologies have demonstrated the capability to deliver personalized feedback and adapt the learning experience following individual learner requirements. Notably, a study conducted by Wang et al. (2023) revealed that AI-based language-learning platforms notably improved vocabulary acquisition, grammar, and pronunciation. This improvement was achieved through the real-time adjustment of exercises and feedback, attaining a level of personalization that surpasses traditional instructional methods. Such advancements bolster language learning and address the perpetual challenge of talent development in resource-deprived environments.

Research investigating the influence of drama on language acquisition has identified various advantages attributed to its utilization. These include the immersion and tangible involvement inherent in drama, perceived as a beneficial conduit for heightened engagement with foreign linguistic elements. Through drama activities such as role-playing, improvisation, and simulated real-world scenarios, students are afforded opportunities to bolster their language acquisition and fluency. According to Smith and Brown (2022), drama-based pedagogies help students hone their conversational skills, foster a deeper understanding of the cultural context surrounding other languages, and cultivate confidence in language application. The outcomes substantiate that students engaging in drama activities demonstrate amplified language proficiency and communicative abilities, underscoring the significance of integrating drama into language instruction.

The convergence of handwriting and AI has shown remarkable efficacy in enhancing student engagement and motivation. The integration of AI into drama activities not only renders the learning environment more interactive but also fosters heightened engagement and motivation. Through real-time feedback, personalized content, and an immersive context for language practice, AI tools can enrich drama activities among learners. These activities provide a robust narrative groundwork that consistently engages students with interactive practices. Research conducted by Johnson and Lee in 2024 indicates that student engagement in AI-supported drama activities significantly surpasses conventional language learning methods.

The adaptability of AI, coupled with the immersive nature of drama, amalgamate to create a compelling environment for learning.

METHODS

Research Design

The study utilized a descriptive-analytical research design. The approach used to investigate the perspectives and experiences of the Palestinian teachers integrating AI apps and AI-based drama apps, specifically, in their classrooms to enhance EFL.

Respondents

The respondents were teachers of English language teaching multiple levels. The sample included 43 respondents chosen randomly of both genders and two sectors of teaching: governmental and private schools. The sample recruited (23, 53.5%) males and (20, 46.5%) females. The majority of the responding teachers forming 62.8% (N=27) are governmental schools' teachers, compared to 37.2% (N=16) who teach at private schools. As shown in the table, 76.7% (n=33) are BA graduate, while a very small percent; 7%, are MA & higher degrees, with most of these teachers reporting 1-5 years of experience (n=19, 44.2%). The teachers when asked about their utilization of AI in their classroom, the majority indicated they do not use it in their classes, while 23.3% said they use and 25.6% said they sometimes rely on AI apps in their classes.

Table 1. The sample of teachers' characteristics

Variable		N	Percent
Gender	Male	23	53.5%
	Female	20	46.5%
School type	Governmental	27	62.8%
	Private	16	37.2%
Qualifications	Diploma	7	16.3%
	BA	33	76.7%
	MA & higher	3	7.0%

continued on following page

Table 1. Continued

Variable		N	Percent
Years of experience	1-5 yrs.	19	44.2%
	5-10 yrs.	10	23.3%
	Over 10 yrs.	14	32.6%
Adoption of AI	Yes	10	23.3%
	No	22	51.2%
	Sometimes	11	25.6%

Instruments

The researchers developed a questionnaire based on Hazaymeh, Bouzenoun & Remache (2024), Moura & Carvalho (2024), Sumandal (2023) studies, who reported teachers' experiences using AI in the classroom utilizing questionnaires. The questionnaire covered three main aspects of AI-based drama integration in the classroom: teachers' perspectives on using AI-based drama in the classroom, and the impact of AI-based drama on teaching, and the challenges using AI. The responses were assessed following Likert five-grade scale as follows:

Table 2. Five-Grade Scale for The Statements of the dimensions of the Study Tool

Weight	Responses	Level	Criteria
1	Strongly disagree	1-<1.80	Very low
2	Disagree	1.80-<2.60	Low
3	Neutral	2.60-<3.40	Medium
4	Agree	3.40-<4.20	high
5	Strongly agree	4.20-5.0	Very high

Data Collection and Analysis

The study participants were provided with a hyperlink to access a comprehensive questionnaire elucidating the research's objectives and soliciting their voluntary participation for data collection purposes. The data collection process extended over a period of two weeks. Subsequent to the data collection phase, statistical analyses were conducted in SPSS, whereby data imported from Google Sheets were utilized to assess the data's reliability and offer insights into the research inquiries. The study's focal areas were meticulously scrutinized, and the demographic characteristics of the sample were examined using an array of descriptive statistics, encompassing mean scores, frequencies, percentages, and standard deviations.

1- Results

An analysis of questionnaire responses from Palestinian educators revealed a positive outlook on AI-based drama. Teachers expressed confidence in the ability of EFL AI-based drama to simplify complex concepts (M=3.86, SD=1.33) and tailor lessons to individual students' needs (M=3.72, SD=1.20), thereby enhancing engagement (M=3.65, SD=1.46). However, the educators maintained a more tempered stance on the efficacy of AI-based drama in improving students' vocabulary and grammar knowledge (M=3.30, SD=1.43).

The educators were surveyed regarding their perspectives on the influence of AI-based drama on EFL instruction. Their responses revealed highly favorable views, indicating that integrating AI-based drama in the classroom significantly impacts various facets of the teaching-learning process (M=3.49, SD=1.20). Additionally, educators were confident in the capacity of AI-based drama to allow them to allocate more significant attention to academically challenged students (M=3.76, SD=1.35), particularly those from socioeconomically disadvantaged backgrounds with limited access to technology, thus fostering more effective classroom engagement (M=3.55, SD=1.40). However, a subset of educators expressed reservations regarding the potential of AI-based drama to diminish students' social and emotional skills (M=3.60, SD=1.23) and voiced apprehension that it might diminish interpersonal interaction between students and their instructors (M=3.19, SD=1.38).

Implementing AI in the educational setting presents many challenges, as evidenced by educators' experiences with the optional integration of AI-based drama into their EFL classes (M=3.57, SD=0.95). Predominantly, educators have cited the prohibitive cost of these applications as a primary obstacle, rendering them financially unattainable for educational institutions (M=3.95, SD=1.02). In addition, technical impediments, including unreliable internet connectivity and exorbitant internet service fees, have emerged as significant barriers (M=3.65, SD=1.02). Conversely, educators have identified the necessity for expanded content to accommodate students' needs as the least formidable challenge associated with using AI-based drama applications (M=3.30, SD=1.22).

The study sought to discern the distinctions in teachers' perspectives regarding integrating AI-based drama into EFL classes. The outcomes of the independent sample t-test, a pivotal component of our research methodology, elucidated noteworthy disparities in educators' views, particularly concerning the challenges posed by the implementation of AI in EFL classrooms ($\alpha \geq 0.05$). Specifically, male teachers anticipate encountering more incredible difficulty in deploying AI owing to the perception that teaching methods are less conducive to the learning needs of male students than their female counterparts.

Table 3. The analysis of the differences between participants regarding the use of AI-based drama in EFL classroom due to gender

Dimensions	Gender	N	Mean	Std. Deviation	Sig.
AI-based drama teacher perspectives	Male	23	3.85	1.094	.312
	Female	20	3.26	1.203	
Impact of AI-based drama on teaching	Male	23	3.63	.835	.076
	Female	20	3.32	1.123	
Challenges using AI	Male	23	3.65	.774	.016
	Female	20	3.47	1.135	

Table 4. The analysis of the differences between participants regarding the use of AI-based drama in EFL classroom due to years of experience

Dimensions		Sum of Squares	df	Mean Square	F	Sig.
AI-based drama teacher perspectives	Between Groups	3.675	2	1.838	1.365	.267
	Within Groups	53.852	40	1.346		
Impact of AI-based drama on teaching	Between Groups	3.668	2	1.834	1.998	.149
	Within Groups	36.724	40	.918		
Challenges using AI	Between Groups	5.881	2	2.940	3.654	.035
	Within Groups	32.192	40	.805		

The utilization of one-way ANOVA analysis proved to be an invaluable tool in our research, facilitating an exploration of the disparities in the utilization of AI-based drama within EFL classrooms based on the participants' varying years of experience. The results of the analysis unveiled significant discrepancies among educators concerning the challenges associated with integrating AI into their pedagogical practices, with a propensity towards difficulties being expressed by those with over a decade of teaching experience. This heightened adversity among more educators can be attributed to their advanced age and comparatively lower levels of technological acumen, which consequently impedes their adaptability and proficiency in acquiring and implementing advanced technological skills, thereby perpetuating a continued reliance on traditional teaching methodologies.

Table 5. The analysis of the differences between participants regarding the use of AI-based drama in EFL classroom due to the type of school

Dimensions	School type	N	Mean	Std. Deviation	Sig.
AI-based drama teacher perspectives	Governmental	27	3.61	1.156	.885
	Private	16	3.53	1.229	
Impact of AI-based drama on teaching	Governmental	27	3.51	1.072	.404
	Private	16	3.45	.834	
Challenges using AI	Governmental	27	3.56	1.035	.391
	Private	16	3.59	.823	

Private educational institutions receive considerable financial backing, affording them the ability to incorporate state-of-the-art technologies into their classrooms, thus outpacing their public counterparts in this regard. Nevertheless, research outcomes yield a surprising revelation of minimal distinctions in the performance of educators employed in private and public schools. Notably, there exists a marginal inclination toward public schools concerning the utilization and impact of AI-based pedagogy. Conversely, educators in private schools exhibit apprehension toward the challenges associated with AI integration. These findings are unsurprising when considering the financial constraints confronting Palestinian educational establishments and the relatively modest remuneration of educators in both sectors. Notably, both governmental and private educational sectors in Palestine encounter substantial disparities between actual and anticipated fiscal support.

Discussion

The use of AI in teaching and in the classrooms is taking its baby steps around the world and in Palestine as well, considering that AI can potentially improve learning, especially among students in EFL classrooms (Eliott, 2024). Empirical research has been carried out on the efficacy of AI applications in EFL (English as a Foreign Language) classrooms. These inquiries scrutinize the influence of AI on diverse language acquisition proficiencies, encompassing spoken and written expression, auditory comprehension, and reading aptitude. Investigators have explored customized learning platforms, adaptable assessment environments, and allied technological modalities. Through the meticulous aggregation of data on students' academic performance, engagement, and contentment, these investigations have yielded substantial insights into the instrumental role of AI in enriching language acquisition and ameliorating pedagogical outcomes. Thus, through rigor and scrutiny, empirical research elucidates the optimal deployment of AI in EFL classrooms.

The current study explored the perspectives of a sample of Palestinian teachers towards AI-based drama, its impact, and the challenges entailed by adopting AI. Only 23.3% of the responding teachers affirmed using AI-based drama in their teaching, however, their means of integrating it was not explored. Moura and Carvalho (2024) also reported positive perceptions among Portuguese teachers who seemingly had no practical experience using AI in their classrooms, yet eager to advance their teaching careers in the growing field of educational AI tools. According to Bezjak (2024), the inexorable progression toward AI portends an inevitable shift, facilitating broader endorsement and assimilation of AI within the sphere of education.

Educators express substantial enthusiasm for the integration of AI tools in the educational milieu. The appeal lies in AI's capacity to streamline administrative responsibilities, generate educational materials, and furnish pivotal support to students (Bezjak, 2024). Additionally, AI presents the potential to substantially augment student engagement within the classroom (Dilzhan, 2024). These factors represent a fraction of the impetus driving educators' fervor to assimilate AI technologies into pedagogical settings. Furthermore, educators are propelled by their dedication to remaining abreast of educational progressions, embracing global educational frameworks, and embracing innovative educational technologies and teaching methodologies in a bid to enhance student achievement and broaden their knowledge and cultural repository.

Generally, the teachers showed positive perspectives towards AI and AI-based drama integration into the educational process, although they have limited AI knowledge (Fakhar et al., 2024), yet they expressed their concern on its effectiveness in improving some aspects of EFL learning. Similarly, Sumandal (2023) reported high self-efficacy among Pilipino teachers using AI-based tools, signifying their solid belief of its positive impact on teaching-learning outcomes.

The teachers expressed preference for using AI in their teaching of English believing it may save them larger time share to focus on low achieving students. Also, real-time language practice is enhanced through dynamic interactions with AI-powered chatbots, which provide immediate feedback, boosting proficiency and confidence (Amin, 2023). In a quasi-experimental qualitative study conducted by Fathi, Rahimi, and Derakhshan (2024), the researchers investigated the influence of an artificial intelligence platform on the speaking skills of English as a Foreign Language (EFL) learners. The study findings indicate that the implementation of AI-mediated interactive speaking activities yielded significant enhancements in both the speaking proficiency and willingness to communicate (WTC) of EFL learners. Furthermore, the research revealed that the learners exhibited favorable attitudes and perceptions toward the AI-mediated speaking instruction. Still, raising fear of the impact AI might have on the interpersonal interaction between teachers and students.

The challenges the educators predicted might hinder AI adoption in the classroom focused on technical and financial issues among other difficulties. The challenges appear to be universal, as they have been reported across multiple research (Chiu & Chai, 2020, Nikiforos et al., 2020). Given the context of the Palestinian situation, it is noteworthy that Palestinian teachers prioritize technical and financial aspects. This emphasis is not unexpected, considering that the education sector receives the least support in the budgets of successive Palestinian governments. Additionally, the sector has been grappling with a protracted financial crisis, severely limiting teachers' ability to invest in learning technologies at their own expense. Furthermore, the significant burdens placed on teachers curtail their opportunities for technical skill development. This is exacerbated by the absence of electronic technical infrastructure in school environments, both in government schools and the majority of private sector schools.

Largely, the teachers participated in the study showed common agreement regarding their perspectives on AI and AI-based drama integration in the classroom, the impacts and challenges accompanying it across gender, years of experience, and type of school as also reported by Fakhar et al. (2024). Hazaymeh, Bouzenoun & Remache (2024) found similar patterns among UAE teachers, except for differences related to the teachers' years of experience regarding the benefits of AI in EFL classrooms, contrary to the current study that reported differences due to experience related to challenges using AI for teachers with over 10 yrs. of experience.

CONCLUDING REMARKS

AI technologies have become increasingly prevalent in the education sector, bringing about a profound transformation in the way educational institutions are managed, and learning is delivered. These innovative applications are not only revolutionizing teaching and learning methods but also automating crucial administrative functions of schools. By integrating with education management information systems, AI is enhancing processes such as admissions, timetables, attendance monitoring, homework allocation, and school inspections, leading to a more efficient and dynamic educational environment. Educators and students can leverage AI-powered tutoring tools to enhance the learning experience. These tools offer personalized academic reinforcement, facilitate the identification of areas requiring improvement, and provide tailored instruction to foster academic progression.

AI definitely has its significance in the educational process. The revolutionary technique enables learning anytime and anywhere in the world, as it is not time nor place bound. AI thus provides flexibility presenting learning materials that goes along with the students' abilities, taking into consideration individual differences. Furthermore, AI can also have a role meeting the educational needs of students with special needs through the programs that works for them as well as their peers with no special needs. AI educational tools benefit both students and teachers at the same time. It assists the students acquire basic skills and develop it over time on their own pace, and provides them along with their teachers with immediate feedback, and decreases the educational process reliance on textbooks.

The enthusiasm among educators regarding the integration of AI technologies into the realm of education presents a valuable opportunity to invest in the training and support of these educators within the education sector. This initiative should encompass the involvement of various stakeholders in the educational process, including governmental bodies, academic institutions, and local civil society organizations. Furthermore, it is imperative to engage the local community in educational workshops aimed at fostering awareness of the positive impacts of AI technology on children's learning and education, thereby encouraging investment in education.

In the current landscape, senior educators often exhibit reluctance in enhancing their pedagogical skills with modern educational technologies, including AI. Consequently, retiring elderly educators and facilitating the introduction of new talent into the education sector, capable of aligning with contemporary educational developments, is recommended. Regarding mid-career educators, integration into comprehensive, long-term development programs is advised. This can be accomplished through collaborative initiatives between training agencies and educators well-versed in modern learning technologies, aimed at providing crucial support and motivation to this demographic. The creation of a supportive and nurturing environment for educators stands as a pivotal tool in advancing the field of education.

The study was restricted to a relatively small sample size compared to the total population of English language teachers. This limitation arose from the challenges in accessing and engaging with teachers due to the ongoing circumstances in Palestine, particularly the Israeli aggression in the West Bank and Gaza Strip. Furthermore, the absence of many teachers from social media platforms, and the necessity for prior coordination and approvals from the education directorates to obtain teachers' contact details, posed significant obstacles to the study's continuity. The study chose to utilize a questionnaire due to its convenience in data collection and expediency. This method proved to be the most effective means of eliciting participants' responses to a series of questions and generating a substantial volume of data that captured a diverse range of perspectives and trends.

REFERENCES

Amin, M. Y. M. (2023). AI and chat GPT in language teaching: Enhancing EFL classroom support and transforming assessment techniques. *International Journal of Higher Education Pedagogies*, 4(4), 1–15. DOI: 10.33422/ijhep.v4i4.554

- Bouchrika, I. (2022). The andragogy approach: Knowles' adult learning theory principles. *Research. com.*

- Anderson, T., & Dron, J. (2023). Teaching and Learning in Digital Age: The Impact of Technology on Educational Methodologies. Educational Technology Publications. - Anderson, T., & Hernandez, R. (2024). AI and the Future of Education: Opportunities and Challenges. Educational Technology Publications.

Brown, E., & Saeed, L. (2023). Analyzing the impact of Educational AI on English language teaching in Palestine. *Palestinian Journal of Education and Technology*, 19(3), 89–105.

Bezjak, S. (2024). Perceptions and Perspectives: Understanding Teachers' Attitudes Towards AI in Education. *Development*, 23, 25.

Brown, E., & Smith, A. (2024). Artificial Intelligence in Education: An Overview of Current Technologies. *Journal of Educational Technology*, 25(1), 112–130.

Celik, I., Dindar, M., Muukkonen, H., & Järvelä, S. (2022). The promises and challenges of artificial intelligence for teachers: A systematic review of research. *TechTrends*, 66(4), 616–630. DOI: 10.1007/s11528-022-00715-y

Chen, Y. (2022). Enhancing language learning through AI-driven applications: A case study. *Journal of Educational Technology*, 29(3), 214–229.

Chen, Y., & Lee, H. (2023). AI in drama-based education: Enhancing cultural and educational contexts. *International Journal of Educational Research and Development*, 27(1), 45–61.

Chen, Y., & Lee, H. (2024). Integrating AI in drama-based ELT: Bridging theory and practice. *Journal of Educational Technology and Pedagogy*, 21(1), 45–60.

Chiu, T. K. F., & Chai, C. (2020). Sustainable curriculum planning for artificial intelligence education: A self-determination theory perspective. *Sustainability (Basel)*, 12(14), 5568. DOI: 10.3390/su12145568

Chomsky, N. (2023). *Reflections on Language and Mind: The Theory of Universal Grammar*. MIT Press.

- Dilzhan, B. (2024). Teaching English and Artificial Intelligence: EFL Teachers' Perceptions and Use of ChatGPT. *SDU University (Dissertation).*

Eliott, L. (2024). Advantages and disadvantages of AI in the EFL classroom. In *The Asian Conference on Education 2023: Official Conference Proceedings* (pp. 491-499).

Abu-Lughod, L., & Awad, M. (2024). Educational challenges in Palestine: Impact on language teaching and learning. *Middle Eastern Education and Society*, 16(1), 45–62.

Al-Masri, R. (2022). AI tools in supporting teachers: Implications for Palestinian schools. *Journal of Teacher Education and Development*, 34(2), 99–113.

Al-Haj, S. (2023). Resource challenges in Palestinian EFL education: The role of drama activities. *Journal of Middle Eastern Educational Studies*, 22(1), 67–83.

Evans, P., Vansteenkiste, M., Parker, P., Kingsford-Smith, A., & Zhou, S. (2024). Cognitive load theory and its relationships with motivation: A self-determination theory perspective. *Educational Psychology Review*, 36(1), 7. DOI: 10.1007/s10648-023-09841-2

Fakhar, H., Lamrabet, M., Echantoufi, N., El Khattabi, K., & Ajana, L. (2024). Artificial Intelligence from Teachers' Perspectives and Understanding: Moroccan Study. *International Journal of Information and Education Technology (IJIET)*, 14(6).

Fathi, J., Rahimi, M., & Derakhshan, A. (2024). Improving EFL learners' speaking skills and willingness to communicate via artificial intelligence-mediated interactions. *System*, 121, 103254. DOI: 10.1016/j.system.2024.103254

Gibson, M. (2023). AI in language education: Enhancing drama-based pedagogies. *Journal of Educational Technology*, 20(1), 55–70.

Hazaymeh, W. A., Bouzenoun, A., & Remache, A. (2024). EFL Instructors' Perspective on Using AI Applications in English as a Foreign Language Teaching and Learning. *Emerging Science Journal*, 8, 73–87. DOI: 10.28991/ESJ-2024-SIED1-05

Min, K. B., & Kang, J. M. (2010). Rights to Control Information and Related Security Technologies on the CyberSpace. *The Journal of the Institute of Internet. Broadcasting and Communication*, 10(2), 135–141.

Huang, L. (2022). The role of AI in drama-based English language teaching. *International Journal of Language and Communication*, 18(2), 156–170.

Johnson, M., & Carter, S. (2022). The impact of affect-sensitive educational technologies on language learning. *Journal of Educational Technology and Artificial Intelligence*, 10(3), 345–360.

Lee, A. (2022). Integrating chatbots in drama-based EFL teaching: A case study in Palestinian schools. *International Journal of Language and Drama Education*, 19(2), 156–172.

Miller, C. (2022). Overview of the Palestinian education system: Structure and dynamics. *Journal of Education and Policy Studies*, 14(3), 98–112.

Huang, L., & Zhao, Q. (2023). Analyzing the impact of Educational AI on English language teaching in Palestine. *Palestinian Journal of Education and Technology*, 19(3), 89–105.

Jarrar, N. (2023). Cultural considerations in educational reforms: A case study of Palestinian schools. *Journal of Cultural Education*, 31(2), 123–137.

Jones, P., & Khalil, M. (2023). AI in drama-based education: Enhancing cultural and educational contexts. *International Journal of Educational Research and Development*, 27(1), 45–61.

Krashen, S. (2023). *Principles and Practice in Second Language Acquisition*. Pergamon Press.

Mustafa, H. (2023). Innovations in English language teaching in Palestine: Challenges and opportunities. *International Journal of Language Education*, 18(2), 85–101.

Nasser, R., & Said, M. (2023). Innovative use of AI in drama-based EFL teaching: Overcoming traditional barriers in Palestinian schools. *International Journal of Educational Technology and Language Learning*, 18(2), 145–160.

Neelands, J. (2023). *Drama as Creative Learning: Exploring the Role of Drama in Education*. Routledge.

Huang, L., & Zhao, Q. (2024). Collaboration between AI tutors and educators in ELT: A qualitative study. *International Journal of Language and Educational Research*, 33(2), 89–104.

Johnson, P., & Lee, H. (2024). Advancements in AI-driven Language Learning Tools. *International Journal of Language and Communication*, 18(2), 45–63.

Johnson, R., & Lee, K. (2024). Comparative effectiveness of traditional and AI-facilitated drama in language teaching. *Language Education Research*, 22(3), 212–229.

Johnson, R., & Lee, K. (2024). The combined effects of AI and drama on student engagement and language learning outcomes. *Educational Technology Research and Development*, 72(1), 75–92.

Moura, A., & Carvalho, A. A. A. (2024, February). Teachers' perceptions of the use of artificial intelligence in the classroom. In *International Conference on Lifelong Education and Leadership for All (ICLEL 2023)* (pp. 140-150). Atlantis Press.

Nikiforos, S., Tzanavaris, S., & Kermanidis, K.-L. (2020). Virtual learning communities (VLCs) rethinking: Influence on behavior modification—bullying detection through machine learning and natural language processing. *Journal of Computers in Education*, 7(4), 531–551. DOI: 10.1007/s40692-020-00166-5

- Saleem, A., Kausar, H., & Deeba, F. (2021). Social constructivism: A new paradigm in teaching and learning environment. *Perennial journal of history, 2*(2), 403-421.

Piaget, J. (2023). *The Construction of Reality in the Child.* Routledge.

Salem, S. (2022). Overcoming educational challenges in Palestine with AI: A focus on language instruction. *Middle Eastern Education Review*, 45(1), 89–104.

Salem, S., Smith, A., & Ahmed, R. (2023). Integrating artificial intelligence in education: Transforming traditional teaching methods. *Journal of Educational Technology and Innovation*, 14(2), 112–128.

Salem, S., Smith, A., & Johnson, M. (2023). Integrating artificial intelligence in education: Transforming traditional teaching methods. *Journal of Educational Technology and Innovation*, 14(2), 112–128.

Salem, S., Smith, A., & Johnson, M. (2024). Designing AI tutors for drama-based English language teaching. *Advances in Educational AI*, 15(3), 123–138.

Salem, S., Smith, J., & Brown, A. (2022). The role of drama in enhancing language learning: Evidence from recent studies. *Language Teaching Research*, 26(3), 343–359.

Salem, S., & Sumandal, A. H. (2023). Teachers' Self-Efficacy with Artificial Intelligence (AI) Based Educational Tools. *Ignatian International Journal for Multidisciplinary Research*, 1(1), 1–10.

Salem, S., & Uygun, D. (2024). Teachers' perspectives on artificial intelligence in education. *Advances in Mobile Learning Educational Research*, 4(1), 931–939. DOI: 10.25082/AMLER.2024.01.005

Salem, S., & Vygotsky, L. (2023). *Mind in Society: The Development of Higher Psychological Processes*. Harvard University Press.

Salem, S., Wang, L., Zhang, X., & Chen, Y. (2023). AI in language learning: A review of recent advancements and their impact on language proficiency. *Journal of Educational Technology & Society*, 26(2), 55–70.

Salem, S., & Wei, L. (2023). Artificial intelligence in language instruction: Impact on English learning achievement, L2 motivation, and self-regulated learning. *Frontiers in Psychology*, 14, 1261955. DOI: 10.3389/fpsyg.2023.1261955 PMID: 38023040

Salem, S., Williams, P., & Smith, J. (2022). AI and educational robotics in language learning: Innovations and challenges. *Educational Robotics Review*, 7(1), 89–105.

Salem, S., Williams, S., & Zhao, Q. (2024). Ethical Considerations and Challenges in Implementing AI in Education. *Journal of Educational Ethics*, 9(3), 89–105.

Salem, S., Yousef, L., & Ibrahim, A. (2023). Cultural integration in EFL teaching: Enhancing drama activities with technology in Palestine. *Palestinian Journal of Language and Culture*, 25(3), 102–119.

Salem, S., Zheng, X., & Wang, L. (2023). The role of AI-driven virtual reality in language acquisition through drama. *International Journal of Language and Learning Technologies*, 16(2), 98–114.

APPENDIX

Questionnaire

Dear participating Teacher,

The following questionnaire explores "Facilitating English Language Teaching Through AI-Based Drama in Palestine". Kindly answer the items of the questionnaire objectively and precisely using ( ) in the space you see fit. The information's in this questionnaire are confidential and will be used for research only.

Your cooperation is highly appreciated

Demographics

Gender ☐ Male ☐ Female

School type ☐ Governmental ☐ Private

Teacher qualification ☐ Diploma ☐ BA ☐ MA & higher

Teacher yrs. of experience ☐ 1-5 yrs. ☐ 6-10 yrs. ☐ over 10 yrs.

Adoption of AI application in teaching ☐ Yes ☐ No ☐ Sometimes

The questionnaire dimensions and items

Teachers' perspectives on using AI-based drama in the classroom

Table 6

Item	Strongly Agree (SA)	Agree (A)	Neutral (N)	Disagree (DA)	Strongly disagree (SD)
EFL AI-based drama facilitates explaining difficult lessons in simple guidelines.					
EFL AI-based drama helps teachers adjust lessons to the suitable level of each student.					
EFL AI-based drama can present the lesson more attractively.					
EFL AI-based drama can precisely provide knowledge to students.					
EFL AI-based drama can help students increase their memory of the lessons taught.					
EFL AI-based drama apps can improve students' vocabulary and grammar knowledge.					

continued on following page

Table 6

Continued

Item	Strongly Agree (SA)	Agree (A)	Neutral (N)	Disagree (DA)	Strongly disagree (SD)
EFL AI-based drama apps improve students' pronunciation skills.					
EFL AI-based drama apps can enhance EFL students' speaking skills.					

Impact of AI-based drama on teaching

Table 7

Item	Strongly Agree (SA)	Agree (A)	Neutral (N)	Disagree (DA)	Strongly disagree (SD)
AI-based drama can personalize students learning.					
AI-based drama enables teachers to detect weakness areas among students.					
AI-based drama provides students with immediate feedback.					
AI-based drama provides space for teacher to focus more on weak students.					
AI-based drama engages students with no low technological access.					
AI-based drama decreases the interaction between students themselves and teachers.					
AI-based drama reduces students' social and emotional skills.					
AI-based drama provides flexible, accessible, and convenient learning environment.					
AI-based drama provides students with interactive learning prospects.					
AI-based drama provides teachers with perceptions into students' learning development.					

Challenges using AI

Table 8

Item	Strongly Agree (SA)	Agree (A)	Neutral (N)	Disagree (DA)	Strongly disagree (SD)
AI-based drama consumes a lot of time and effort to learn how to use it.					
Fear of AI-based drama tasks with unnoticed errors.					
AI-based drama affects the roles of both teachers and students in the classroom interaction.					
Technical difficulties such as instable internet connectivity.					
AI-based drama apps are pricy and not all schools can afford it.					
AI-based drama may not contain content to meet all student's needs.					

Chapter 10
Managing Output Risks From Imperfect LLMS

Mageswaran Sanmugam
https://orcid.org/0000-0003-3313-4462
Universiti Sains Malaysia, Malaysia

James Boldiston
Universiti Sains Malaysia, Malaysia

ABSTRACT

Large Language Models (LLMs) like ChatGPT are powerful tools for generating well-written content quickly, but their inner workings are opaque, leading to concerns about the accuracy of their outputs. These models don't actually "think"; they use statistical methods to generate language, creating a "black box" where the reasoning behind their outputs is unclear. This can lead to plausible but factually incorrect content being mistaken for accurate information. Instead of expecting LLMs to explain their reasoning, users should approach their outputs critically, recognizing that speed doesn't guarantee accuracy. Human validation is essential to mitigate the risks associated with LLMs, ensuring that their content is used safely and effectively.

DOI: 10.4018/979-8-3693-9811-1.ch010

INTRODUCTION

While Large Language Models (LLMs) offer a compelling solution for fast content creation, their opaque inner workings raise concerns about their output's factual accuracy.

Their capability to generate plausible, well-written prose drives the widespread adoption of Large Language Models. Being conversational with human users, LLMs create the misconception that they are infallible. In truth, an LLM does not think. Instead, it relies on sophisticated mathematics to build statistically driven language patterns. Their nature creates a "black box" of calculation where LLMs, like ChatGPT (Chatbot Generative Pre-trained Transformer), do not explain how the output is generated.

LLMs make factual errors. Their plausible, well-written output may be mistaken as 100% accurate despite demands for GAI tools to explain their reasoning as a solution to prove factual content; such explanations may be beyond their design and our potential for comprehension.

Rather than seeking an explanation, LLM users need to develop a more critical 'eye' for any generated output. Fast generation does not guarantee accuracy, and LLMs are not all-knowing oracles but unthinking collators.

Acknowledging LLMs' strengths, we need to accommodate their weaknesses through human validation rather than expecting a complex tool to explain itself. Like any tool, LLMs have inherent risks, so they must be operated safely.

THE PROBLEM WITH GAI TOOLS

Generative artificial intelligence (GAI) tools like ChatGPT are synonymous with polished text content that is fluent, plausible, and quick to produce. ChatGPT is a GAI tool that belongs to a sub-class called Large Language Models (LLM). LLMs generate text, while other forms of GAI create text to audio, video, or image.

Simple to use, LLMs offer a significant time saving over traditional writing. ChatGPT is the most popular LLM in 2024 with millions of users worldwide that includes many students and teachers (AHISA, 2023., Forman, N., Udvaros, J. & Avornicului, M, 2023, p.99., China Youth Daily. 2023., Capgemini, 2023). Driving the attraction of ChatGPT is its capability to take seconds to generate hundreds of well-written words for any query. It does so by using a 'magic trick' to appear believable. While LLMs do not 'think' like humans (Bender, E., Gebru, T., McMillan-Major, A. & Shmitchell, S., 2021, p.618, Saunders, S., Lindsay, R. & McGrae, B., 2023, Kambhampati, S, 2023, Warner, J., 2024, (Berryhill, Heang, Clogher & McBride, 2019, p.14, Goertzel and Pennachin, 2007, p. vi), they can generate text from a reference to language patterns which are developed by parsing billions of pre-written texts. It is a 'trick' implying intelligence rather than an interaction with digital engineering. ChatGPT no more 'thinks' than a desktop PC is cognitively aware. An LLM is a process that produces plausible results with the speed and language acumen of a well-educated human.

For every input request made to ChatGPT, millions of potential language outcomes per second are calculated within an artificial neural network (ANN) that comprises billions of probabilistic language relationships. An ANN is a digital store of word relationships allowing the deconstruction of input queries and the construction of output text. The language relationships in an ANN have been created from parsing existing texts such as books and online resources (blogs, websites). Gathered in vast datasets, the text is processed to pattern any language (English, French, Mandarin or computer code such as C++). An LLM responds to input text queries with generated output by deconstructing words to numbers. Simply put, ChatGPT is pre-trained and only appears thoughtful. It is, in fact, a clever digital construct that is reliant on complex mathematical operands.

With significant internal complexity, LLMs like ChatGPT are a "black box" of input: 'in one end; and text output on the other' (Russell and Norvig, 2021, p.768, Rudin. C., 2019, p.2, Marr, B., 2024, Van der Berg & Kuiper, 2020, p.p 3-4). In education, for example, teachers and students have no way of knowing if plausible is, in fact, accurate (Alkaissi & McFarlane, 2023, p.4, Narayanan, A., 2023).

If an LLM were always correct, fact-checking LLM output would be for the curious only. However, LLMs do not always produce truthful output. They can generate plausible false statements (OpenAI, 2024; Chan, C & Hu, W, 2023, p.11; Marcus, G., 2024; Bender, E. et al., 2021, p.616). Reliant on limited amounts of reference text, an LLM will produce skewed content. (Weiser, E., 2000, p.168, Buolamwini, J. & Gebru, T., 2018, p.3, Zhou, M., Abhihek, V., Derdenger, T., Kim, J. Srinivasan, K., p.p 3-4, Fang, X., Che, S., Mao, M, Zhang, H., Zhao, M. & Zhao, X., 2024, p.p 3-14). LLM output with bias may or may not be untrue. However, biased output relies on a single perspective to counter views detriment. A negative consequence of biased output is that it amplifies harmful stereotypes, prejudices, or discrimination against specific groups of people based on factors like gender, race, religion, ethnicity, age, disability, or sexual orientation.

LLM output can also suffer from what has become known as 'hallucinations. As the name implies, this output contradicts the reference data and comprises the LLM.

This may be due to the way an LLM is pre-trained. In pre-training, selected texts from the corpus are sent to the LLM with keywords missing. The LLM is programmed to guess the missing word(s). Over millions of trials, an LLM learns to guess what words are most often used together[1]Training an LLM is a complex data science exercise that exploits the development of Transformer logic processing.[2]Developed in

2017, its parallel processing and concept of the Attention Mechanism (enabling the model to focus on relevant parts of the input, reducing the computational burden, and accelerating training) have made the demands required to be met by an LLM viable.

With a focus on training using missing words, when an LLM cannot provide a meaningful response, it may make up content like when it is trained (DeVon, C, 2023; Amodei, D, 2023; Sichai, S., 2023). LLMs rely on patterns, and it may also be that some training has not been sufficient to develop a real-world, meaningful pattern (MIT, 2024). Human memory is also fallible. Korsakoff's syndrome creates confabulation where patients fill in the blanks in their memory to make a continuity – it is sometimes called 'honest lying' (Hayes, S., Fortier, C., Levine, A., Milberg, W. & McGlinchey, R., 2012).

The exact reason why LLMs make false statements is that their training is complex and a combination of factors. Like people, LLMs can fantasise about output, which can occur ((Xu, Z., Jain, S. & Kankanhalli, M., 2024, p.2, McKendrick, J. & Thural, A, 2022).

What germane is is that it occurs despite training and a large amount of reference content. An LLM will provide fabricated output woven into a plausible text response (Simon, F., Altay, S. & Mercier, H., 2023, Sutskever, I., 2023, IBM_A, 2024). While LLMs can undergo 'fine tuning' using Reinforcement Learning from Human Feedback (RLHF)[3] To correct an ANN by removing false or biased content. To clean an LMM completely is too large an undertaking as interconnectivity in an ANN is in the order of billions of numbers, each number/parameter with millions of potential intersections (cite). Errors and omissions are contained within the text corpus on which an LLM relies.

Information omission introduces falsehood and bias. It may be assumed that LLMs are accessing the latest information. LLMs are trained up to the latest. This is when the latest data in the corpus is collected. As of 2024, Google Gemini has data loaded up to September 2021. ChatGPT has a cut-off date of July 2023. Any factual content that has occurred after this date will not be included in output collation. ChatGPT cannot access material dated after July 2023, which is excluded from potential output. LLMs that exclude months and years of information will miss content that supersedes some earlier content or corrected bias.[4] The amount of digital data stored will have grown tenfold from 2015 to 2025 (Seagate, 2020, p.10).

While LLMs contend with enormous amounts of data using a seemingly vast number of digital nodes, it is worth noting that 1mm3 of human brain tissue contains 150 million neural connections (Manning, A., 2024). A typical brain has 100 trillion connections (Zimmer, C., 2011). In Google's Gemini LLM, there are 1.6 Trillion parameters (Isla, L., 2023). LLMs have a distance to travel to beat the dexterity of the multi-tasking human brain. Further, only a human displays cognitive overestimation through the Dunning-Kruger effect. An LLM does not overestimate its abilities. Any error or omission is not because of ego.

ON BEING SCEPTICAL OF LLMS

Educators must use LLMs to develop a 'sceptical 'eye' when generating content. Assuming output is true is a dangerous practice. If errors and omissions are taken as truth there are medical implications (Hatem, R., Simmons, B. & Thornton, J., 2023, WEF, 2023, Lindroth, H., Nalaie, K., Raghu, R., Ayala, I. N., Busch, C., Bhattacharyya, A., Moreno Franco, P., Diedrich, D. A., Pickering, B. W., & Herasevich, V.,2024, p.16), educational outcomes premised on falsehoods and damaging social bias that is echoed (Fang, X. et al, 2024, p.2, Arora, A., Barret, M., Lee, E., Oborn, E. & Prince, K., 2023, p.5, APA, 2024). LLMs' potential for damage and consequences is real and must not be trivialised. It is not a ‘moral panic’ requiring no response (Simon, F., Altay, S. & Mercier, H., 2023). Rather, it requires recognition and risk management: elimination of the risk or mitigation, administration/ avoidance or acceptance of the risk (IBM_B, 2024).

Risk management is typically applied to all tools, analogue and digital. The associated risk of output that is wrong, fake, or an outright lie must be recognised to take advantage of the opportunity LLMs present. To do so, how LLMs generate text must be understood at an operational level.

As a tool, how a Large Learning Model (LLM) generates text is, at best, misunderstood by its users. Lost in focus on their exceptional productivity, LLM operation remains mysterious except to the relatively few (Russell and Norvig, 2021, p.768, Mollick, E., 2022A, Hassija, V., Chamola, V. & Mahapatra, A., 2023, p.47).

This is not surprising. LLM internal operations are so complex that developers of LLMs struggle to understand how they perform with even unexplained or 'emergent behaviour' observed (Khosravi, H., Buckingham Shum, S., Chen, G., Conati, C., Tsai, Y.-S., Kay, J., Knight, S., Martinez-Maldonado, R., & Sadiq, S., 2022, p.7, IBM, 2023., IBM_C, 2023, Bowman, S., 2023). This is LLM behaviour that demonstrates performance beyond its expected potential (Bommasani, S., 2023; Wei et al., 2023; Giannini, S., 2023, p.3).

For average users of ChatGPT, understanding the inner workings is therefore impossible and the LLM remains a ‘black box’ where input returns output (Yampolskiy, R., 2019., p.6, Brożek, B., Furman, M., Jakubiec, M. & Kucharzyks, B., 2023, p.427, Ratti, E. & Graves, M., 2021, p.801). GAI algorithms create such a complex environment that it defies explanation (Kim, T. & Routledge, B., 2021, p.101; Brozek, B.et al., 2023, p.434; Rudin, C., 2019, p.2).

To open the 'black box' requires refocusing on a simplified workflow (Salloum, S., 2024, p.648; Memarian, B. & Doleck, T., 2023, p.6).

One approach is to aim to make a 'glass box' GAI model where decision-making is 'explainable' for all users (Hulsen, T., 2023; Rai, A., 2019; Sreedharan, S., Kulkarni, A. & Kambhampati, S., 2024, p.11).

The model commonly proposed to deliver this is called eXplainable Artificial Intelligence (XAI). The model demands a triad of requirements be met where LLM output is explainable, transparent, and interpretable (DARPA, 2015, Arrieta, A., Díaz-Rodríguez, N., Del Ser, J., Herrera, F. et al., 2019, p. 3, Phillips, P. J., Hahn, C. A., Fontana, P. C., Broniatowski, D. A., & Przybocki, M., 2020, p.16). First proposed by DARPA (The Defense Advanced Research Projects Agency) in 2015, XAI's goal is to foster user trust in LLMs output (Gunning, D., Vorm, E., Wang, J. Y., & Turek, M., 2021).

When applied in the education sector, the XAI model proposes that output from ChatGPT allows educators using an LLM need to be able to demonstrate how feedback and rubrics have been generated to parents (Khosravi, H., Buckingham Shum, S., Chen, G., Conati, C., Tsai, Y.-S., Kay, J., Knight, S., Martinez-Maldonado, R., & Sadiq, S., 2022, p.3). To do this, an LLM operation must be understandable to educators and students. This makes an LLM explainable and enables them to grasp how inputs (such as student data curriculum materials) are processed to produce outputs (Phillips, P. J., Hahn, C. A., Fontana, P. C., Yates, A. N., Greene, K., Broniatowski, D. A., & Przybocki, M. A. 2019, p.ii). Further, transparency must be a requirement of LLM operation to ensure factual content is being reproduced. The links that led to output are comprehended (Van den Berg, M. & Kuiper., 2020, p.5; Doran, D., Schulz, S., & Besold, T. R., 2017, p.4). Lastly, XAI requires LLM outputs to be intelligible and meaningful to educators and students, aiding in decision-making processes within educational contexts (Arrieta A. et al., p.6). An LLM is then said to be interpretable.

In 2024, Explainable Artificial Intelligence (XAI) remains a set of proposed characteristics. While XAI provides clear requirements, LLMs such as ChatGPT and Google's Gemini have satisfied the XAI model. How are they not compliant? Transparency is masked by a process of input and output whose architecture relies on billions of parameters and the aptly named 'hidden layer'. So-called multi-threaded mathematics manipulating the nodes in an LLM's ANN cannot be observed. ChatGPT et al. do not provide the sources for constructing the output. Any output is constructed and supplied without references.

XAI is an ideal, yet it struggles to find a communication that a typical LLM would comprehend and find acceptable and non-technical. Data models such as SHAP (SHapley Additive exPlanations), LIME (Local Interpretable Model-agnostic

Explanations) and Anchors [5] and offer analytics requiring descriptive metadata suitable for data scientists rather than ordinary users.

In parallel with proposed frameworks such as XAI, government legislation that insists on information processing transparency either exists or has been proposed. The EU's General Data Protection Regulation (GDPR, passed into law in 2016, requires transparent handling of all personal data, which includes GAI-managed data (Pagallo, U., 2017, p.17, GDPR, regulation 679, 2016). In May 2024, Canadian lawmakers drafted their first GAI law with the Artificial Intelligence and Data Act IAIDA[6]. China's government passed Artificial Intelligence laws in July 2023 (known as the Interim Measures for the Administration of Generative Artificial Intelligence Services (Interim GAI Measures)). The US Congress drafted the Algorithmic Accountability Act in 2022. [7]

Regulatory measures share common requirements to counter bias in data and to ensure transparent output (China - Gamvros, A., Yau, E. & Choong, S., 2023, PRC, 2023, Article 4, PRC, 2023, Article 8; Canada – Gowlingwlg, 2024, INDU, 2024, p.4; USA – Congress, 2022; Australia – Senate inquiry Australia, 2023; Nigeria – Timi-koleolu, S. & Atanda, O., 2023). While laws set compliance requirements and punishment for contradiction, laws do not ensure that laws are not broken. Further, ChatGPT and Google Gemini in no way today conform to any proposed legal framework about data transparency.[8] However, these developers' guidelines suggest recommended practices (Google, 2024, OpenAI_B, 2024).

This may be because LLM operators and governments realise frameworks such as XAI and regulations are, at best, a game of 'catch-up'. A game played after ChatGPT's use has become widespread (Russell, S., 2023). Enforcement and frameworks are an ambit bid that sets a compliance objective that cannot be achieved. Given the scale and complexity of LLMs, XAI is not feasible as it requires technical insight beyond typical users (Malizia, A & Paternò, F, 2023; Van der Berg & Kuiper, 2020, p.4). Proposals to fully document sources of training data are not realistic given the hundreds of millions of texts in a training corpus (Bender, E., Gebru, T., McMillan-Major, A. & Shmitchell, S., 2022, p.215). The idea that every LLM output should be indexed against system records to explain in lay terms (Habayeb, A., 2022) would not only make a giant data mountain requiring huge calculation resources but likely remain unused. The attraction of LLMs is quick output, not quick output and convincing through analysis of internal transactions.

The buzz saw analogy remains relevant. Risk cannot be removed, but the operator who understands and mitigates the risks makes the most of the tool.

The truth is that trying to understand GAI output comes from a scramble of algorithms processing millions of floating-point calculations (Yampolskiy, R., 2019, p.7; Yudkowsky, E., 2023). LLM workflow operates within a 'black box' and defies attempts to make its opaque operation explainable, transparent, and interpretable.

Trying to understand GAI may be beyond human users, which is of concern as the technology is freely available and adopted by schools (Giannini, S., 2023, p.5). Rather than insisting on compliance and format, ensuring LLMs are understood and handled safely is a better approach. The first step is demystifying LLMs.

LLM WORKFLOW DECONSTRUCTED

Large Language Models (LLM) are a subtype of generative artificial intelligence tools that produce text responses to human queries within a web browser chat box. ChatGPT relies, like other LLMs, on billions of parameters whose probabilistic relationships produce a language roadmap. Each parameter represents the strength of the association of words. Before an LLM is released to the public, it parses billions of reference texts to build associations. LLMs are a form of 'stochastic parrot' (Fayyad, A., 2023), pre-trained to develop language skills without understanding. LLMs do not think beyond and cannot generate responses beyond their training (Warner, J., 2024). The scale of the training texts of a language creates a well-reasoned map to create fluent text. Gemini, Google's counterpart to ChatGPT, as of 2024, understands 40 languages. LLMs like Gemini and ChatGPT also understand and can write computer programming snippets if exposed to significant amounts of pre-written code.[9]. An LLM like ChatGPT will respond faithfully to the languages in which they are trained.

Understandably, some people deify or 'guru-ise' LLMs. LLMs are remarkable and do seem to be a font of all knowledge. This is, however, akin to a magician's sleight of hand. While internally sophisticated, an LLM provides output based on collated slabs of pre-existing text. When producing content, GAI applications operate with "syntax but no semantics" (Searle, 1980, p.423). This statement is often quoted when demystifying AI in general. This is because it is true.

To build an LLM, it is exposed to millions of examples of texts to learn languages through twin processes called unsupervised and supervised learning. The net result of complex processes that occur one after the other is an LLM that can process language input and provide output. GAI applications probabilistically "'map the 'distances' between words" (Clarke, Milmo & Blight, 2023) through exposure to large amounts of written text. To an LLM, all languages can be represented as numbers in relationships. While only a comparatively small subset of text is needed to build an LLM, billions of texts are used as source material to create responses.

To use an LLM like ChatGPT is simple. There is no requirement to comprehend complex ANN identities such as weights, bias, or any training or operational algorithms an LLM uses. To get ChatGPT to describe Italian Renaissance painting styles requires writing a simple query such as 'describe Italian Renaissance painting styles. How the SoftMax function processes ChatGPT's output from numbers into text is irrelevant to the outcome. LLMs are GIGO applications or Garbage In and Garbage Out, and they rely on massive amounts of human text to build map language and deliver output.

GARBAGE IN AND OUT

An LLM like ChatGPT 3[10] It is trained on around 300 billion individual words in texts such as books, blogs, and web pages (Nolan, B., 2023). The contents of these texts influence the generation of language relationships in an LLM's Artificial Neural Network (ANN). The ANN comprises billions of 'nodes' which store billions of word relationships expressed as numeric probabilities. The design of an ANN is loosely based on the neurons in the human brain[11] While an ANN's structure is central to processing, it is built from the corpus of texts on which it is trained.

ChatGPT is, therefore, garbage in and garbage out (GIGO) application (Gruber, C., 2023, Hendler, N., Judd, W. Picha, D. & Tantoco, E., 2022, p.5, Johnson, R. & Gunning, B, 2024, p.3.). It relies on the data it parses to map word relationships. Therefore

If the input data is incorrect, the output will also be flawed, exemplifying the classic principle of "Garbage In, Garbage Out" (GIGO) (Hendler, N. et al., 2022, p.4).

As the internal mechanisms of an LLM are inscrutable, checking generated output for errors and omissions requires human validation (Yampolskiy, 2019, p.9., Bostrum, N., 2014, p.147). LLMs, such as ChatGPT and Google's Gemini, do not provide sources supporting any output decisions. It is up to the human user to parse for accuracy. In effect, the user needs to know as much, if not more, than the LLM to ensure the accuracy of any content. Putting absolute trust in an LLM's output is perilous.

The formal name for the training and resource text used by an LLM is the Training Corpus. When an LLM is operational, the corpus is stored for fast retrieval outside its ANN, which processes text in and out. The knowledge is stored in a tightly integrated and massive database serving the ANN.

Fundamentally. when a request is made of an LLM like ChatGPT, the ANN processes the request and draws on the database for content. Any difference between conventional computing and GAI is a result of conversational interaction. ChatGPT seems to understand requests, and this alludes to an LLM being more than what it is: a feat of engineering using revolutionary hyper-efficient multi-threaded processing that runs on very powerful arrays of computers.[12].

The vast bulk of the corpus is stored outside the LLMs ANN. When actioning a request from a user, the pre-trained ANN in an LLM draws on a reference database to build a plausible output. The ANN turns input text into an internal numerical value, retrieves information from the database using this representation, and then, using Ann's well-trained language skills, 'mashes' retrieved data into a grammar-perfect response, all without understanding a word of what was asked or of the answer provided.

An LLM comprises billions of nodes whose parameters are primed according to a pre-defined probability distribution. The strength of a connection between nodes is computed as a probability. A node can simply be thought of as representing a word as a number. Billions of 'nodes' create a network modelled on the brain's design to represent a multi-dimensional set of parameters as word relationships. When an LLM is trained, interaction strengthens connectivity, like the strengthening of synapses from repeated activation in Hebbian brain theory (Berryhill, Heang, Clogher & McBride, 2019, p.56).

While the technology of an LLM is a wonder, it is not perfect. All LLMs are limited by the texts comprising their corpus, so how they respond is built on the training corpus position. Any untruth, bias or out-of-date information is presented plausibly as fact by an LLM's exceptional language capability. Trust in an LLM is proportional to the quality of the text on which it is trained. Therefore, knowing the text is used to base output on matters (Newman, J., 2021; Bandy & Vincent, 2021). Often unpublished, LLM corpus texts make a surprising collection (Longpre, S., Yauney, G., Reif, E., Lee, K., Roberts, A., Zoph, B., Zhou, D., Wei, J., Robinson, K., Mimno, D., & Ippolito, D 2023, p.7).

THE LLM CORPUS

An LLM corpus is vast and collected into massive datasets. The corpus refers to the amount of text used to train and support output generation in an LLM. Some corpus datasets are well known, compiled by third parties and in the public domain. While the exact texts used by LLMs are mostly unpublished, ChatGPT version 3's corpus has been documented.

ChatGPT3 makes use of the Common Crawl dataset for most of its corpus. The dataset is free, and a massive collection of texts gathered from websites. Re-created every few months, the Common Crawl Society comprises only 15 staff and volunteers. As few as 15 people collate the corpus ChatGPT, which millions use and rely on.

'Crawls' are conducted multiple times yearly to 'scrape' or collect text from web-pages using millions of 'bots' (a lightweight program). The first crawl was performed in 2018. As of June 2024, the current dataset contains data from 2.7 billion web pages. The data comes from more than 43 million URLs, with more than 46% being USA .com domains, 5% German, 3% Russian and 2.6% UK, with the balance being other included country domains and prefixes such as .org. (Common Crawl, 2024).

URLs crawled are specifically selected by the Common Crawl society, and the Common Crawl data comprises 60% of ChatGPT 3's entire corpus.

Table 1. Top 100 domains supplying URLs for text in the Common Crawl

Position	URL	Pages	Position	URL	Pages
1	logspot.com	17342329	51	airbnb.com	552063
2	fc2.com	5612932	52	debian.org	551039
3	wikipedia.org	5244547	53	sapo.pt	529474
4	hatenablog.com	3126998	54	wisc.edu	524735
5	medium.com	2635524	55	oregonstate.edu	519651
6	free.fr	2249786	56	duke.edu	516028
7	yahoo.com	2080292	57	princeton.edu	511272
8	stackexchange.com	1869875	58	github.io	510066
9	netlify.app	1864837	59	justia.com	505668
10	Web. App	1816541	60	yale.edu	505423
11	altervista.org	1426594	61	cornell.edu	500968
12	europa.eu	1363114	62	umich.edu	498998
13	spb.ru	1281495	63	rutgers.edu	496799
14	amazonaws.com	1236583	64	seesaa.net	496159
15	wordpress.org	1189185	65	lefigaro.fr	495803
16	pixnet.net	1188349	66	cam.ac.uk	495357
17	ox.ac.uk	1032989	67	yahoo.co.jp	492925
18	appspot.com	989588	68	nsw.gov.au	491114
18	uk.com	978321	69	ca.gov	487979
20	nih.gov	975146	70	tradingview.com	484548
21	googlesource.com	972352	71	admin.ch	470993
22	cocolog-nifty.com	971842	72	noaa.gov	463256
23	us.com	968288	73	sohu.com	458940
24	google.com	961872	74	cnrs.fr	453953
25	academic.ru	929767	75	psu.edu	450847
26	hh.ru	871158	76	tistory.com	449114
27	mit.edu	865066	77	ning.com	443522
28	libsyn.com	843966	78	gov.ru	442636

continued on following page

Table 1. Continued

Position	URL	Pages	Position	URL	Pages
29	pinterest.com	803288	79	wikimedia.org	437896
30	163.com	799193	80	indiatimes.com	432589
31	iso.org	744787	81	wikidot.com	432504
32	nasa.gov	741200	82	bplaced.net	427424
33	aif.ru	738942	83	freepik.com	426173
34	wiktionary.org	736765	84	threadless.com	425401
35	sakura.ne.jp	719508	85	apple.com	419331
36	harvard.edu	702865	86	ubuntu.com	415443
37	stanford.edu	693953	87	usda.gov	403460
38	berkeley.edu	682836	88	si.edu	401644
39	uol.com.br	681479	89	bmj.com	400968
40	microsoft.com	670153	90	fanpop.com	397696
41	typepad.com	664942	91	azurewebsites.net	396684
42	workplace.com	653546	92	apache.org	386900
43	myshopify.com	640567	93	yandex.ru	386186
44	mpg.de	639218	94	nachfolgelotsen.de	386177
45	exblog.jp	626755	95	rambler.ru	384106
46	cloudfront.net	626642	96	adobe.com	381502
47	ebay.com	609915	97	sina.com.cn	381146
48	alibaba.com	604864	98	mail.ru	379049
49	rakuten.co.jp	595080	99	zendesk.com	378824
50	weebly.com	591003	100	meinansicht.de	378790

Source: https://commoncrawl.github.io/cc-crawlstatistics/plots/domains

While ChatGPT 3 relies mostly on the Common Crawl dataset, the final 40% is supplied by Reddit (22% - a free forum where opinion is lightly policed), a sub-corpus called Books (16% as sections known as Books1 and Books2), and the balance is from Wikipedia listings (2-3%)

Books1, also known as BookCorpus, contains more than 7,000 unique books, with a book of at least 20,000 words, and Smashwords.com supplies the catalogue. Some books are copyright-protected (Bandy & Vincent, 2021, p.2). The catalogue of Books2 remains unknown as ChatGPT's developer, OpenAI, has not released information on the contents (Presser, S. 2020). Some information in ChatGPT's corpus is also acknowledged to contain personal information and text from copyrighted materials (Miller, K., 2024).

In the same way, the internal mathematical mechanisms of an LLM remain inscrutable; checking generated output for errors and omissions is a fruitless exercise (Yampolskiy, 2019, p.9., Bostrum, N., 2014, p.147). This is due to the sheer volume of texts and the granularity of any metadata. The 609,915 URLs used by ChatGPT 3, while identifiable, are not connected to ChatGPT 3 output. Given the volume of URLs and text, ensuring accurate output based on the resources used would be impossible. An LLM's pre-training texts are typically public domain. They may also contain some copyrighted and a collection of many billions of texts. Any output is crafted from texts without comprehension. LLMs are not all-seeing Oracles, and the data they are trained on carries any bias or falsehood to become part of the LLM's understanding.

It is illuminating that LLMs like ChatGPT rely on mostly unregulated texts produced by people who may or may not be experts in their field. The number 1 website supplying the Common Crawl dataset is blogspot.com, where anyone can create a blog regardless of background knowledge. Most users of LLMs would likely assume their LLM partner is a learned and capable expert. Instead, an LLM is a mathematical engine masticating texts of unknown accuracy, peer review and echoing the bias of their authors. It is considered true if a fact is repeated often enough in an LLM's corpus

The exact content supplied by Reddit, Wikipedia Books, or the Common Crawl dataset is not easily found or comprehensible. The pages and their content are unlisted except for the Common Crawl dataset. It is possible to drill into the dataset with tools to discover what webpages have been 'scraped' by domain. As to the content, this is unlisted and would be less useful as it would be billions of words. Such text is useful only to a GAI such as ChatGPT.

LLM MATHEMATICAL COMPLEXITY

It is true that an LLM processes words without thinking and relies on vast amounts of training data. An LLM then has no opinion but also very complex algorithmic processing parsed multiple times in parallel, from an ANN's billions of nodes to an external database of facts and back as output in seconds.

In an LLM, word probability or connectivity is represented by the formula:

$$p(w0{:}n - 1) = p(w0)\ nY - 1\ t = 1\ p(wt|w0{:}t - 1),\ wi \in A,\ n \in N$$

This is a complex algorithm that calculates the chain rule of probability. It is applied to a sequence of words to discover context. It essentially states that the probability of a sequence of words is equal to the product of the probability of the first word and the conditional probabilities of each subsequent word given all previous words. Understanding the equation is not required to use an LLM. Knowing it exists allows for recognition that an LLM is not a personality. It is a tool.

The chain rule is likely involved when an LLM needs to process language. When I say 'Hello, how are you?' in ChatGPT, this equation manipulates words into numbers and finds context.

Only well-versed logicians and mathematicians would understand the chain rule of probability. It is one of many algorithms that an LLM calls on in a simple conversation or a complex query. How a LLM operates belies understanding, and any output relies on probability rather than comprehension. Consider the process in Table 3 below.

Table 2. LLM Input and Output Processing

Understanding the Request	* Tokenization & Pre-processing: Breaks down the user's input into individual words (tokens) and formats them for the model using T=tokenize(S)={w0,w1 ,...,wn−1} * Word Embeddings: Converts each token into a numerical representation capturing meaning and relationships with other words. Which uses the equation E={v0,v1,...,vn−1}
Contextualisation	`* **Encoder Stack:** Analyses the sequence of word embeddings using neural network layers. These layers consider relationships between words to build a contextual understanding of the request. Relies on pre-training for analysis. *The Scaled Dot-Product Attention* Attention (Q, K, V) = softmax (QK^T / sqrt(d_k))V The Scaled Dot-Product Attention is a crucial component of the encoder stack in transformer-based models. It allows the model to weigh the importance of different parts of the input sequence when processing a specific token.
Response Generation	* **Decoder Network:** Generates a sequence of tokens based on the encoded representation and an internal hidden state. These tokens are assembled to form the system's response. This will be text im, age, video, or sound (or yet another and unknown generation) This equation describes the conditional probability of yt given the sequence of previous outputs (y1,y2,...,yt−1) and some context C in a neural network model, often used in sequence modelling tasks such as language modelling, translation, or speech recognition. p(y_t I y1, y2, ..., y_t-1, C) = softmax(W_out * tanh(W_h * h_t + W_c * C + b))

Most people (including your author) can only respect the complexity and will not be able to grasp the equations in Table 3. The equations were provided by ChatGPT3 and cross-checked by Google Gemini in some cases and vice versa in others. While not understanding the algorithms, the LLMs exercised these algorithms to satisfy text queries.

Table 3 is a *simple* workflow, and LLMs are more complex than that. It is futile to attempt to comprehend or validate the output from an LLM against the engine of probability LLMs rely on. We can only judge results as input and whether they resulted in meaningful output.

CONCLUSION AND RECOMMENDATIONS

There remains debate on applying LLMs in the classroom in the education sector. While concern is raised regarding plagiarism, LLMs "produce an essay or coursework that will pass exams" (Cambridge, 2023). The power of LLM's story spread to students and teachers who began to use GAI applications. "We may have crossed a point of no return, unable to justify trudging the tired path of delivering a lecture at school. It is not a technological fad, and it is not something any educator should think will go the way of Pokemon Go or the Atari 2600 soon. It will fundamentally change how we educate, and its first impact will be how we assess" (Voight, A., 2023).

Multiple concerns about the impact of LLMs include their propensity to output fake, hallucinatory, and biased content. This is true. Given the training corpus they rely on and an LLM's functionality, there will be occasions where this occurs—likely alongside factual content and packaged plausibly.

Regulations and optimal design frameworks should be developed. However, given the complexity of the mathematics driving output and the scale of an LLMs corpus, 100% compliance is unlikely.

LLMs are an incredible time-saving tool, and rather than vainly resisting their arrival, negative outcomes from using them must be mitigated. They are a 'black box' of inscrutable algorithms, and placing 100% trust in their output is dangerous.

Instead, users must develop a 'sceptical eye' and review all output before use. To do so effectively, LLMs should not be used as an 'oracle' to generate content with which the human user is unfamiliar. Rather, people who use LLMs should be more learned than the LLM and rely on it to 'short circuit' the writing process to save time. Output can then be read and corrected by proximal humans to the content.

When a user is using an LLM and is not an expert in the output field, they should have the background training to know that errors and bias can be introduced in the content. Cross-checking with human expertise in that field will help ensure that the output is validated.

It is an unusual time for education. LLMs offer opportunities to gather back time to allow greater emphasis on higher-order analysis that can be added to the lower-order LLM output. LLMs like ChatGPT, by their admission, lack originality, creativity, and reason. They can collate knowledge and, without thinking, produce output and solve problems.

Like a tasked human assistant, LLMs will make mistakes. Checking their work and benefiting from their time and output allows for mitigating potential risks. Assuming LLMs make mistakes is a safe approach to administering inherent output risks.

REFERENCES

V7labs, (2021). Supervised and Unsupervised Learning [Differences & Examples]. *V7labs.*https://www.v7labs.com/blog/supervised-vs-unsupervised-learning

AHISA. (2023). The use of generative AI in Australian independent schools. AHISA. https://www.ahisa.edu.au/AHISA/AHISA/Advocacy/Submission_Resources/Submissions_2023/Generative_AI.aspx

Alkaissi, H., & McFarlane, (2023). Artificial Hallucinations in ChatGPT: Implications in Scientific Writing. Cureaus. DOI: DOI: 10.7759/cureus.35179

Amodei, D. (2023). Chatbots sometimes make things up. Is AI's hallucination problem fixable? *AP.* https://apnews.com/article/artificial-intelligence-hallucination-chatbots-chatgpt-falsehoods-ac4672c5b06e6f91050aa46ee731bcf4

APA. (2024). Addressing equity and ethics in artificial intelligence. *APA.*https://www.apa.org/monitor/2024/04/addressing-equity-ethics-artificial-intelligence

Arora, A., Barret, M., Lee, E., Oborn, E. & Prince, K., (2023). Risk and the future of AI: Algorithmic bias, data colonialism, and marginalisation. *Information and Organisation (33, 3).* .DOI: 10.1016/j.infoandorg.2023.100478

Arrieta, A. B., Díaz-Rodríguez, N., Del Ser, J., Bennetot, A., Tabik, S., Barbado González, A., Garcia, S., Gil-Lopez, S., Molina, D., Benjamins, V. R., & Chatila, R. (2019). Explainable Artificial Intelligence (XAI): Concepts, Taxonomies, Opportunities and Challenges toward Responsible AI. *Information Fusion*, 58, 15–36. DOI: 10.1016/j.inffus.2019.12.012

Bender, E., Gebru, T., McMillan-Major, A., & Shmitchell, S. (2022). On the Dangers of Stochastic Parrots: Can Language Models Be Too Big? *FAccT '21: 2021 ACM Conference on Fairness, Accountability, and Transparency.* DOI: DOI: 10.1145/3442188.3445922

Bommasani, S. (2023). The Unpredictable Abilities Emerging From Large AI Models. *Quanta.*https://www.quantamagazine.org/the-unpredictable-abilities-emerging-from-large-ai-models-20230316/

Bostrom., (2014). Superintelligence: Paths, Dangers, Strategies. Oxford University Press. ISBN 978 0 19 967811 2

Bowman, S. (2023). Even the scientists who build AI cannot tell you how it works. Vox. https://www.vox.com/unexplainable/2023/7/15/23793840/chat-gpt-ai-science-mystery-unexplainable-podcast

Brożek, B., Furman, M., Jakubiec, M., & Kucharzyks, B. (2024). 2023. The black box problem was revisited. Real and imaginary challenges for automated legal decision-making. *Artificial Intelligence and Law*, 32(2), 427–440. DOI: 10.1007/s10506-023-09356-9

Buolamwini, J., & Gebru, T. (2018). Gender Shades: Intersectional Accuracy Disparities in Commercial Gender Classification. *Proceedings of Machine Learning Research*, 81, 1–1. https://www.media.mit.edu/publications/gender-shades-intersectional-accuracy-disparities-in-commercial-gender-classification/

Cambridge, (2023). ChatGPT (We need to talk). University of Cambridge. https://www.cam.ac.uk/stories/ChatGPT-and-education

Capgemini, (2023). Half of secondary school teachers believe that the benefits of generative AI in education outweigh the risks. Capgemini. https://www.capgemini.com/news/press-releases/half-of-secondary-school-teachers-globally-believe-that-benefits-of-generative-ai-as-an-educational-tool-outweigh-the-risks/

Chan, C., & Hu, W. (2023). Students' voices on generative AI: Perceptions, benefits, and challenges in higher education. *International Journal of Educational Technology in Higher Education*, 20(1), 43. DOI: 10.1186/s41239-023-00411-8

China Youth Daily & China Youth School Media. (2023). More than eighty per cent of college students surveyed have used AI tools. China Youth Daily. https://edu.cnr.cn/sy/sytjB/20231120/t20231120_526491878.shtml

Congress, (2022). H.R.6580 - Algorithmic Accountability Act of 2022. *Congress of the United States*.https://www.congress.gov/bill/117th-congress/house-bill/6580

DARPA. (2016). Explainable Artificial Intelligence (XAI) (Archived). DARPA. https://www.darpa.mil/program/explainable-artificial-intelligence

DeVon, C. (2023). AI chatbots can 'hallucinate' and make things up—why it happens and how to spot it. *CNBC*. https://www.cnbc.com/2023/12/22/why-ai-chatbots-hallucinate.html

Doran, D., Schulz, S., & Besold, T. R. (2017). What Does Explainable AI Really Mean? A New Conceptualization of Perspectives. Wright State University. https://arxiv.org/abs/1710.00794

Edelsen, L. (2023). Google's Bard A.I. chatbot is being trained by contractors who say they're overworked, underpaid, stressed and scared. Fortune. https://fortune.com/2023/07/12/google-bard-ai-chatbot-trained-contractors-overworked-underpaid-stressed-scared/

Fang, X., Che, S., Mao, M., Zhang, H., Zhao, M., & Zhao, X. (2024). Bias of AI-generated content: An examination of news produced by large language models. *Scientific Reports*, 14(1), 5224. DOI: 10.1038/s41598-024-55686-2 PMID: 38433238

Fayyad, U. (2023). Stochastic Parrots Or Intelligent Assistants? Forbes. https://www.forbes.com/sites/forbestechcouncil/2023/06/06/stochastic-parrots-or-intelligent-assistants/?sh=562777665966

Forman, N., Udvaros, J., & Avornicului, M. (2023). ChatGPT: A new study tool shaping the future for high school students. *International Journal of Advanced Natural Sciences and Engineering Research*, 7(4), 95–102. DOI: 10.59287/ijanser.562

Gamvros, A., Yau, E., & Choong, S. (2023). China finalises its Generative AI Regulation. Norton Rose Fulbright. https://www.dataprotectionreport.com/2023/07/china-finalises-its-generative-ai-regulation/

Garante, (2023). Personal data is collected unlawfully; no age verification system is in place for children. *Garante*.https://www.garanteprivacy.it/home/docweb/-/docweb-display/docweb/9870847#english

GDPR. (2016). What does the General Data Protection Regulation (GDPR) govern? *GDPR*.https://commission.europa.eu/law/law-topic/data-protection/reform/what-does-general-data-protection-regulation-gdpr-govern_en?

Giannini, S. (2023). Generative AI and the future of education. UNESCO. https://unesdoc.unesco.org/ark:/48223/pf0000385877

Golda, A., Mekonen, K., Pandey, A., Singh, A., Hassija, V., Chamola, V., & Sikdar, B. (2024). Privacy and Security Concerns in Generative AI: A Comprehensive Survey. *IEEE Access : Practical Innovations, Open Solutions*, 12, 48126–48144. DOI: 10.1109/ACCESS.2024.3381611

Google. (2024). Responsible AI practices. Google. https://ai.google/responsibility/responsible-ai-practices/

Gowlingwlg, (2023). The artificial intelligence and data act (AIDA). *Gowling WLG*. https://gowlingwlg.com/en/insights-resources/articles/2023/artificial-intelligence-and-data-act-aida/

Gruber, C. (2023). GIGO: The Production Function of AI's Market Impact. Medium. https://medium.com/@gruberai/gigo-the-production-function-of-ais-market-impact-97a724b1650b

Gunning, D., Vorm, E., Wang, J. Y., & Turek, M. (2021). DARPA's explainable AI (XAI) program: A retrospective. AI & Logic, Advance online publication. DOI: 10.1002/ail2.61

Habayeb, H. (2022). Explainable AI Isn't Enough; We Need Understandable AI. Techopedia. https://www.techopedia.com/explainable-ai-isnt-enough-we-need-understandable-ai/2/34671

Hassija, V., Chamola, V., Mahapatra, A., Singal, A., Goel, D., Huang, K., Scardapane, S., Spinelli, I., Mahmud, M., & Hussain, A. (2024). Interpreting Black-Box Models: A Review on Explainable Artificial Intelligence. *Cognitive Computation*, 16(1), 45–74. DOI: 10.1007/s12559-023-10179-8

Hatem, R., Simmons, B., & Thornton, J. (2023). A Call to Address AI "Hallucinations" and How Healthcare Professionals Can Mitigate Their Risks. *NIH.*https://www.ncbi.nlm.nih.gov/pmc/articles/PMC10552880/

Hayes, S., Fortier, C., Levine, A., Milberg, W., & McGlinchey, R. (2012). *NIH.* https://www.ncbi.nlm.nih.gov/pmc/articles/PMC3955262/

Hendler, N., Judd, W. Picha, D. & Tantoco, E., (2022). Inherent GIGO Problems with Medical AI. *Journal of Artificial Intelligence and Soft Computing Techniques.*

Hulsen, T. (2023). Explainable Artificial Intelligence (XAI): Concepts and Challenges In Healthcare. *MDPI.*https://www.mdpi.com/2673-2688/4/3/34

IBM. (2023). What is explainable AI? IBM. https://www.ibm.com/topics/explainable-ai

IBM_A (2023). What are AI hallucinations? IBM. https://www.ibm.com/topics/ai-hallucinations

IBM_B, (2024). What is risk mitigation? *IBM.* https://www.ibm.com/topics/risk-mitigation

IBM_C (2024). What is explainable AI? IBM. https://www.ibm.com/topics/explainable-ai

INDU. 2024. The Artificial Intelligence and Data Act (AIDA). *House of Commons' Standing Committee on Industry and Technology (INDU.*https://www.ourcommons.ca/Content/Committee/441/INDU/Brief/BR12711885/br-external/CanadianLabourCongress-e.pdf

Islas, L. (2023). Gemini, Google's new Artificial Intelligence, is explained in 5 key facts. Merca2.0. https://www.merca20.com/gemini-googles-new-artificial-intelligence-explained-in-5-key facts/#:~:text=Gemini%201.0%201%20Parameters%3A%201.6%20trillion%202%20Training,date%3A%20December%202023%204%20Execution%3A%20In%20the%20cloud

Johnson, R., & Gunning, B. (2024). How does data science impact publication? *Geoconvention 2024*.https://geoconvention.com/wp-content/uploads/abstracts/2024/106445-how-does-data-science-impact-publication.pdf

Kambhampati, S. (2023). Can LLMs Really Reason and Plan? *ACM*.https://cacm.acm.org/blogcacm/can-llms-really-reason-and-plan/

Khosravi, H., Buckingham Shum, S., Chen, G., Conati, C., Tsai, Y.-S., Kay, J., Knight, S., Martinez-Maldonado, R., & Sadiq, S. (2022). Explainable Artificial Intelligence in Education. Elsevier. Computers and Education: Artificial Intelligence 3 (2022)

Kim, T. & Routledge, B., 2022. Why a Right to an Explanation of Algorithmic Decision-Making Should Exist: A Trust-Based Approach. *Business Ethics Quarterly 32:1 (pp. 75–102)*. DOI:DOI: 10.1017/beq.2021.3

Lindroth, H., Nalaie, K., Raghu, R., Ayala, I. N., Busch, C., Bhattacharyya, A., Moreno Franco, P., Diedrich, D. A., Pickering, B. W., & Herasevich, V. (2024). Applied artificial intelligence in healthcare: A review of computer vision technology application in hospital settings. *Journal of Imaging*, 10(4), 81. DOI: 10.3390/jimaging10040081 PMID: 38667979

Longpre, S., Yauney, G., Reif, E., Lee, K., Roberts, A., Zoph, B., Zhou, D., Wei, J., Robinson, K., Mimno, D., & Ippolito, D. (2023). A Pretrainer's Guide to Training Data: Measuring the Effects of Data Age, Domain Coverage, Quality, & Toxicity. Cornell. https://doi.org//arXiv.2305.13169DOI: 10.48550

Malizia, A., & Paterno, F. (2023). Why Is the Current XAI Not Meeting the Expectations? *ACM*.https://acmwebvm01.acm.org/magazines/2023/12/278147-why-is-the-current-xai-not-meeting-the-expectations/fulltext

Manning, A. (2024). Researchers publish largest-ever dataset of neural connections. *Harvard Gazette*. https://news.harvard.edu/gazette/story/2024/05/the-brain-as-weve-never-seen-it/

Marcus, G. (2024). Will AI Destroy Us? - AI Virtual Roundtable. Coleman Hughes. https://www.youtube.com/watch?v=xkQSiS8hpZA

Marr, B. (2023). The 15 Biggest Risks Of Artificial Intelligence. Forbes. https://www.forbes.com/sites/bernardmarr/2023/06/02/the-15-biggest-risks-of-artificial-intelligence/?sh=72f2c8502706

McKendrick, J., & Thural, A. (2022). AI Isn't Ready to Make Unsupervised Decisions. *HBR*.https://hbr.org/2022/09/ai-isnt-ready-to-make-unsupervised-decisions

Mollick, E. (2022). ChatGPT Is a Tipping Point for A. *HBR*.https://hbr.org/2022/12/chatgpt-is-a-tipping-point-for-ai

Murgia, M. (2023). Generative AI exists because of the transformer. Financial Times. https://ig.ft.com/generative-ai/

Narayanan, A. (2023). Disinformation Researchers Raise Alarms About A.I. Chatbots. New York Times. https://www.nytimes.com/2023/02/08/technology/ai-chatbots-disinformation.html

Nolan, B. (2023). Google researchers say they got OpenAI's ChatGPT to reveal some of its training data with just one word. Business Insider. https://www.businessinsider.com/google-researchers-openai-chatgpt-to-reveal-its-training-data-study-2023-12#:~:text=The%20AI%20model%20powering%20ChatGPT,or%20570%20GB%2C%20of%20data

NOYB. (2024). ChatGPT provides false information about people, and OpenAI can't correct it. *NOYB*.https://noyb.eu/en/chatgpt-provides-false-information-about-people-and-openai-cant-correct-it

Open, A. I. (2024). Limitations. *OpenAI*.https://openai.com/blog/chatgpt

Open, A. I. _A, (2024). Improving language understanding with unsupervised learning. *OpenAI*.https://openai.com/research/language-unsupervised?ref=superpower-chatgpt-extension

Open, A. I. _B, (2024). How should AI systems behave, and who should decide? *Open_AI*.https://openai.com/index/how-should-ai-systems-behave/

Pagallo, U. (2017). Automation to autonomous systems: A legal phenomenology with problems of accountability. In C. Sierra (Ed.). Proceedings of the twenty-sixth joint Conference in artificial intelligence (issue 26th international joint conference on artificial intelligence (IJCAI) (pp. 17–23). DOI: 10.24963/ijcai.2017/3

Perrigo, B. (2023). Exclusive: OpenAI Used Kenyan Workers on Less Than $2 Per Hour to Make ChatGPT Less Toxic. Time. https://time.com/6247678/openai-chatgpt-kenya-workers/

Phillips, P. J., Hahn, C. A., Fontana, P. C., Yates, A. N., Greene, K., Broniatowski, D. A., & Przybocki, M. A. (2019). Four Principles of Explainable Artificial Intelligence. NIST Interagency/Internal Report (NISTIR) 8312. https://nvlpubs.nist.gov/nistpubs/ir/2021/NIST.IR.8312.pdf

PRC. (2023). Interim Measures for the Management of Generative AI Services. People's Republic of China. https://www.cac.gov.cn/2023-07/13/c_1690898327029107.htm

Rai, A., (2020). Explainable AI: from black box to glass box. *Journal of the Academy of Marketing Science Volume 48*, p.p 137–141. DOI: 10.1007/s11747-019-00710-5

Ratti, E. & Graves, M., (2021). Explainable machine learning practices: opening another black box for reliable medical AI. *AI and Ethics (2022) 2:801–814.* https://doi.org/DOI: 10.1007/s43681-022-00141-z

Rudin, C. (2019). Stop Explaining Black Box Machine Learning Models for High Stakes Decisions and Use Interpretable Models Instead. National Library of Medicine. https://pubmed.ncbi.nlm.nih.gov/35603010/

Russell, S. (2023). AI likely to spell end of traditional school classroom, leading expert says. Guardian. https://www.theguardian.com/technology/2023/jul/07/ai-likely-to-spell-end-of-traditional-school-classroom-leading-expert-says

Russell, S., & Norvig, P. (2021). *Artificial Intelligence: A Modern Approach* (4th ed.). Pearson.

Safety Summit, A. I. (2023). Capabilities and risks from frontier AI. Department of Science, Innovation & Technology. https://assets.publishing.service.gov.uk/media/65395abae6c968000daa9b25/frontier-ai-capabilities-risks-report.pdf

Saunders, S., Lindsay, R., & McGrae, B. (2023). A spotlight on Generative Artificial Intelligence (GAI) Technology. *University of Liverpool.*https://www.liverpool.ac.uk/media/livacuk/centre-for-innovation-in-education/staff-guides/generative-artificial-intelligence-technology/generative-artificial-intelligence-technology.pdf

Schaffer, A., Miranda, B., & Koyejo, S. (2023). Are Emergent Abilities of Large Language Models a Mirage? NeuralIPs. https://proceedings.neurips.cc/paper_files/paper/2023/file/adc98a266f45005c403b8311ca7e8bd7-Paper-Conference.pdf

Seagate, (2020). Rethink data. Seagate. https://www.seagate.com/content/dam/seagate/migrated-assets/www-content/our-story/rethink-data/files/Rethink_Data_Report_2020.pdf

Senate Inquiry Australia. (2023). Inquiry into the use of generative AI in the education system. *University of Melbourne*.https://about.unimelb.edu.au/__data/assets/pdf_file/0032/396446/UoM-Submission-Inquiry-into-Generative-AI-in-Education-FINAL.pdf

Sichai, S. (2023). Is artificial intelligence advancing too quickly? What AI leaders at Google say. *60 Minutes*.https://www.cbsnews.com/news/google-artificial-intelligence-future-60-minutes-transcript-2023-04-16/

Simon, F., Altay, S., & Mercier, H. (2023). Misinformation reloaded? Fears about the impact of generative AI on misinformation are overblown. *HKS*.https://misinforeview.hks.harvard.edu/article/misinformation-reloaded-fears-about-the-impact-of-generative-ai-on-misinformation-are-overblown/

Sreedharan, S., Kulkarni, A., & Kambhampati, S. (2024). Explainable Human-AI Interaction: A Planning Perspective. *SLAIML*.https://link.springer.com/book/10.1007/978-3-031-03767-2

Sutskever, I. (2023). OpenAI Is Working to Fix ChatGPT's Hallucinations. *IEEE Spectrum*.https://spectrum.ieee.org/ai-hallucination

Techcrunch, (2023). Poland opens privacy probe of ChatGPT following GDPR complaint. TC. https://techcrunch.com/2023/09/21/poland-chatgpt-gdpr-complaint-probe/

Techcrunch_A, (2024). ChatGPT is violating Europe's privacy laws, Italian DPA tells OpenAI. *TC*.https://techcrunch.com/2024/01/29/chatgpt-italy-gdpr-notification/

Timi-koleolu, S., & Atanda, O. (2023). Artificial intelligence in Nigeria: legal and regulatory guidance. Pavestones. https://pavestoneslegal.com/artificial-intelligence-in-nigeria-legal-and-regulatory-guidance/

Van den Berg, M., & Kuiper, O. (2020). Xai in the financial sector. Hogeschool Utrecht. https://www.researchgate.net/publication/344079379_XAI_in_the_Financial_Sector_A_Conceptual_Framework_for_Explainable_AI_XAI

Vaswani, A., Shazeer, N., Parmar, N., Uszkoreit, J., Jones, L., Gomez, A. N., Kaiser, L., & Polosukhin, I. (2017). Attention is All You Need. In *Proceedings of the Neural Information Processing Systems (NIPS) conference*

Voight, A. (2023). Schools right now face a choice - fight the wave of ChatGPT, or surf it. Sydney Morning Herald. https://www.smh.com.au/national/schools-right-now-face-a-choice-fight-the-wave-of-chatgpt-or-surf-it-20230118-p5cdle.html

Warner, J. (2024). ChatGPT Cannot Teach Writing. Insider Higher ED. https://www.insidehighered.com/opinion/blogs/just-visiting/2024/01/22/arizona-state-announces-plan-give-education#

WEF. (2023). How will generative AI impact healthcare? *WEF*.https://www.weforum.org/agenda/2023/05/how-will-generative-ai-impact-healthcare/

Wei, J., Tay, Y., Bommasani, R., Raffel, C., Zoph, B., Borgeaud, S., Yogatama, D., Bosma, M., Zhou, D., Metzler, D., Chi, E. H., Hashimoto, T., Vinyals, O., Liang, P., & Dean, J. (2022). Emergent Abilities of Large Language Models. Transactions on Machine Learning Research. https://arxiv.org/pdf/2206.07682.pdf

Weiser, W. (2000). Gender Differences in Internet Use Patterns and Internet Application Preferences: A Two-Sample Comparison. Cyberpsychology & behavior (3,2). https://www.researchgate.net/publication/250899760_Gender_Differences_in_Internet_Use_Patterns_and_Internet_Application_Preferences_A_Two-Sample_Comparison

Yampolskiy, R., (2019). Unexplainability and Incomprehensibility of AI. International Journal of Machine Consciousness (2009 Vol. 1 Issue 1 - 2014 Vol. 6 Issue 2 https://doi.org/DOI: 10.1142/S2705078520500150

Yudowsky, E. (2023). Will AI Destroy Us? - AI Virtual Roundtable. Coleman Hughes. https://www.youtube.com/watch?v=xkQSiS8hpZA

Zhou, M., Abhihek, V., Derdenger, T., & Kim, J. $ Srinivasan, K., (2024). Bias in Generative AI. https://doi.org//arXiv.2403.02726DOI: 10.48550

Zimmer, C. (2011). 100 Trillion Connections: New Efforts Probe and Map the Brain's Detailed Architecture. *Scientific American*. https://www.scientificamerican.com/article/100-trillion-connections/#:~:text=Join%20together%20100%20billion%20neurons%E2%80%94with%20100%20trillion%20connections%E2%80%94and,a%20human%20brain%2C%20capable%20of%20much%2C%20much%20more

ENDNOTES

1 This is called Unsupervised Learning and is 90% of the language training an LLM will typically receive (OpenAI_A, 2024). Some texts in the corpus receive limited filtering to remove metadata and target subjects and then elected the LLM parses texts with selected words in sentences masked. The LLM uses Backpropagation, a mathematical algorithm. to compare the LLM's prediction with the actual word masked. Any error between the prediction and the actual

word is calculated to adjust an LLM's probabilistic relationship parameters for these words. These adjustments aim to improve the LLM's ability to predict missing words based on context in future encounters. Unselected texts for sentence masking are statistically massaged to produce a map of word relationships. A training algorithm in the LLM analyses co-occurrence patterns within a dataset, tracking how often words appear together. The algorithm builds a richer understanding of word relationships within the language by analysing these co-occurrence patterns across the entire dataset. A feature is the creation of word embeddings. These are numerical representations of words where words with similar meanings have close relationships. These word embeddings and other statistical features derived from the unselected text become the LLM's "diet" during training. So, a massive dataset of 300 billion words is only part selected for pre-training, and the remainder is then statistically analysed and applied later as a synthesised summary. Once complete, Supervised Learning occurs. In Supervised Learning, humans or other language models are crucial in providing labelled examples to train the model. Supervised training uses marked-up or labelled texts to fine-tune an LLM. While simplified, in supervised learning, machines learn from data with labels, like teachers showing students right and wrong answers. This helps them sort information and make predictions. (V7labs, 2023). In supervised training, the LLM learns from carefully curated datasets. These datasets can be tailored for specific tasks, like identifying emotions in text or structured to direct the model's behaviour in ways that align with what developers value and what users expect. (AI Safety Summit, 2023).

2 Transformers analyse the entire input, from sentences to articles, considering all its elements together, not just isolated words. (Murgia, M., 2023). The change from previous ANNs to the Transformer model allows for the fast parallel processing of data. The key function is the Self-Attention mechanism, which allows for processing text in "different positions of a single sequence to compute a representation of the sequence" (Vaswani A. et al.., 2017, p.2). In plain words, the Self-Attention mechanism dynamically adjusts its focus across different segments of the input sequence, efficiently capturing contextual relationships within the data. Unlike traditional ANNs, the Transformer architecture breaks down processing into identical building blocks. Each block uses self-attention and standard neural networks to understand complex relationships within text data (Golda A. et al., 2024).

3 Google and OpenAI use cheap human labour to edit pre-training supervised learning. This is Reinforcement Learning with Human Feedback (RLHF) – a process that corrects LLM text responses. ChatGPT Kenyan text editors "could

take home up to $2 per hour if they met all their targets." (Perrigo, B., 2023). "It is worth remembering that these systems are not the work of magicians — they are the work of thousands of people and their low-paid labour." (Edelsen, L., 2023).

4 Some GAIs have exits to the Internet when processing requests to embed searched-up information. This information is not part of the ANN and will be NLP processed rather than included in any generative process. This is problematic as the content is raw and may add to the burden of falsehood and bias.

5 SHAP, LIME, and Anchors are all separate techniques used for explaining the predictions of machine learning models, but they can be complementary: SHAP (SHapley Additive exPlanations): This method calculates how much each feature contributes to a specific prediction. It uses game theory concepts to credit features based on their influence on the model's output. SHAP provides global explanations, meaning it helps understand the importance of features across the entire dataset. LIME (Local Interpretable Model-agnostic Explanations): This technique explains individual predictions. It works by creating a simpler model around a specific data point and analysing how that model arrives at the prediction. LIME is model agnostic, meaning it can be used with any type of machine learning model. Anchors: Anchors are a method of explanation related to LIME. They express explanations as easy-to-understand "IF-THEN" rules. These rules highlight the conditions on a specific set of features sufficient to cause a particular prediction. Anchors provide local explanations similar to LIME, but they are more interpretable and reusable

6 As of July 2024, the ACT remains unpassed

7 As of July 2024, the ACT remains unpassed.

8 Complaints regarding data privacy have been filed against LLM operators. For example, this occurred in Italy in March 2023 (Garante, 2023), Austria in April 2024 (NOYB, 2024) and Poland in August 2023 (Techcrunch, 2023). Violations were filed with their respective countries' data protection authorities. To date, these complaints have not been acted on. In January 2024, Italy's data protection authority further flagged that ChatGPT was not compliant with the EU's General Data Protection Laws, which is an ongoing complaint (Techcrunch_A, 2024).

9 Common languages such as C++, JSON, Python, Java, and JavaScript can be generated to provide useful programming snippets.

10 The current free-to-use version of ChatGPT in 2024 is version 3.5

11 Warren McCulloch and Walter Pitts developed the initial concept of the node in 1943. They described mathematically an artificial neuron network with a "character of nervous activity, neural events and the relations among them can be treated using propositional logic" (McCulloch & Pitts 1943, p. 99). Their artificial 'neuron' was called a Perceptron.

12 All LLMs rely on a type of artificial neural network called Transformer architecture. GAI applications rely on Transformer architecture. Before Transformers, language was processed sequentially. Transformer model can also "efficiently handle large inputs and outputs such as images, audio and video. Making generation less sequential" (Vaswani, A., Shazeer, N., Parmar, N., Uszkoreit, J., Jones, L., Gomez, A. N., Kaiser, L., & Polosukhin, I., 2017., et al. 2017, p.10). The key function is the Self-Attention mechanism, which allows for processing text in "different positions of a single sequence to compute a representation of the sequence" (Vaswani A. et al., 2017, p.2). In plain words, the Self-Attention mechanism dynamically adjusts its focus across different segments of the input sequence, efficiently capturing contextual relationships within the data. Without the development of Transformer processing, a tool such as ChatGPT would be snail-like in responsiveness.

Chapter 11
Redefining Learning Experiences With Virtual and Augmented Reality:
Creating Engaging Learning Environments Through Immersive Technologies

Seema Yadav
https://orcid.org/0000-0002-8477-5895
Bhopal School of Social Sciences, India

ABSTRACT

The integration of immersive technologies such as Virtual Reality (VR), Augmented Reality (AR), and Extended Reality (XR) is reshaping educational paradigms by offering immersive, interactive, and contextually rich learning environments. These technologies are pivotal in cultivating critical 21st-century skills like critical thinking, self-directed learning, and teamwork, bridging theoretical knowledge with practical application. This article synthesizes insights from various studies highlighting the transformative potential of VR and AR in education. Despite their promising benefits, challenges such as cost, accessibility, and pedagogical integration remain critical considerations. By fostering an environment conducive to their ethical and impactful deployment, educational institutions can prepare students for success in a digitally-driven world. The article concludes by advocating for a balanced approach that leverages technological advancements while prioritizing thoughtful implementation to maximize educational outcomes.

DOI: 10.4018/979-8-3693-9811-1.ch011

INTRODUCTION

The advancement of technology in actual life these days will inevitably affect education. Virtual reality (VR), augmented reality (AR), and mixed reality (MR) are examples of Extended Reality (XR) technologies, which are currently thought to be extremely promising technical aids for education (Alnagrat et al., 2022). Virtual reality (VR) and augmented reality (AR) are the two primary categories of parallel realities that are represented by the term "immersive technology." Both put the user in a completely or partially virtualised setting (Sadek et al., 2023). Immersive technology, encompassing Virtual Reality (VR) and Augmented Reality (AR), is revolutionizing education by creating highly engaging and interactive learning environments. These technologies transport students into simulated worlds where they can explore complex concepts, practice skills, and experience scenarios that would be impossible or impractical in a traditional classroom (Tom Dieck & Han, 2022). For example, VR can take students on virtual field trips to historical sites or inside the human body, while AR can overlay digital information onto the physical world, enhancing real-world interactions with educational content. This level of immersion not only captivates students' attention but also enhances their understanding and retention of material through experiential learning. By making abstract concepts tangible and fostering active participation, immersive technology is poised to transform educational practices and significantly improve learning outcomes (Pears & Konstantinidis, 2022). The integration of Virtual Reality (VR) and Augmented Reality (AR) into educational environments represents a transformative shift in how learners engage with content.

The world has changed, and it will continue to change at an ever-increasing rate due to the pervasiveness of AI and the promise of immersive technologies, which blur the lines between the actual and virtual worlds (Aerts, 2024). As digital technologies have developed, a variety of interface technologies have emerged (Partarakis & Zabulis, 2024). Nations all around the world looked for creative ways to adjust to the new realities of distance learning. Innovative teaching and learning techniques have been adopted by online education systems globally. Animating content with Augmented Reality/Virtual Reality (AR/VR) technologies is one of the most effective ways to include students and improve learning outcomes (Giri & Sharma, 2024). With the use of immersive augmented reality (AR) surroundings, dermatological students may now explore well-known virtual environments and learn new things through simulated scenarios, all while applying their abilities to a familiar context (Muralidharan et al., 2024). The fields of augmented reality (AR) and virtual reality (VR) have significantly grown in the last few years. These state-of-the-art tools open up new scientific vistas for dermatology researchers, educators, proceduralists, and patients. While AR adds virtual aspects to a user's perspective of their real surroundings, VR allows users to interact with varied simulated settings through natural head and hand movements, facilitating engaging human encounters (Muralidharan et al., 2024). With its ability to create immersive and engaging learning environments, extended reality (XR) technology which includes virtual reality (VR), augmented reality (AR), and mixed reality (MR) has the potential to completely transform education (Asif Zaman et al., 2024). Extended reality (XR) technology allows students to investigate ideas and phenomena that might be difficult or impractical to encounter in traditional classroom settings by imitating real-world contexts (Asif Zaman et al., 2024).

One of the key elements affecting a country's progress is education. Teachers must always be looking for fresh and innovative ways to provide material to students and encourage their participation in the learning process. Immersion technology has a lot of potential to help with the educational process. Because immersive technology facilitates higher learning outcomes, it may raise student achievement (Nastiti et al., 2024). The development of multimedia and computer technologies has led to an increased emphasis on technology-enhanced learning in the classroom (Nastiti et al., 2024). Immersions in technology can promote innovative thinking, inquisitiveness, and skill development (Taggart et al., 2023). Technologies involving virtual and augmented reality (VR/AR) have enormous potential in several medical domains. The introduction of a new line of medical-enhanced VR/AR (MER) headsets creates new opportunities for applications in urology and other medical fields to increase accessibility for all patients. With their deep immersion and controller-free design, these cutting-edge headsets provide a fresh take on VR/AR interaction (Zattoni et al., 2023). Modern technology has the power to transform medical processes, communication strategies, and urology practices, all of which could ultimately benefit patients' experiences with their urological conditions (Zattoni et al., 2023). VR/AR is new because it immerses users in virtual interfaces, enabling them to interact with their perceived environment. The experience of literally existing in a nonphysical environment is known as immersion (Muralidharan et al., 2024).

The field of information and communication technologies (ICT) and human-computer interaction (HCI) are undergoing a shift as a result of the introduction of novel design paradigms that are changing how people engage with digital data (Partarakis & Zabulis, 2024). Virtual reality (VR), mixed reality (MR), and augmented reality (AR) are examples of immersive technologies that are thought to be the foundation of contemporary digital experiences. In order to create intelligent, user-centered information systems, reasoning on knowledge sources is supported by knowledge representation in a machine-interpretable format (Partarakis & Zabulis, 2024). Virtual reality (VR) and augmented reality (AR) have become potent instruments for expanding the boundaries of medical treatment (Zattoni et al., 2023). Developments in virtual reality (VR), augmented reality (AR), and visual effects

(VFX) enable students to take part in interactive odysseys that immerse them in real-life circumstances, accelerating the retention of knowledge and the learning of practical skills (Giri & Sharma, 2024). As the commercial potential of these technologies becomes apparent, virtual reality (VR) and augmented reality (AR) are experiencing rapid growth (Taggart et al., 2023).

Immersion technologies such as virtual tours, virtual laboratories, augmented reality books and games, and 3D modeling are used in elementary school classrooms. Immersion technologies help improve the quality of education by making it more dynamic, flexible, and engaging for students; they also increase learning opportunities and aid in the better understanding of difficult concepts; they also encourage students' motivation and interest in learning, memory and focus, creativity, and activity, especially in younger students. It was discovered that utilizing immersive technology in the classroom greatly increases a modern specialist's toolbox (Shvardak, 2023). When compared to conventional approaches, the use of immersive technology greatly enhanced student learning. The foundation for creating educational applications was established from thirteen learning theories. To increase the range of applications focused on education, they recommended that promising but constrained new technologies be developed further. Immersion technologies for education are still in their infancy, and there are still many issues to be resolved, such as user preparation and potential privacy and security concerns (Nastiti et al., 2024).

Virtual reality and other immersive technologies allow for more involvement with the medium and accurately experienced training experiences (Rosendahl et al., 2024). With multi-perspective movement observation, 360° movies leverage the training potentials of traditional video technology without requiring significant creative work. Therefore, 360° videos are at least as appropriate as traditional training videos for providing visual assistance during instruction (Rosendahl et al., 2024). Instructors who wish to use such devices' capabilities to access VR and/or AR experiences will need to get around any limitations that certain schools may impose. This problem is solved by HMDs with integrated screens, although many schools may not be able to buy them given their present price (Taggart et al., 2023). The constant innovation may help teachers stay current and promote the use of cutting-edge technology in the classroom, but the constant pressure to share and keep up with the rapidly changing technologies will probably lead to professional pressures and even burnout (Taggart et al., 2023).

The rapid advancement of AI and immersive technologies like AR and VR is significantly transforming education by enhancing the learning experience. These technologies create engaging and interactive learning environments, allowing students to explore complex concepts in a more practical and immersive manner. In fields like dermatology and urology, AR and VR provide realistic simulations that improve training and patient care. The adoption of immersive technologies in classrooms, such as virtual tours and 3D modeling, makes education more dynamic and flexible, increasing student motivation and understanding. Despite the benefits, challenges remain, including the need for user preparation, privacy concerns, and the high cost of equipment. Additionally, educators face pressure to continuously update their knowledge and integrate new technologies, which can lead to professional stress. However, the potential of immersive technologies to transform education and training across various fields is immense, promising more effective and engaging learning experiences.

INTRODUCTION TO VIRTUAL REALITY AND AUGMENTED REALITY (VR AND AR) AND IMMERSIVE TECHNOLOGIES

In recent years, there has been tremendous advancement in the fields of virtual reality (VR) and augmented reality (AR). New scientific vistas are opened up for dermatology researchers, educators, proceduralists, and patients thanks to these state-of-the-art tools. While augmented reality (AR) adds virtual components to a user's perspective of their real surroundings, virtual reality (VR) technology allows users to interact with diverse simulated environments through natural head and hand movements (Muralidharan et al., 2024). While immersive technologies and artificial intelligence (AI) are not new, their quick revolution has surprised us. The world has gone crazy over the metaverse and generative AI in particular since OpenAI's ChatGPT 3.5 was released. Even the most reputable AI researchers are being surprised by the rise in chatbots and multimodal generative AI technologies that rely on LLMS (long language model) advancements (Aerts, 2024). In actuality, virtual reality (VR) systems and other remote learning strategies offer a lifelike virtual environment that mimics the real world. Virtual reality (VR) can be used in educational settings that are challenging for students to physically access and experience, which makes it a useful tool for fostering creative learning and the emergence of diverse learning styles. Students can interact with realistic environments because to VE (Alnagrat et al., 2022). Technologies related to virtual, augmented, extended, and mixed reality are constantly evolving and integrating into society in new ways. These technologies range from adding layers to the physical world to fully submerging a user in a virtual one (Wagner & Alharthi, 2023).

Integration of AR and VR is the foundational technology for the much-anticipated metaverse, erasing boundaries between actual and virtual worlds. For these methods to work, AR and VR technology compatibility is essential. Virtual reality (VR) and augmented reality (AR) have different uses, advantages, and disadvantages. All technologies, though, aim to expand reality in the same way (Kang et al., 2024a).

Through the use of technology, virtual reality, and augmented reality can help students achieve more since they provide them with superior learning outcomes. Additionally, students' interest, motivation to study, and favorable attitude toward using technology in the classroom are strengthened and increased by immersive technology (Nastiti et al., 2024). Technology needs to act as an enabler, spearheading initiatives to surmount obstacles in order to provide a user-centric experience that is global (Nagtilak, 2024). The term virtual reality (VR) refers to technology that connects people to different simulated settings through natural head and hand movements, hence enabling immersive human experiences. With augmented reality (AR), users may see and hear their surroundings while also adding a virtual overlay (Muralidharan et al., 2024). Virtual reality (VR) offers the freedom to evaluate research results using both quantitative and qualitative approaches in a variety of study designs, from case studies to randomized control trials. It's important to think about when and how VR fits into a certain curriculum when utilizing the technology for teaching and pedagogical improvement (Muralidharan et al., 2024).

In recent years, significant advancements in VR and AR have opened new opportunities for various fields, including dermatology and education. AR overlays virtual elements onto real-world environments, while VR immerses users in simulated settings through natural movements. These technologies, integral to the metaverse, enhance learning outcomes, student interest, and motivation by providing engaging and interactive experiences. Despite their benefits, VR and AR come with distinct uses, advantages, and challenges. The rapid evolution of AI and immersive technologies, exemplified by the rise of generative AI and chatbots, continues to surprise even seasoned researchers. For effective integration, it's crucial to consider how these technologies fit within specific educational curricula and to ensure they provide a user-centric experience.

THEORETICAL FOUNDATIONS FOR VIRTUAL REALITY AND AUGMENTED REALITY

This education revolution has its origins in years of intensive study and research in the multimedia and animation fields. Technologies like AR and VR have become essential components in the creation of this new educational paradigm. Animation, augmented reality, and virtual reality have all advanced rapidly thanks in large part to technological advancements (Giri & Sharma, 2024). Increasing the chances that students have to build knowledge is one way to improve learning. This becomes a crucial factor to take into account when providing learners with appropriate materials, their surroundings, and an appropriate social setting. Offering beginner learners a simple starting point and progressively more complex assignments as time goes on is essential for the efficient use of technology (Nastiti et al., 2024). Constructivism, which holds that students actively create their understanding of the world via experiences and interactions with their surroundings, provides one theoretical basis for the use of digital immersive technology in education (Tang, 2024). Students may participate, explore, learn, and become more motivated thanks to immersive technologies like virtual reality, which offers a 3D immersive interactive environment. This is an important aspect of education (Nastiti et al., 2024). With its ability to promote student enthusiasm, active learning, and involvement, virtual reality is an extremely important teaching tool in the constructivist educational paradigm (Nastiti et al., 2024). Increasingly important are the ethical issues, technological literacy education, accessibility, interoperability, user trust, environmental sustainability, and regulatory frameworks along with the seamless integration of technology (Partarakis & Zabulis, 2024). When viewed through a human-centered lens, these technologies can potentially dynamically adjust to the various and changing needs of users. New technologies for information processing, information visualization, and interaction are all made possible by the simultaneous integration of artificial intelligence (AI) with ICT (Partarakis & Zabulis, 2024). Providing high-quality data creates the illusion of engaging with fictitious objects in an imaginary setting. Because users are fully submerged in the virtual environment, just like in the real world. In particular, virtual reality games can cause users to lose their sense of time, space, and reality and give them the impression that they are a part of the environment (Alnagrat et al., 2022).

The education revolution, driven by advancements in multimedia, animation, AR, and VR, is rooted in years of intensive research. These technologies are now essential in modern educational paradigms, enhancing learning by providing immersive, interactive environments. By adopting constructivist principles, where students build understanding through experiences and interactions, AR and VR boost engagement, motivation, and active learning. Effective use of technology involves gradually increasing task complexity to support learner development. Ethical considerations, technological literacy, accessibility, and regulatory frameworks are crucial for the successful integration of these technologies, which can dynamically adapt to users' evolving needs through the synergy of AI and ICT.

BENEFITS OF IMMERSIVE LEARNING WITH VIRTUAL REALITY AND AUGMENTED REALITY

The use of immersive technologies, such as augmented reality, in certain learning contexts is also beneficial to students' learning outcomes (Nastiti et al., 2024). Immersion technologies, such as virtual reality, are very useful tools (Nastiti et al., 2024). Recognizing these technologies' revolutionary power over distant learning experiences, educational institutions around the world embraced them. Globally, educators have come to rely heavily on animation and AR/VR technology as indispensable teaching aids. This broad acceptance demonstrated the technology's general applicability in raising student engagement and creating a supportive learning environment (Giri & Sharma, 2024). Immersion technology still has a long way to go in the future. For immersive technology to completely support the learning process, a great deal of progress is still needed (Nastiti et al., 2024).

Medical education can benefit from the pandemic-era simulation scenarios made possible by immersive technologies (Nastiti et al., 2024). Programs for physical education in preschools can make use of virtual reality and other immersive technology. They can improve students' focus and engagement, make it easier for them to integrate both new and existing knowledge, and support the growth of cognitive structures and cognitive psychological activities (Nastiti et al., 2024). Collaborative learning is facilitated when technologies, such as immersive technology, are included in instructional procedures. The term "collaborative learning" is an approach in which students collaborate in small groups to create things, concepts, ideas, phenomena, or applications that involve more people. Teachers adopt collaborative learning approaches because interactions between learners and the teacher-student relationship are two important ways that knowledge is created (Nastiti et al., 2024).

Immersion technologies that facilitate novel learning features or techniques are those that can offer profound, original, interactive, and captivating learning experiences that improve learning on both cognitive and emotional levels and boost learning motivation (Nastiti et al., 2024). AR encourages interaction between the user, digital content, and actual physical things in the real world by fusing immersive techniques with real-world visuals. As a result, an augmented reality format aims to provide a more realistic user experience. The learner could feel closer to their current surroundings as a result. Users using VR and AR can see, feel, interact with their surroundings, and receive real-time feedback (Muralidharan et al., 2024). Nearly every industry is using AR and VR technologies, including advertising, education, training, entertainment, and therapy. Specifically, a great deal of research has been done on using these technologies in teaching. Training is a type of instruction where students practice a particular ability or gain better knowledge by repeating the procedure (Kang et al., 2024a).

The integration of extended reality (XR) technology into education is crucial for revolutionizing conventional teaching methods and generating immersive, captivating, and productive learning environments (Asif Zaman et al., 2024). With the help of XR, students may interact with and explore virtual worlds, simulations, and three-dimensional content, which helps them understand difficult ideas and theories more naturally and experientially. XR improves student motivation, engagement, and information retention by providing an interactive, multisensory learning environment that encourages a better comprehension of the material (Asif Zaman et al., 2024). XR technologies allow for real-time collaboration among users by enabling them to interact with shared virtual objects. In order to improve their cooperation and communication skills, students can work together to solve problems, carry out virtual experiments, or develop collaborative projects. The shortcomings of conventional teaching techniques can potentially be addressed by incorporating XR into education (Asif Zaman et al., 2024).

In order to achieve sustainability in education through teamwork, they specifically look into how digital immersive technology can support critical thinking and self-directed learning in English language learners. They also evaluate how digital immersive experiences can improve learning outcomes by looking at how they can support critical thinking abilities and self-directed learning practices (Tang, 2024). Using digital immersion technology to teach English as a second language has grown in popularity in recent years. With the use of 3D simulations, educational games, and virtual and augmented reality, these technological innovations provide students with a dynamic and engaging learning environment. However, the relationship between using digital immersive technology, critical thinking, and teamwork is still unknown (Tang, 2024). Augmented reality (AR) provides learners with context-sensitive, real-time assistance by superimposing digital information onto the physical environment in vocational education. Augmented Reality (AR) enhances the learning process by enabling students to learn through hands-on equipment maintenance simulations and interactive instructional overlays (Partarakis & Zabulis, 2024). Augmented Reality (AR) unifies the physical and digital learning realms by expanding reality through instructive digital overlays, so creating a bridge between theory and practice (Partarakis & Zabulis, 2024). Its capacity to produce immersive, human-centered learning experiences is what gives XR applications for vocational education their added value. XR technologies encourage active engagement and information retention by imitating real-world work environments, which engages learners at a deeper level. Immersion-based, performance-based evaluations are used in vocational education in addition to regular exams. Extended Reality (XR) applications allow educators to evaluate students' application of abilities in real-world situations in addition to their theoretical understanding (Partarakis & Zabulis, 2024).

Immersive technologies, such as augmented and virtual reality, have been widely embraced by educational institutions worldwide for their ability to enhance learning outcomes and engagement. These technologies are especially useful in distance learning and have proven beneficial in fields like medical education and physical education for young children. They support collaborative learning by enabling students to work together in interactive and realistic environments, thus fostering cognitive and emotional development. The integration of extended reality (XR) into education is transforming traditional teaching methods by creating immersive and productive learning environments. XR technologies allow students to explore virtual worlds and simulations, enhancing their understanding of complex concepts and theories. This interactive, multisensory approach improves motivation, engagement, and information retention. Despite their benefits, immersive technologies still require significant progress to fully support the learning process. Ethical considerations, technological literacy, accessibility, and regulatory frameworks are essential for successful integration. Additionally, the use of these technologies in teaching English as a second language and other fields shows promise, though the relationship between digital immersion, critical thinking, and teamwork needs further exploration.

CHALLENGES AND LIMITATIONS OF VIRTUAL REALITY AND AUGMENTED REALITY

Pre-service instructors had no prior experience with VR or AR technology, despite their awareness of them. Their perceptions and self-assurance on utilizing cutting-edge ICT in the classroom were improved by the intervention. Pre-service teachers can assess their digital skills through brief interventions and create an action plan to improve them (Taggart et al., 2023). It is clear that pre-service teachers believe there is value in VR and/or AR technologies. It is reasonable to anticipate that they would encounter similar implementation challenges as other ICTs in schools. These are mostly related to long-term implementation, a lack of embedding time, and the rate of change in software and hardware development in the context of rising teacher demand in an educational climate driven by neoliberalism and outcomes (Taggart et al., 2023). It is important to understand that immersive technology depends not only on its users but also on the way it was designed, how it was implemented, and how it was integrated into the classroom. Since they are involved in the resources, information, and effective tools required to accomplish the intended results, teachers play a critical role in the implementation of this technology (Nastiti et al., 2024).

Notwithstanding the potential benefits of VR or AR for research, it is important to be aware of any possible ethical issues with capacity, permission, privacy, and confidentiality (Muralidharan et al., 2024). One of the most important benefits of XR in education is personalized learning. Personalized feedback and content catered to each student's strengths and limitations can be provided via adaptive learning systems, which are also capable of evaluating students' progress. Students who get challenges and content that are tailored to their learning styles and abilities are more engaged and motivated (Asif Zaman et al., 2024). Universities and other educational institutions have a history of being late adopters and implementers of digital technology for teaching and learning. The use of online meeting platforms, which were first introduced in the business sector, and learning management systems are examples of digital innovation in higher education that haven't extended very far beyond these first uses (Aerts, 2024).

Pre-service teachers, despite being aware of VR and AR technologies, had no prior experience using them. However, brief interventions improved their perceptions and confidence in using advanced ICT in the classroom. They recognized the value of VR and AR but anticipated similar implementation challenges as other ICTs, such as long-term adoption, lack of embedding time, and rapid changes in technology. Successful integration of immersive technology depends on its design, implementation, and integration into the classroom, with teachers playing a crucial role. Ethical concerns, such as capacity, consent, privacy, and confidentiality, are important to consider when using VR or AR in education. One major benefit of XR technology is personalized learning, which provides tailored feedback and content based on individual students' strengths and weaknesses, increasing engagement and motivation. Although universities and educational institutions have historically been slow to adopt digital technologies for teaching, recent developments like online meeting platforms and learning management systems show progress in this area.

VIRTUAL REALITY AND AUGMENTED REALITY (AR AND VR) IN VARIOUS ACADEMIC FIELDS

In dermatology, VR/AR offers potentially innovative technological solutions for learning, changing behavior, and forming habits. Financial factors may have an impact on how accessible technology is, as VR/AR research is still a costly endeavor (Muralidharan et al., 2024). It is encouraging to see the use of VR and AR in dermatology as digital health techniques expand. New digital technologies, however, need to be handled carefully and with an awareness of the risks. In this work, we address VR and AR in dermatology by giving a general review of the technology, describing useful procedures, and going over several real-world cases (Muralidharan et al., 2024). Students can collaborate and work as a team while using digital immersive technologies. Through the use of digital platforms, students can collaborate on tasks, discuss linguistic concepts, and provide feedback to one another. Such collaborative learning fosters independent learning and helps students develop critical interpersonal and collaboration skills. Students may get access to real-world experiences that they might not otherwise have because of digital immersion technologies (Tang, 2024). The integration of virtual, augmented, mixed, and extended reality technologies into the distant learning model created new avenues for research on the effectiveness of these tools. The research found that virtual, augmented, extended, and mixed reality technologies can improve knowledge attainment and retention, are effective at delivering content in a realistic and applicable way, and that users enjoyed the learning process and were motivated to use them to learn the material all without being a perfect solution or a complete replacement for traditional learning modalities (Wagner & Alharthi, 2023). Using immersive technologies to teach mathematics to students has several benefits, including: complete immersion in the learning process, the "presence effect," visibility and brightness of sensations, interactivity, facilitation of understanding and simplification of perception, use of visual and auditory channels of perception, the capacity to scale, rotate, assemble, and disassemble complex objects into their constituent parts, the potential for gamification of education, and the ability to repeat information and train skills (Kramarenko & Kochina, 2023). In the modern world, using emerging technologies is growing more widespread, and as media integration becomes more complex, so does the use of virtual reality technology. VR/AR technology is a rapidly developing field of science and technology that is essential to the growth and advancement of society. Its use in modern teaching and learning has made it a vital component. The creation of highly skilled animators is the key to the growth of the cultural and creative industries, the cornerstone of national soft power, and the response to the pressing talent shortage in both society and the animation industry (Fu, 2021).

Augmented reality (AR) in vocational education provides learners with context-sensitive, real-time assistance by superimposing digital information onto the physical world. AR enhances the educational experience by facilitating learning by doing through interactive instructional overlays and hands-on equipment maintenance simulations (Partarakis & Zabulis, 2024). In dermatology, VR, and AR offer innovative solutions for learning and habit formation, though their accessibility is affected by high costs. The expanding use of these technologies in dermatology requires careful handling and risk awareness. They provide detailed overviews, useful procedures, and real-world cases. Digital immersive technologies enable collaborative learning, allowing students to work together on tasks, discuss concepts, and provide feedback, fostering independent learning and essential interpersonal skills. AR, specifically in vocational education, enhances learning by overlaying digital information onto the physical world, providing real-time, context-sensitive assistance, and facilitating hands-on simulations and interactive instructional overlays.

DESIGNING VIRTUAL REALITY AND AUGMENTED REALITY (VR AND AR) LEARNING EXPERIENCES

Through collaborative learning, students can hone other abilities including reflection, critical thinking, teamwork, and participatory learning (Nastiti et al., 2024). Digital immersion technology fosters the growth of 21st-century skills including collaboration, critical thinking, and digital literacy all of which are essential for success in the contemporary world. This is consistent with the 21st-century learning paradigm, which highlights the need to prepare pupils for the demands of the digital age (Tang, 2024).

Experiences with AR and VR have their advantages and disadvantages. Augmented Reality (AR) offers a high degree of realism and exceptional accessibility without isolating users from the outside world by fusing the actual and virtual worlds. However, compared to VR, AR may be less immersive and its utility may vary based on the device's screen size and resolution. VR, on the other hand, is suitable in a variety of sectors because of its exceptional immersion and capacity to portray endless virtual worlds (Kang et al., 2024a). The spectrum of activities in the physical area is severely limited by VR's full separation from reality, which can also lead to motion sickness. Through integration, we were able to overcome the shortcomings of each technology and provide new advantages. We were able to verify through usability testing that consumers found the combined AR/VR technology to be both entertaining and helpful (Kang et al., 2024a). Virtual reality (VR) isolates users from reality, which drastically reduces the variety of activities possible in real space and increases the risk of motion sickness. By integrating the various technologies, we were able to overcome their shortcomings and provide new advantages. We were able to verify through usability testing that the combined AR/VR technology was both practical and engaging for users (Kang et al., 2024a).

Extended reality (XR) technologies can improve student motivation, engagement, and retention by providing immersive and interactive experiences. They can provide individualized learning pathways, accommodate different locations, and suit a variety of learning styles (Asif Zaman et al., 2024). Furthermore, XR can close the knowledge gap between theory and practice by enabling students to use what they have learned in real-world, authentic circumstances. Even with XR's growing appeal and promise in the classroom, there are still issues that need to be resolved (Asif Zaman et al., 2024). Over the next ten years, Extended Reality (XR) has the potential to completely transform education. Students' comprehension, retention, motivation, and engagement are all improved by XR's individualized, dynamic, and captivating learning experiences. Historical settings and virtual surroundings can be used to teach subjects like physics and history. XR facilitates practical learning and skill practice through virtual field excursions and simulations (Asif Zaman et al., 2024).

Adopting AI and immersive technology for dentistry education's future demands thoughtful deliberation and the creation of pedagogies that are still frequently mired in the world of analog lectures. New educational models and modalities in online and offline learning environments will require experimentation with and evaluation of digital pedagogies and innovation. These will probably change and grow as new digital tools and technology allow for more tutor-student, peer, tutor-machine, and learner-machine connection and involvement (Aerts, 2024). The development of the requisite cognitive skills may depend more on problem-solving and individual learning than on teamwork. Nonetheless, self-directed learning, teamwork, and critical thinking can all be promoted by digital immersive technology. For example, challenging scenarios requiring creative thinking and problem-solving skills may be presented via interactive educational games (Tang, 2024). Digital immersion technology allows for customized learning experiences that are beneficial to students. By using virtual worlds that teachers create that replicate real-life circumstances, students can practice language skills in context. It can help kids develop their critical thinking skills to make them use language to solve difficulties (Tang, 2024).

Collaborative learning with digital immersion technology enhances skills such as reflection, critical thinking, teamwork, and participatory learning. It aligns with the 21st-century learning paradigm, promoting essential skills like digital literacy, collaboration, and critical thinking for the digital age. AR and VR each have distinct advantages and disadvantages. AR offers realism and accessibility without isolating users but may lack the full immersion of VR. VR provides exceptional immersion and limitless virtual worlds but can cause motion sickness and isolate users from reality. Combining AR and VR technologies can mitigate their individual shortcomings and offer practical and engaging experiences, as verified through usability testing. XR technologies improve student motivation, engagement, and retention by offering immersive, interactive experiences and individualized learning pathways. They bridge the gap between theory and practice, allowing students to apply learned concepts in real-world scenarios. XR's potential to transform education over the next decade includes teaching subjects through virtual environments and facilitating practical learning through simulations. The integration of AI and immersive technology in fields like dentistry requires thoughtful consideration and the development of new pedagogies that blend digital and traditional learning methods. Digital immersive technology fosters self-directed learning, teamwork, and critical thinking by presenting challenging scenarios and interactive educational games. Customized learning experiences using virtual worlds help students practice language skills and develop critical thinking by solving real-life problems.

FUTURE DIRECTIONS AND EMERGING TRENDS IN THE APPLICATION OF VIRTUAL REALITY AND AUGMENTED REALITY

Our lives have been profoundly changed by artificial intelligence and will continue to be so in the future. It is becoming more prevalent across a wider range of contexts and businesses, and it is generating novel approaches to a number of pressing problems. However, it still has a number of issues that hinder its implementation in applications meant to improve system speed and user interactions. The widespread application of AI and its intimate connections to industry and society have produced benefits and raised productivity. It will surely have an impact on the existing social structure and pose moral dilemmas (Asif Zaman et al., 2024).

Virtual reality (VR) decreased the frequency of itching but did not lessen its intensity. Again, despite the modest study sizes, VR use demonstrated therapeutic promise that warrants additional investigation (Muralidharan et al., 2024). In addition to fostering positive attitudes and dispositions towards innovation, curiosity, and the benefits of developing ICT skills, a brief introduction to VR and AR technologies may allow pre-service teachers to assess their self-efficacy regarding digital capabilities (Taggart et al., 2023). The specialized requirements of dermatology clinical and research teams are met by VR/AR apps through effective storyboarding, user-driven design, and interdisciplinary teamwork. We offer a methodical approach to their design, team structure, and assessment in medical education, procedures, research in dermatology, and habit development techniques (Muralidharan et al., 2024). By connecting people to different simulated settings through natural head and hand movements, virtual reality (VR) refers to technology that aims to enable immersive human experiences. Users can see, hear, and interact with an actual world while adding a virtual overlay using augmented reality (AR) (Muralidharan et al., 2024). VR and AR have a wide range of potential applications in dermatology and have been employed in a variety of contexts, including dermatological procedures (Muralidharan et al., 2024). In the process of using immersive technology in the classroom, teachers are also vital players in the transition and training phase. Immersion technology, like augmented reality, provides more efficient teaching representations, removing the need for non-immersive usage groups to obtain representations from other tools or communication strategies (Nastiti et al., 2024).

The development of immersive technological methods in medical education, including virtual reality, can be used to teach competencies that would be challenging to teach systematically during clinical rotations (Nastiti et al., 2024). The application of extended reality technology will grow in the future. Technologies like virtual reality (VR) and augmented reality (AR) will be used by people for work, leisure, and a variety of other purposes. Applications for medical augmented reality have developed, spanning from picture-guided surgery to medical education (Taghian et al., 2023). The incorporation of technology has emerged as a key driver for transformative learning experiences in the constantly changing field of education. In particular, the revolutionary developments in Augmented Reality (AR) and Virtual Reality (VR) coupled with computer animation-based teaching have emerged as a promising catalyst for project-based learning systems (Giri & Sharma, 2024).

With a variety of study designs, including case reports and randomized control trials, virtual reality (VR) offers the freedom to evaluate study outcomes using both quantitative and qualitative approaches. It's important to think about when and how VR fits into a certain curriculum when utilizing the technology for teaching and pedagogical improvement (Muralidharan et al., 2024). It seems hopeful that VR and AR-based digital health techniques are being more widely used in dermatology. However new digital technologies need to be used carefully, keeping possible risks in mind. We cover VR and AR in dermatology in this paper by giving a general overview of the technology, laying down doable stages, and going over a number of real-world instances (Muralidharan et al., 2024). Because hybrid AR/VR technologies can get around obstacles like real-world challenges, risk factors, space and time constraints, and training expenses, they could be beneficial when used in fire drill training (Kang et al., 2024a). The capacity to generate lifelike virtual worlds in any given space is a step forward for the combination of AR and VR and shows great promise for improving training and teaching across a range of subjects. As technology develops, more sophisticated virtual simulations that offer compelling and useful learning experiences will be created (Kang et al., 2024a).

Virtual reality (VR), augmented reality (AR), and mixed reality (MR) are examples of extended reality (XR) technology that has the potential to completely transform education by offering engaging and dynamic learning environments. Establishing trust among all stakeholders is crucial to guarantee the successful deployment and adoption of XR in education. They offer a thorough framework for raising training and educational levels of confidence in XR technology (Asif Zaman et al., 2024). Important factors including openness, confidentiality, moral usage, research, user education, cooperation, accessibility, ongoing development, risk reduction, and long-term viability are all covered by the framework. Educational institutions may create a trusting environment that supports the successful integration of XR into the classroom and enhances student engagement and learning results by putting this approach into practice (Asif Zaman et al., 2024).

XR enables tailored and flexible learning experiences that accommodate various learning preferences and individual requirements. Additionally, it makes remote and distance learning possible, increasing access to education and removing constraints related to geography. Additionally, XR technology fosters teamwork, creativity, critical thinking, and problem-solving abilities all necessary for the workforce of the twenty-first century (Asif Zaman et al., 2024). Teachers can create a generation of learners who are prepared for the future and who possess the knowledge and abilities necessary to succeed in a quickly changing digital world by incorporating XR into the educational system (Asif Zaman et al., 2024). Students, parents, instructors, and administrators are among the stakeholders who need to have faith in the moral application, privacy protection, security, and general effectiveness of XR technology.

To guarantee that XR technology is accepted and utilized to its greatest extent, it is imperative to allay their worries and build confidence (Asif Zaman et al., 2024).

The future of dentistry education will involve the adoption of AI and immersive technology, which calls for innovative pedagogies that are not too far from the analog lecture world. In both online and offline learning environments, new educational models and modalities will require the testing and assessment of innovative digital pedagogies. These will probably develop along with new digital tools that become available, allowing for peer, tutor-student, tutor-machine, and learner-machine connection and participation (Aerts, 2024). Students can identify areas for growth by using interactive feedback and virtual and augmented reality to spark their curiosity and enthusiasm for learning. Digital immersion technology has the potential to foster teamwork and collaboration among students, inspiring them to strive towards a shared goal. By providing students with the skills and knowledge they need to prosper in the twenty-first century, digital immersion technology has the potential to make a substantial contribution to the attainment of sustainability in education (Tang, 2024). Hardware performance must be at an elevated level to enhance the integration of AR and VR technologies. Outstandingly high resolution and low latency are anticipated to make up for this integrated technology's drawbacks and improve immersive experiences. AI agent characters that can converse with users in AR/VR mixed environments may be created in the future to deliver interesting and productive instruction without the need for specialists (Kang et al., 2024b).

Artificial intelligence (AI) has profoundly impacted various industries and contexts, presenting novel solutions to pressing problems and enhancing productivity. However, its widespread adoption poses challenges such as ethical concerns and system integration issues. In dermatology, VR and AR applications have shown potential in learning, behavior change, and habit formation, though financial and ethical considerations remain. Virtual reality (VR) has demonstrated therapeutic potential, reducing the frequency of itching in studies. VR and AR technologies can significantly enhance medical education by offering immersive experiences and improving self-efficacy among pre-service teachers. Effective storyboarding, user-driven design, and interdisciplinary teamwork are crucial for developing VR/AR applications in dermatology. Teachers play a vital role in integrating immersive technologies into classrooms, using AR to provide efficient teaching representations. In medical education, VR helps teach complex competencies. The adoption of XR technologies, including VR and AR, is expected to grow, with applications ranging from surgery to project-based learning systems. These technologies enable immersive and dynamic learning environments, fostering student engagement and retention. Combining AR and VR technologies can overcome individual limitations, offering practical and engaging experiences. XR technologies can bridge the gap between theory and practice, providing tailored learning experiences that accommodate different learning styles and facilitate remote learning. Establishing trust among stakeholders is crucial for the successful deployment of XR in education. AI and immersive technology will play a significant role in the future of education, including dentistry. Innovative pedagogies and digital tools will enhance tutor-student and peer interactions. Digital immersion technology fosters teamwork, collaboration, critical thinking, and problem-solving skills, preparing students for success in the digital age.

PROFESSIONAL DEVELOPMENT OF THE EDUCATORS FOR USING VIRTUAL REALITY AND AUGMENTED REALITY IN THE CLASSROOM

In order to raise pre-service teachers' understanding and evaluation of VR and AR's potential to help learning, teacher education providers should think about incorporating immersive experiences with these technologies (Taggart et al., 2023). VR experiences can be used by pre-service teachers and those in charge of assisting them to assess their current level of digital proficiency and create an action plan to update and/or grow those skills as needed (Taggart et al., 2023). Today's pre-service teachers, who represent the future teaching workforce, must have access to immersive, creative, and pedagogically robust learning experiences. These should be designed and guided by knowledgeable teacher educators, particularly considering the projected rapid growth of VR and AR technologies and their significant potential for enhancing education. It is essential that today's pre-service teachers, who will make up tomorrow's teaching workforce, have access to similarly immersive, creative, and pedagogically robust learning experiences that are created and facilitated by knowledgeable teacher educators. This is especially true if the high growth projections for VR and AR are to be believed, as well as the potential educational gains of these technologies (Taggart et al., 2023).

Additionally, it is critical that the pre-service teachers join a field that is prepared to support these new educators by offering free access to current and dependable ICT resources, as well as support from school administrators and a welcoming environment that encourages ICT use throughout the institution (Taggart et al., 2023). Community members must be made aware of practical considerations in light of the growing number of small-scale VR/AR studies in dermatology. It also demands that rules, regulations, and best practices be developed in the future (Muralidharan et al., 2024). With applications in a variety of industries, including education, training, healthcare, and defense, the need for XR is growing in both daily life and business. AR and VR are also included in the concept of XR technology. Although augmented reality is limited in space, it is grounded in reality and allows for realistic and natural interaction. While virtual reality (VR) is based on a virtual environment and offers a high level of immersion and flexibility in terms of space, it differs from reality and is not easy to use for extended periods of time (Kang et al., 2024a).

By bridging the knowledge gap between theory and practice, XR enables students to apply their learning in real-world, authentic circumstances. Even with XR's growing appeal and promise in the classroom, there are still issues that need to be resolved. These include the price and availability of XR technology, the necessity of sufficient professional development and training for teachers, and the creation of efficient pedagogical strategies for incorporating XR into the curriculum (Asif Zaman et al., 2024). The utilization of digital immersion technology holds promise in promoting critical thinking, self-directed learning, and teamwork—all essential components in attaining sustainability in education. Legislators may help guarantee that children have access to technology by establishing laws that support collaboration, curriculum development, and teacher preparation (Tang, 2024).

To enhance pre-service teachers' understanding and evaluation of VR and AR's potential in education, teacher education providers should incorporate immersive experiences with these technologies. These experiences allow pre-service teachers to assess their digital proficiency and create plans to improve their skills. It is crucial that these future educators have access to innovative and pedagogically robust learning experiences facilitated by knowledgeable educators, especially given the projected growth and educational benefits of VR and AR. Pre-service teachers need to enter a field that supports these technologies with access to reliable ICT resources, administrative backing, and a welcoming environment that encourages the use of ICT. As VR/AR studies in fields like dermatology increase, community awareness and the development of rules, regulations, and best practices are necessary. The demand for XR technology, encompassing AR and VR, is growing across various sectors such as education, training, healthcare, and defense. While AR provides realistic and natural interaction grounded in reality, VR offers high immersion in a virtual environment but is not always practical for prolonged use. XR technology bridges the gap between theory and practice, enabling students to apply learning in real-world scenarios. However, challenges like the cost and accessibility of XR technology, the need for professional development for teachers, and the creation of effective pedagogical strategies remain. Digital immersion technology promotes critical thinking, self-directed learning, and teamwork, which are vital for sustainable education. Legislators can support this by establishing laws that facilitate collaboration, curriculum development, and teacher preparation, ensuring that students have access to these transformative technologies.

CONCLUSION

The integration of immersive technologies such as VR, AR, and XR into education holds transformative potential for enhancing teaching and learning experiences. These technologies offer immersive, interactive, and contextually rich environments that foster critical 21st-century skills, including critical thinking, self-directed learning, and teamwork. By bridging the gap between theoretical knowledge and practical application, they enable students to engage deeply with the content and apply their learning in real-world scenarios. However, the successful implementation of these technologies requires a multifaceted approach. Teacher education providers must ensure that pre-service teachers have access to immersive, creative, and pedagogically robust learning experiences. This includes facilitating their assessment of digital proficiency and providing opportunities for skill development. Additionally, the education sector must support these new educators with reliable ICT resources, administrative backing, and a conducive environment for ICT integration. The broader adoption of VR and AR in various fields, including dermatology and other specialized areas, underscores the need for community awareness and the development of comprehensive rules, regulations, and best practices. Despite the promising prospects of XR technologies in education, challenges such as cost, accessibility, and the need for professional development and effective pedagogical strategies must be addressed. Legislative support can play a crucial role in ensuring equitable access to these technologies, promoting collaboration, curriculum development, and teacher preparation. By fostering an environment that supports the ethical and effective use of immersive technologies, educational institutions can prepare a generation of learners equipped with the skills and knowledge necessary to thrive in a rapidly evolving digital world. The path forward involves not only embracing technological advancements but also ensuring their thoughtful integration into the educational landscape to maximize their potential benefits.

REFERENCES

Aerts, C. (2024). Dental education: Not immune to AI and immersive technologies. *Faculty Dental Journal*, 15(1), 14–17. DOI: 10.1308/rcsfdj.2024.7

Alnagrat, A., Che Ismail, R., Syed Idrus, S. Z., & Abdulhafith Alfaqi, R. M. (2022). A Review of Extended Reality (XR) Technologies in the Future of Human Education: Current Trend and Future Opportunity. *Journal of Human Centered Technology*, 1(2), 81–96. Advance online publication. DOI: 10.11113/humentech.v1n2.27

Fu, L. (2021). Research on the Teaching Model of Animation Professional Class Based on AR/VR Technology and 5G Network. *Wireless Communications and Mobile Computing*, 2021(1), 1–10. DOI: 10.1155/2021/1715909

Giri, D., & Sharma, J. (2024). TRANSFORMATIVE EFFECTS OF ANIMATION, VIRTUAL REALITY, AND AUGMENTED REALITY IN EDUCATION - A STUDY. *ShodhKosh: Journal of Visual and Performing Arts*, 4(2SE), 237–246. DOI: 10.29121/shodhkosh.v4.i2SE.2023.563

Kang, H., Yang, J., Ko, B. S., Kim, B. S., Song, O. Y., & Choi, S. M. (2024a). Integrated Augmented and Virtual Reality Technologies for Realistic Fire Drill Training. *IEEE Computer Graphics and Applications*, 44(2), 89–99. DOI: 10.1109/MCG.2023.3303028 PMID: 37585326

Kang, H., Yang, J., Ko, B. S., Kim, B. S., Song, O. Y., & Choi, S. M. (2024b). Integrated Augmented and Virtual Reality Technologies for Realistic Fire Drill Training. *IEEE Computer Graphics and Applications*, 44(2), 89–99. Advance online publication. DOI: 10.1109/MCG.2023.3303028 PMID: 37585326

Kramarenko, T. H., & Kochina, O. S. (2023). The use of immersive technologies in teaching mathematics to vocational students. *Journal of Physics: Conference Series*, 2611(1), 1–14. DOI: 10.1088/1742-6596/2611/1/012006

Muralidharan, V., Tran, M. M., Barrios, L., Beams, B., Ko, J. M., Siegel, D. H., & Bailenson, J. (2024). Best Practices for Research in Virtual and Augmented Reality in Dermatology. *The Journal of Investigative Dermatology*, 144(1), 17–23. DOI: 10.1016/j.jid.2023.10.014 PMID: 38105083

Nagtilak, P. S. A. (2024). Survey on Gesture-Based Virtual Keyboard and Mouse. *Interantional Journal of Scientific Research in Engineering and Management*, 08(01), 1–13. DOI: 10.55041/IJSREM28414

Nastiti, K. A., Santoso, H. B., Junus, K., Ahmad, M., & Purwandari, E. P. (2024). The Effect of Immersive Technology on Enhancing Student Learning:A Systematic Literature Review. [Jurnal Ilmu Pengetahuan Dan Teknologi Komputer]. *JITK*, 9(2), 208–217. DOI: 10.33480/jitk.v9i2.4775

Partarakis, N., & Zabulis, X. (2024). A Review of Immersive Technologies, Knowledge Representation, and AI for Human-Centered Digital Experiences. *Electronics (Basel)*, 13(269), 1–19. DOI: 10.3390/electronics13020269

Pears, M., & Konstantinidis, S. (2022). The Future of Immersive Technology in Global Surgery Education. *Indian Journal of Surgery*, 84(S1), 281–285. DOI: 10.1007/s12262-021-02998-6 PMID: 34230785

Rosendahl, P., Müller, M., & Wagner, I. (2024). A 360° video as visual training support for independent movement acquisition—Benefit evaluation with the TAM. *German Journal of Exercise and Sport Research*, 54(3), 383–392. Advance online publication. DOI: 10.1007/s12662-023-00930-6

Sadek, O., Baldwin, F., Gray, R., Khayyat, N., & Fotis, T. (2023). Impact of Virtual and Augmented Reality on Quality of Medical Education During the COVID-19 Pandemic: A Systematic Review. In *Journal of Graduate Medical Education* (Vol. 15, Issue 3). DOI: 10.4300/JGME-D-22-00594.1

Shvardak, M. V. (2023). IMMERSIVE TECHNOLOGIES IN PRIMARY EDUCATION. *Scientific Herald of Sivershchyna. Series: Education.Social and Behavioural Sciences*, 2(11), 227–239. DOI: 10.32755/sjeducation.2023.02.227

Taggart, S., Roulston, S., Brown, M., Donlon, E., Cowan, P., Farrell, R., & Campbell, A. (2023). Virtual and augmented reality and pre-service teachers: Makers from muggles? *Australasian Journal of Educational Technology*, 39(3), 1–16. DOI: 10.14742/ajet.8610

Taghian, A., Abo-Zahhad, M., Sayed, M. S., & Abd El-Malek, A. H. (2023). Virtual and augmented reality in biomedical engineering. *Biomedical Engineering Online*, 22(76), 1–25. DOI: 10.1186/s12938-023-01138-3 PMID: 37525193

Tang, F. (2024). Understanding the role of digital immersive technology in educating the students of english language: Does it promote critical thinking and self-directed learning for achieving sustainability in education with the help of teamwork? *BMC Psychology*, 12(144), 1–14. DOI: 10.1186/s40359-024-01636-6 PMID: 38481260

Tom Dieck, M. C., & Han, D. (2022). The role of immersive technology in Customer Experience Management. *Journal of Marketing Theory and Practice*, 30(1), 108–119. DOI: 10.1080/10696679.2021.1891939

Wagner, P., & Alharthi, D. (2023). Leveraging VR/AR/MR/XR Technologies to Improve Cybersecurity Education, Training, and Operations. *Journal of Cybersecurity Education Research and Practice*, 2024(1), 1–26. DOI: 10.32727/8.2023.32

Zaman, A., Abir, M. R., & Mursalin, S. (2024). Extended reality in education and training: Enhancing trustworthiness. *International Journal of Science and Research Archive*, 11(1), 1705–1720. Advance online publication. DOI: 10.30574/ijsra.2024.11.1.0206

Zattoni, F., Carletti, F., Randazzo, G., Tuminello, A., Betto, G., Novara, G., & Dal Moro, F. (2023). Potential Applications of New Headsets for Virtual and Augmented Reality in Urology. *European Urology Focus*, xxxx, 10–14. DOI: 10.1016/j.euf.2023.12.003 PMID: 38160172

KEY TERMS AND DEFINITIONS

Critical Thinking: The ability to analyze, evaluate, and synthesize information to make reasoned judgments and decisions, essential for solving complex problems and understanding diverse perspectives in various contexts.

Digital Literacy: The ability to effectively find, evaluate, use, and create digital information and media, encompassing skills in information literacy, technological proficiency, and critical thinking applied in a digital environment.

Educator Readiness: The preparedness of teachers and instructors to effectively integrate new technologies, such as VR and AR, into their teaching practices through training, professional development, and pedagogical adaptation.

Ethical Use: The responsible and conscientious application of technologies, including VR and AR, to ensure privacy, fairness, transparency, and societal benefit while minimizing potential harms and risks.

Immersive Technologies: These encompass Virtual Reality (VR), Augmented Reality (AR), and Extended Reality (XR), technologies that create immersive, interactive, and multi-sensory experiences by blending digital content with the real world.

Pedagogical Integration: The systematic incorporation of VR and AR technologies into educational curricula and teaching methodologies to enhance learning outcomes, engage students, and bridge theoretical knowledge with practical application.

Regulatory Frameworks: Legal and policy guidelines established by governments or institutions to govern the development, deployment, and use of technologies like VR and AR, ensuring compliance with ethical standards and promoting safe and equitable access.

Teamwork: Collaborative effort among individuals to achieve a common goal, involving effective communication, coordination, and mutual support, crucial for fostering innovation and achieving shared objectives.

Chapter 12
The Role of Using ChatGPT in Improving the Creative Writing of English Language Students From the Perspectives of Instructors and Postgraduate Students

Khaled N. Yaseen
An-Najah National University, Palestine

Khaled Dweikat
https://orcid.org/0000-0002-4848-3756
Al-Quds Open University, Palestine

ABSTRACT

This study investigates the role of ChatGPT in enhancing the creative writing skills of English language students from the perspectives of instructors and postgraduate students at Palestinian universities. Using a mixed-methods research design, the researchers employed both quantitative and qualitative descriptive analytical approaches, including questionnaires and interviews for data collection. The study population consisted of 89 instructors and 169 postgraduate students. Quantitative data were analyzed using SPSS V.23, while qualitative data were analyzed themati-

DOI: 10.4018/979-8-3693-9811-1.ch012

cally. The findings indicate that both instructors and doctoral students have positive perspectives on ChatGPT's role in improving creative writing. The study also found a high level of implementation of ChatGPT as an AI tool in higher education for creative writing enhancement. No statistically significant differences were observed among participants' perspectives.

INTRODUCTION

In light of the rapid global changes and developments in the fields of technology and education, technology has been proven to play a significant role, where the comprehensive definition of technology must be distinct from a reference to the crucial rule of technology in the educational process. The successful and effective teaching process in the current era is wider than conveying educational content to learners traditionally using conventional methods. Ahmadi & Razan (2018) claimed that essential learning tools provided by technology help facilitate cooperative learning and offer exciting choices for developing language skills through experimentation.

During the last few decades, the role of technology has become more effective to the extent that it has greatly impacted our daily lives. In the field of education, Information and Communication Technology (ICT) in general and Artificial Intelligence (AI) in particular have been proven to be effective in identifying students' needs and familiarizing teachers with the best methods to achieve them, prompting feedback and helping teachers assess students' work and tasks (Celik, Dindar, Muukkonen, & Järvelä, 2022). Ratheeswari (2018) argued that ICT is promoted as a catalyst for change. This change involves exchanging information, teaching methods, learning approaches, scientific research, information communication technologies, and a change in working conditions. Furthermore, ICT contributes to presenting different teaching content attractively and allows learners to learn and apply 21st-century skills. Ishaq et al. (2020) stated that the use of ICT is highly associated with enhancing students' competencies, where students with ICT skills can improve their essential skills and learning by performing different tasks effectively, such as preparing assignments and classroom activities and planning their lessons effectively. Accordingly, the question of using Artificial Intelligence applications and tools to accomplish educational tasks has been raised by many specialists around the world. One major objective of raising this question is to help teachers and educators find new methods and modes of content delivery that can replace the traditional methods of teaching and learning as well.

The modern efficiency of technology in the educational industry is represented through the global spread of Artificial Intelligence in different academic institutions, where it is globally promoted in the shadow of the recent appearance of technological applications such as machine translation, text editors including spelling checkers, paraphrasing, grammar checkers, citation generators, and ChatGPT. Fahimirad & Kotamjani (2018) stated that Artificial Intelligence is promoted as a useful tool in educational instruction by providing solutions to the challenges and difficulties in the learning process. AI then has the potential to motivate students, help them learn content, and achieve educational goals. Besides, AI is featured for supporting learning, automatic grading, constructive feedback, and learning from mistakes.

Within the field of English language skills, writing is one of the most important skills that cannot be separated from the successful and comprehensive plans in global educational institutions. In other words, writing takes the highest priority in teaching languages, where writing is regarded as a tool for communication, sharing ideas, critical thoughts, and comments through the formation of meaningful sentences, paragraphs, reports, and other academic writing productions. Rushkoff (2013) commented that educational and professional success is highly associated with effective writing, according to teachers, parents, and students. As a result, employees and students with good writing skills are highly valuable, promotable, and recommended.

Although writing has many advantages for learners in expressing different contents easily, it still needs to be considered an obstacle for learners in various educational institutions. Fareed, Ashraf, & Bilal (2016) concluded that the problems of ESL learners in writing are highly associated with the structure of the content, vocabulary, spelling, punctuation, word form and word order, spoken expressions, contracted forms, cohesion, and repetition of ideas, in addition to the possible impact of the first language.

Creative writing is generally defined as the art of building words together in a systematic, syntactic, and coherent way of expressing ideas, thoughts, and expressions. In other words, creative writing content is all about collecting different thoughts, views, contributions, and beliefs to formulate a new innovative product. Sundt (2018) stated that the increased knowledge in various domains and fields of study stresses the importance of employing creative writing as an upturn in classrooms. However, students in different educational institutions face challenges in building creative writing skills. These challenges cannot be understood without a reference to the core of writing problems, such as students' mistakes, uniqueness of ideas, and lack of interest in the topics that they are writing about. Tirumalesh (1991) suggested that lack of proficiency, crowdedness of students in writing classrooms, and unqualified teachers are obstacles that limit knowledge of English writing. In other words, the crowdedness of students in writing classes negatively affects learning writing effectively, where it is an obstacle to sharing ideas and thoughts needed for writing between parts of the learning process in educational institutions.

Based on the introduction above, recent technology has made a prominent contribution to creative writing. Jiang, Li, Han, & Yang (2019) argued that technology plays a significant role in creative writing skills when it allows pupils and educators to cooperate and communicate with each other to facilitate product prototyping, cooperative projects, and joint learning. Besides, Veiga & Andrade (2021) stated that technology helps students understand complex knowledge and fosters motivation and engagement. Similarly, Williams & Beam (2019) added that using technology in EFL classrooms encourages students' participation and involvement in writing assignments and enhances peer collaboration and social interaction. Hence, the process of introducing the so-called ChatGPT has been a revolutionary step towards reshaping the educational status all over the world.

The Birth of ChatGPT

In April 2023, UNESCO released an introductory guide to ChatGPT, an AI tool that quickly gained 100 million users within two months of its launch. This guide highlights ChatGPT's capabilities and potential uses in the educational sector. ChatGPT, which stands for "Generative Pre-trained Transformer," is a language model that allows natural and conversational interaction with computers or tablets. Developed by OpenAI, it can generate coherent, human-like text based on large datasets, predicting the next word in a sentence to produce relevant content. With 300 billion words in its system and 175 billion parameters, ChatGPT is known for its user-friendly interface and effectiveness in generating original written content based on user prompts. The guide emphasizes its utility in data analysis and content generation despite limitations in interpretation due to its reliance on statistical analysis of online texts (UNESCO, 2023).

ChatGPT, based on the GPT-3.5 architecture, is a machine learning model designed to perform various natural language processing tasks, such as answering questions and generating formal letters, essays, jokes, poetry, stories, and exam responses. This versatile tool can produce multiple outputs for similar inputs in a short time (Liu et al., 2021). However, there are concerns that over-reliance on ChatGPT might diminish users' critical thinking skills and the originality of their writing (Iskender, 2023). Future versions of ChatGPT are expected to offer enhanced functionalities, including advanced data interpretation and more sophisticated writing capabilities. In essence, ChatGPT represents a significant advancement in AI technology, with broad applications and notable benefits, particularly in education and content creation.

Application of ChatGPT in Education

Despite being a recent development, ChatGPT has already been widely applied in different institutions for different functions in higher education. Due to its ability to generate and assess information, ChatGPT can play a variety of roles in teaching and learning processes. Together with other forms of AI, ChatGPT has the potential to improve the educational process for students. To achieve this, ChatGPT can be used as a standalone tool, or it can be integrated into other systems and platforms used by higher education institutions (HEIs). ChatGPT can perform many samples or technical tasks, including basic research, calculations, proofreading, paraphrasing, and translation.

ChatGPT could be incorporated and used to augment teaching and learning by providing alternative ways of articulating the educational content, especially the writing content. In other words, users of ChatGPT can receive different responses and different styles of writing related to the chosen topic that they are looking for. Besides, ChatGPT has the potential to develop arguments and claims for discussion. Moreover, ChatGPT is highly regarded as a problem-solving tool and a reliable guide for students who search for information to complete tasks and assignments. It provides students with essential support and advice needed for learning specific concepts.

Moreover, ChatGPT plays a prominent role in personal tutoring and design, providing teachers and students with personalized feedback by providing rubrics for assessments. Employing ChatGPT can also be useful for exploring and interpreting data and motivating users, as it offers games and competitions to extend learning through quizzes and exercises. Additionally, educators are provided with students' current knowledge and level by interacting with ChatGPT in a tutorial-type dialogue and then asking ChatGPT to give a summary of the student's current knowledge to participate in debates and classroom activities with their teacher for assessment.

ChatGPT is a beneficial AI tool that can be used effectively for learning. Rudolph, Tan, & Tan (2023) argued that ChatGPT helps learners master problem-solving skills by generating scenarios related to different topics. Moreover, ChatGPT encourages learners to receive personal tutoring, where using ChatGPT in learning enables learners to concentrate on their lessons. Rudolph, Tan, & Tan (2023) stated that uncountable advantages stem from applying ChatGPT in education. On the one hand, ChatGPT provides resources related to learning and teaching.

On the other hand, ChatGPT is a good choice for students who prefer individual learning. Lin (2022) commented that ChatGPT enhances and supports independent learning. In addition, ChatGPT encourages the practicality of learning languages and practice between learners. In addition, Su & Yang (2023) commented that ChatGPT facilitates learners in overcoming their obstacles by providing constructive feedback, quizzes, and questions. Furthermore, ChatGPT enables learners to chat with the application. Thus, this kind of conversation improves learners' level of language proficiency. Moreover, talking with ChatGPT positively affects learners' acquisition of the four basic skills, improves grammar accuracy, and increases learners' vocabulary size.

Furthermore, ChatGPT serves schools by enabling students to improve their writing skills. In other words, ChatGPT suggests well-written samples of writing with good syntactic and grammatical choices and corrections. Besides, ChatGPT develops students' styles of writing and critical thinking. ChatGPT is also an effective AI tool used to enhance reading skills by providing summaries, clarifications, and explanations for complex reading comprehension. Thus, students find reading texts easy to comprehend and analyze. ChatGPT is applicable in higher schools, where it facilitates the way that students are involved with topics in literature, physics, mathematics, and other subjects. ChatGPT can provide students with explanations, step-by-step procedures, and solutions.

Consequently, ChatGPT helps higher school students enhance their problem-solving skills and get properly involved in the learning process. The application of ChatGPT in universities develops different types of writing, enabling students to generate summaries, comprehend the main points of texts, and organize their thoughts in writing. Besides, it clarifies current research topics with critical analyses and interpretations (Kasneci et al., 2023).

Application of ChatGPT in Writing Courses

ChatGPT is a global trend that can be used to perform a wide array of tasks, write essays, and provide advice. Using ChatGPT helps learners in generating texts that are similar to texts written by humans. Buriak et al. (2023) stated that the factors that make ChatGPT amazing are that it has a human-like feel, where the promoted structure cannot be differentiated from human writings. Chen (2023) argued that ChatGPT has a human-like competence in writing. Furthermore, ChatGPT can serve academic writing by providing different components of research papers. Alkaissi & McFarlane (2023) stated that ChatGPT could generate research ideas and hypotheses, write previous studies, and design research experiments and manuscripts. Alkaissi & McFarlane (2023) also commented that ChatGPT contributes to shortening and reducing the time needed for submission. Also, it could assist in the writing process by providing an initial draft of research papers, and it helps researchers come up with ideas and new titles related to their interests. Also, Salvagno, Taccone, & Gerli (2023) argued that ChatGPT helps to write the methodology of the study, justify the sample size, and explain data analysis methods. Van Dis, Bollen, Zuidema, van Rooij, & Bockting (2023) claimed that ChatGPT improves academic training because it offers a vast amount of feedback related to the study. Consequently, this will have a positive effect on the researcher's experience and skills.

Challenges of Using ChatGPT in Creative Writing Courses

Although ChatGPT is regarded as a trustworthy AI resource in educational institutions, especially in creative writing courses, some obstacles and challenges are represented by using ChatGPT as a resource in creative writing courses. Dwivedi (2023) indicated that using ChatGPT is highly associated with privacy concerns, misinformation, and a lack of accuracy and deep meaning. Halaweh (2023) commented that plagiarism and academic honesty are major issues caused by the use of ChatGPT in writing. Also, Iskender (2023) stated that the overuse of ChatGPT weakens critical thinking ability and reduces originality and novelty. Cotton, Cotton, & Shipway (2023) suggested that ChatGPT is the main factor for plagiarism and cheating.

Nevertheless, the effective role of ChatGPT cannot cancel or reduce humans' knowledge.

Furthermore, the effective role of ChatGPT cannot cancel or reduce humans' knowledge and capability. Nevertheless, Cotton, Cotton, & Shipway (2023) suggested a need for activating the use of ChatGPT, where universities, language departments, and writing educators must take into consideration the responsible and ethical issues related to the application of AI tools with suitable strategies, such as developing policies and providing training. Van Dis, Bollen, Zuidema, van Rooij, & Bockting (2023) claimed that there are no consensus opinions towards using ChatGPT in writing. In other words, some people still believe that ChatGPT can be employed as a co-author for creative writing conducted by human scholars and researchers by providing samples of writing judiciously and responsibly.

In contrast, others believe that ChatGPT will harm rather than improve the writing process. However, using ChatGPT is highly associated with concerns that may affect the writing and research process. Liebrenz, Schleifer, Buadze, Bhugra, & Smith (2023) stated that ChatGPT is highly associated with plagiarism, academic honesty, and cheating. Bibi & Atta (2024) commented that ChatGPT contains errors and inaccurate data. Also, ChatGPT provides illogical and incorrect connections of ideas and claims.

Statement of the Problem

The authors of this chapter, who have more than ten years of experience as EFL teachers, have observed that many students struggle with practicing creative writing. In light of the diversity of controversial thoughts and attitudes towards using

ChatGPT in education, especially in English language departments, the researchers observed that educational institutions, university instructors, and postgraduate students are concerned about the growing use of technology in education that is highly represented by using ChatGPT since this kind of technology will probably affect the learning and teaching process in light of evaluation, assessment, and involvement in accomplishing tasks required by the university. This concern comes from the unfamiliarity of using ChatGPT. On the other hand, the researchers observed that the students tend to use ChatGPT to help them accomplish their creative writing tasks since they find it an effective tool to help them save time and effort and perform their creative writing tasks. Besides, they find ChatGPT a solution to solve their writing challenges in terms of choosing topics, brainstorming, and forming coherent writing products that are free of mistakes. Consequently, efforts and claims are made by specialists and academic scholars through conducting different research topics on Artificial Intelligence and ChatGPT and its domains to examine the effectiveness, positives, and possible threats to the learning process. Hence, the current study aimed to investigate the role of using ChatGPT in improving the creative writing of English language students from the perspectives of instructors and postgraduate students in Palestinian universities.

Questions of the Study

Based on the study objectives and its problem, the current study sought to address the following questions:

1. What is the role of using ChatGPT to improve the creative writing of English language students from the postgraduate student's perspectives in Palestinian universities?
2. What is the role of using Chat GPT to improve the creative writing of English language students from the instructors' perspectives in Palestinian universities?
3. Are there statistically significant differences at ($\alpha \leq 0.05$) in the role of using ChatGPT in improving the creative writing of English language students between the perspectives of instructors and postgraduate students in Palestinian universities?

Hypothesis of the Study

The current study sought to test the following main hypothesis:

There are no statistically significant differences at ($\alpha \leq 0.05$) in the role of using Chat GPT in improving creative writing of English language students from the perspectives of instructors and postgraduate students in Palestinian universities.

Significance of the study

The significance of the study stems from the following factors: First, it sheds light on the role of using ChatGPT in improving the creative writing of English language students from the perspectives of instructors and postgraduate students in Palestinian universities. Scholars in educational fields confirmed that Artificial Intelligence, with its newborn GPT, is a big revolution and real transformation in the world of education since it helps save time and effort and accomplish different tasks comprehensively, effectively, and in a well-written style, especially the functions that are related to learning the English language. Second, in light of the widespread use of ChatGPT in educational institutions and international propaganda, ChatGPT is regarded as an effective tool to facilitate tasks and school homework. Thus, the study examines the role of using ChatGPT in improving the creative writing of English language students from the perspectives of instructors and postgraduate students in Palestinian universities. The results of this study will encourage instructors and doctoral students in Palestinian universities to activate the role of ChatGPT in creative writing tasks. Fourth, this study is hoped to be the first in the Palestinian context to target the use of ChatGPT and thus add to the literature review of this field. Other researchers might find this study helpful for their research endeavors.

LITERATURE REVIEW

Despite ChatGPT's relatively recent emergence, researchers around the globe are actively engaged in exploring its impact on various educational aspects across different fields. Although this current field is still unknown to everyone, researchers are highly interested in studying the possible uses and functions, advantages, disadvantages, and threats of using ChatGPT in education and educational institutions.

Tossell et al. (2024) explored the impact of AI on student learning and engagement, focusing on the potential of AI to provide personalized learning experiences. Aina & Joshua (2024) investigated the challenges and opportunities of implementing AI-driven learning tools in educational settings. Hidayatullah (2024) evaluated the effectiveness of digital learning platforms in supporting student engagement and academic performance. Arora & Mehta (2024) examined the role of AI in educational assessments, focusing on the accuracy and fairness of AI-driven grading systems. Strobl et al. (2024) investigated the use of AI in personalized learning, evaluating the benefits and limitations of adaptive learning technologies. Mabuan (2024) explored the potential of gamification in education, examining how game-based learning can motivate and engage students. Gill et al. (2024) analyzed the impact of digital storytelling on student learning and engagement. Niloy et al. (2024) investigated the effectiveness of online earning platforms in supporting student learning during the COVID-19 pandemic. Bibi & Atta (2024) examined the role of digital tools in supporting collaborative learning and peer feedback. Zhu et al. (2024) evaluated the potential of AI-driven language learning applications to improve language proficiency. Curry et al. (2024) explored the use of virtual reality in education, highlighting its potential to create immersive and interactive learning experiences. Finally, Datoo & Siddiqui (2024) investigated the impact of digital health tools on student well-being and mental health.

Ali et al. (2023) explored the use of AI chatbots in educational settings, evaluating their effectiveness in providing personalized learning support. Alneyadi & Wardat (2023) investigated the impact of mobile learning applications on student engagement and academic performance. Bin-Hady et al. (2023) examined the challenges and benefits of blended learning, focusing on the integration of online and face-to-face instruction. Lund & Wang (2023) explored the potential of adaptive learning technologies to tailor educational content to individual student needs. Firat (2023) evaluated the effectiveness of flipped classroom models, where students engage with instructional content outside of class and participate in interactive activities during class time. Meniado (2023) investigated the use of learning analytics to track student progress and inform instructional decisions. Mhlanga (2023) explored the digital divide in education, examining how disparities in access to technology impact student learning outcomes. Thompson et al. (2023) assessed the role of digital tools in supporting collaborative learning and peer feedback. Kayalı et al. (2023) examined the impact of digital storytelling on student engagement and learning. Sallam (2023) investigated the use of AI-driven language learning applications, evaluating their effectiveness in improving language proficiency. Ali & OpenAI (2023) explored the potential of AI-generated content to support personalized learning and provide instant feedback. Dergaa et al. (2023) analyzed the role of digital health tools in supporting student well-being and mental health. Imran & Almusharraf (2023) investigated the effectiveness of online discussion forums in fostering critical thinking and deep learning. Marchandot et al. (2023) examined the impact of AI-driven assessment tools on grading accuracy and feedback quality. Cotton et al. (2023) explored the use of virtual laboratories in science education, evaluating their potential to enhance hands-on learning experiences. Mohammed et al. (2023) investigated the challenges and benefits of online exams, focusing on issues of security and academic integrity. Tiwari et al. (2023) evaluated the use of AI in personalized learning pathways, examining how adaptive technologies can support individual student needs. Abdaljaleel et al. (2023) explored the role of digital credentials and badges in recognizing and validating student learning achievements. Hong (2023) investigated the impact of digital note-taking tools on student learning and retention. Algaraady & Mahyoob (2023) analyzed the use of

AI in educational content creation, evaluating its potential to support personalized learning experiences. Deshpande et al. (2023) examined the role of AI in supporting inclusive education, focusing on tools designed to assist students with disabilities. Lin X. (2023) investigated the impact of digital formative assessment tools on student learning outcomes. Matthews & Volpe (2023) explored the use of AI-driven tutoring systems, evaluating their effectiveness in providing personalized learning support.

Rahman et al. (2022) analyzed the impact of digital literacy on student performance, identifying key factors that influence successful online learning experiences. Zhai (2022) investigated the role of big data in educational analytics, exploring how data-driven insights can enhance teaching and learning processes. Aydın & Karaarslan (2022) evaluated the implementation of blockchain technology in education, discussing its potential to improve data security and transparency. Mageira et al. (2022) assessed the effectiveness of gamification in educational settings, examining how game-based learning can motivate and engage students. Lin (2022) explored the integration of virtual reality in education, highlighting its potential to create immersive and interactive learning experiences. Castillo-González et al. (2022) investigated the challenges and opportunities of remote learning in higher education, focusing on the experiences of both students and educators. Hutson et al. (2022) examined the role of social media in education, analyzing how platforms like Facebook and Twitter can support collaborative learning and information sharing. Dewi et al. (2021) examined the effectiveness of online learning platforms during the COVID-19 pandemic, providing insights into student engagement and learning outcomes. Seo et al. (2021) focused on the use of AI in educational assessments, evaluating the accuracy and fairness of automated grading systems. Esling & Devis (2020) explored the potential of artificial intelligence in education, highlighting the benefits and limitations of AI-driven learning tools. Pedró (2020) investigated the transformative effects of digital learning environments on traditional educational paradigms, emphasizing the shift towards more personalized learning experiences. Lindner et al. (2019) delved into the integration of technology in academic settings, focusing on the impact and challenges of implementing digital tools in classrooms.

Disadvantages of using ChatGPT

Contrary to the various advantages of ChatGPT, which covers different disciplines and fields, using ChatGPT entails some disadvantages, according to some educators and researchers around the globe. For instance, Curry, Baker, & Brookes (2024) argued about the contribution of ChatGPT in discourse analysis, where ChatGPT is not effective and not a reliable solution to challenges in analyzing discourses. Despite ChatGPT being designed to generate human-like text, ChatGPT still needs to meet the standards of human analysts in analyzing utterances and situations implying ambiguous meanings. Besides, Datoo & Siddiqui (2024) think that using ChatGPT is highly associated with the death of the author, where overusing ChatGPT leads to a lack of producing texts in educational institutions. Consequently, this will lead to the absence of human contributions and the spread of concerns related to the control and manipulation of human behavior.Matthews & Volpe (2023) argued that ChatGPT should be used in higher education sectors despite its key concerns. The study uses Turing's imitation game to investigate 16 education academics' perceptions of two pairs of texts written by either ChatGPT or a human. The process of investigation involves five themes: voice, word choice, structure, task achievement, and flow. Accordingly, the research concludes that human writing skills are not easy to define and differentiate from texts generated by AI. Accordingly, that motivates educational institutions to redouble efforts to interrogate the adequacy and efficiency of their assessment design. Lin X. (2023) claimed that ChatGPT may lead to the loss of originality and novelty as long as scholars and students rely on it completely concerning plagiarism problems, ethics, and academic integrity. Also, it does not give students a space to share their thoughts and claims. Thus, involvement and participation are negatively affected in educational institutions.

Deshpande, Murahari, Rajpurohit, Kalyan, & Narasimhan (2023) argued that ChatGPT promotes toxicity, represented by the spread of incorrect stereotypes, harmful dialogue, and hurtful opinions about a group of people. That could be defamatory and harmful to individuals or groups of people in different ways.

Hong (2023) claimed that ChatGPT, as a revolutionary online application, has brought prominent worries to the field of education in educational institutions. These worries were highly observed among educators who rely on writing in the process of assessment and evaluation since ChatGPT helps offer opportunities to parts of the learning and teaching process to enhance second or foreign language teaching and assessments and researchers' personalized learning experiences. Already & Mahyoob (2023) commented that ChatGPT successfully detects surface-level errors in writing. Still, ChatGPT is not a good choice for analyzing errors in writing related to deep structure and pragmatics. Therefore, ChatGPT can be applicable and reliable for some issues that are related to superficial-level errors. This limits the role of ChatGPT in writing in comparison with the human role in writing in light of deep meanings and complexity.

Double-edged merits of using ChatGPT

Seo, Tang, Roll, Fels, & Yoon (2021) stated that the impact of AI is viewed as a double-edged sword for teachers and students. In other words, teachers and students see that systems of artificial intelligence are effective in terms of enhancing the quantity and quality of communication, providing just-in-time, personalized support for large-scale students, and improving social interaction to improve the feeling of connection. On the other hand, ChatGPT may lead to misguidance and misunderstanding, putting limits on the ability to learn independently and a loss of privacy. These claims are seen as obstacles for teachers and students in artificial intelligence communication.

Tlili et al. (2023) claimed that there is positive enthusiasm for using ChatGPT in educational settings. They referenced the necessity of cautious voices in approaching it in the educational industries in light of different issues such as educational transformation, response quality, usefulness, personality and emotion, ethics, cheating, honesty and truthfulness of ChatGPT, privacy, misleading, and manipulation.

Malinka, Perešíni, Firc, Hujňák, & Januš (2023) discussed the positive and negative impact of ChatGPT on university education, especially in the field of computer security-oriented specialization. As for the positives of ChatGPT, it is an effective tool for discussing and solving academic problems and challenges and speeding up the learning process. On the other side, the negativity of ChatGPT is highly represented by the students' misuse of ChatGPT in the university educational context, such as cheating.

Kasneci et al. (2023) argued the advantages and challenges of educational applications of large language models from student and teacher perspectives since this kind of technology is classified as an introduction to innovation despite the controversial views. On one hand, the large-level model plays a big role in enhancing students' involvement and interaction and personalizing learning experiences. On the other hand, the use of large language models is necessary to meet privacy, security, and environmental, regulatory, and ethical requirements in light of connection guidance, human monitoring, and critical thinking.

Zhai (2022) stated that ChatGPT is the main factor in changing learning objectives, learning activities, and the process of assessment and evaluation. In addition, ChatGPT will help write academic papers that are informative, systematic, coherent, and partially accurate. Besides, it reduces the level of effort and professional knowledge needed from the researcher during the writing process. It focuses on enhancing students' creativity and critical thinking rather than general skills.

Wardat, Tashtoush, AlAli, & Jarrah (2023) argued that the effective role of ChatGPT in teaching mathematics is to provide math learners with basic knowledge and valuable topics. ChatGPT also serves branches of mathematics by providing comprehensive assistance and instructions. However, there are obstacles related to the use of ChatGPT in teaching mathematics. In other words, learning and teaching mathematics is highly associated with deep analysis and understanding of equations and mathematical tricks. This is what limits the role of ChatGPT in teaching mathematics and other pure sciences.

Gill et al. (2024) discussed how ChatGPT improves the education sector. Basically, ChatGPT helps design instructional content, offers suggestions and solutions for learners, and acts as an online educator for students in educational institutions by answering questions, interpreting claims, and analyzing data. In contrast, the effective performance of ChatGPT is measured differently according to the subject areas. In other words, ChatGPT is noticed as effective and useful in certain fields of study, but it is recognized as a source of inaccuracy, superficial outputs, and mistakes.

Niloy, Akter, Sultana, Sultana, & Rahman (2024) stated that the use of ChatGPT affects the creative writing abilities of university students in terms of originality, accuracy, content presentation, and elaborations. The study's results revealed a detrimental and destructive association between ChatGPT use and university students' creative writing abilities. Also, the study suggested that effective integration of ChatGPT should be thoughtful and highly associated with caution.

Bibi & Atta (2024) argued aspects of learners' involvement with ChatGPT through a mixed-method approach. The study revealed that there are students who would prefer to use ChatGPT as a source for creating content in English. On the other side, a group of students demonstrated that ChatGPT suits different writing choices and styles.

Zhu et al. (2024) stated that ChatGPT affects art creation by providing ideas and creative blocks. However, there are ethical concerns related to the loss of authorship, ownership, and authenticity of artists. In other words, ChatGPT may have a disastrous effect on art and artists, who ChatGPT will completely replace. Consequentially, this will affect human culture, heritage, and identity.

METHODOLOGY

Study Design

The mixed-methods research design was employed to meet the study's objectives. Both quantitative and qualitative descriptive analytical approaches were chosen for their suitability and effectiveness in achieving these goals and addressing the study's questions.

Population and Sample

The study population included English instructors and postgraduate students from the five participating universities. The reason for choosing the five participating universities over other universities in Palestine is that the participating universities offer master's in teaching English as a foreign language (TEFL) . Thus, the targeted sample of the study is available to achieve the study purpose. Also, choosing the five participating universities offers different and comprehensive opinions and claims related to the study, since the topic is newly known in Palestine. The sample, however, was selected using convenience sampling. Besides, 20 instructors and 20 students from the selected universities were interviewed. Tables 1 & 2 show the demographic characteristics of the two samples.

Table 1. Distribution of Instructors Sample According to Study Independent Variables

Variable	Class	Frequency	Percentage %
Gender	Male	50	56.2
	Female	39	43.8
Qualification Degree	MA	53	59.6
	PhD	36	40.4
Years of Experience	Less than 5years	20	22.5
	6-10 years	23	25.8
	More than 10 years	46	51.7
University	An-Najah National University	49	55.1
	Birzeit University	14	15.7
	Arab American University	16	18.0
	Palestine Ahliya University	7	7.9
	Hebron University	3	3.4
Total		**89**	**100.0**

Table 2. Distribution of Students Sample According to Study Independent Variables

Variable	Class	Frequency	Percentage %
Gender	Male	36	21.8
	Female	129	78.2
Academic Year	First-year	26	15.8
	Second- year	62	37.6
	Third- Year	77	46.7

continued on following page

Table 2. Continued

Variable	Class	Frequency	Percentage %
University	An-Najah National University	102	61.8
	Birzeit University	16	9.7
	Arab American University	31	18.8
	Palestine Ahliya University	7	4.2
	Hebron University	9	5.5
Total		**165**	**100.0**

Instruments

To meet the study's objectives, the researcher developed a 51-item questionnaire, which was distributed electronically to instructors and students via email, social media accounts, and official university accounts. The questionnaire had two sections: the first collected demographic data from respondents, such as gender, qualification degree, years of experience, and university for instructors, while for students, it gathered gender, academic degree, and university. The second section comprised four domains: attitudes toward using ChatGPT, the pros and cons of implementing ChatGPT in teaching writing, and the role of ChatGPT in shaping the future of creative writing. Additionally, the qualitative instrument included interviews with 20 instructors and 20 postgraduate students from Palestinian universities, featuring six open-ended questions. These interviews were conducted through WhatsApp, Messenger, email, and voice recordings.

The Questionnaire

The questionnaire was sent to the instructors and postgraduate students via an electronic link. It was split into four domains. Each domain comprised items that contribute to answering the study questions. The items in the first domain revolve around the attitudes of instructors and postgraduate students towards using ChatGPT. The items in the second domain revolve around the pros of implementing ChatGPT in teaching writing. The items in the third domain revolve around the cons of implementing ChatGPT in teaching writing. The last domain rotated around the role of ChatGPT in shaping the future of creative writing. The scores of responses to each item were calculated according to a five-point Likert scale for the first three domains, in which Strongly Agree = 5 points, Agree =4 points, Undecided = 3 points, Disagree = 2 points, and Strongly Disagree = 1 point. The researchers relied on several references to construct the questionnaire of the study (For example,

Abdaljaleel, et al., (2023); Aina & Joshua (2024); Algaraady & Mahyoob (2023); Arora & Mehta (2024); Cotton & Shipway (2023); Deshpande, et al., (2023). Thus, the contents of the studies were comprehensive and informative to the role of using ChatGPT in improving the creative writing of English language students from the perspectives of instructors and postgraduate students in the Palestinian universities.

The Interviews

The interview instrument consisted of six questions relevant to the study's subject. The interview instrument targeted 20 instructors and 20 postgraduate students from five universities. They were chosen and interviewed to exchange their personal experiences and express their thoughts on using ChatGPT to improve the creative writing of English language students. Interviewing instructors and postgraduate students was done through their personal social media sites, such as WhatsApp, Messenger, and official university accounts. The researcher could not interview the instructors and the postgraduate students face-to-face due to the political situation and the war in Gaza. The researcher recorded the instructors' and postgraduate students' conversations and collected notes, classifying them following the research questions after he obtained their consent to the recording and the publication of their responses. The researchers designed two forms of interview questions related to the role of using ChatGPT in creative writing. One of them was designed to be asked to the instructors in the Palestinian universities, and the other one targeted the postgraduate students in the Palestinian universities. The reason for designing two forms of interview questions is that the researchers sought to investigate the role of using ChatGPT as an AI tool for improving the creative writing of English language students in Palestinian universities. The questions of the interview for the instructors and postgraduate students in Palestinian universities related to using ChatGPT in creative writing were as follows:

The instructors' interviews

1. Have you ever used ChatGPT in your teaching? For what? How often?
2. In your opinion, what are the possible positives/advantages of ChatGPT on learning and teaching writing?
3. What are the possible negatives/disadvantages of ChatGPT on learning and teaching writing?
4. To what extent does ChatGPT affect creative writing?

    5- Are you in favour of using ChatGPT as a supplementary tool in teaching writing?

6-Do you advise other colleagues to use ChatGPT when teaching writing courses?

The Students' interviews

1. What do you know about ChatGPT ? Have you ever used it? For what?
2. In your opinion, what are the possible positives or advantages of ChatGPT on learning and teaching writing?
3. What are the possible negatives or disadvantages of ChatGPT on learning and teaching writing?
4. To what extent does ChatGPT affect creative writing?
5. Are you in favour of using ChatGPT as a supplementary tool in teaching writing?
6. Do you advise other students to use ChatGPT while learning?

Validity and Reliability of the Study Instruments

To ensure the validity of the questionnaire, it was rated by a jury of experts in the fields of teaching English, information technology, and artificial intelligence in the Faculty of Engineering at An-Najah National University. The respondents' comments and the jury's suggestions were taken into consideration to modify and improve the questionnaire's content and wording by omitting, adding, or rephrasing items, bringing the number of items to 51.

Reliability of the Questionnaire

The reliability of the questionnaire was calculated using the Cronbach Alpha formula, and the following tables illustrate the results:

Table 3. Cronbach alpha test for the instructors' study tools

Domains	**Items**	**Cornbach Alpha**
Attitudes toward using ChatGPT	10	0.785
Pros of Implementing ChatGPT in teaching writing	24	0.955
Cons of Implementing ChatGPT in Teaching Writing	6	0.872
Role of ChatGPT in Shaping Future of Creative Writing	11	0.875
Total	**51**	**0.948**

Results in Table (3) illustrate that the Cronbach Alpha coefficient of the instructors' tool ranged from (0.785) to (0.955) for the domains of (Attitudes toward using ChatGPT) and (Pros of Implementing ChatGPT in teaching writing). Cronbach Alpha coefficient for the total degree was (0.948). These values are acceptable for the study.

Table 4. Cronbach alpha test for the postgraduate students' study tools

Domains	Items	Cronbach Alpha
Attitudes toward using ChatGPT	10	0.744
Pros of Implementing ChatGPT in Teaching Writing	24	0.901
Cons of Implementing ChatGPT in Teaching Writing	6	0.826
Role of ChatGPT in Shaping Future of Creative Writing	11	0.825
Total	**51**	**0.931**

Results in Table (4) illustrate that the Cronbach Alpha coefficient of the students' tool ranged from 0.744 to 0.901 for the domains of (Attitudes toward using ChatGPT) and (Pros of Implementing ChatGPT in teaching writing). Cronbach Alpha coefficient for the total degree was 0.931. These values are acceptable for the study. To evaluate the results of Cronbach Alpha, the following scale is used as demonstrated in the following table:

Table 5. Cronbach alpha Internal Consistency

Cronbach alpha	Internal Consistency
$\alpha \geq 0.9$	Excellent
$0.7 \leq \alpha < 0.9$	Good
$0.6 \leq \alpha < 0.7$	Acceptable

Data Analysis

The researcher used the following scale to represent the estimation level of the sample responses. To limit the length of the cells on the five-point Likert scale (minimum and maximum limits, the range has been calculated (5-1=4), divided by the correct length of the cell (4/5=0.80) and added to the lowest value of the scale (1) to reach the highest limit of this cell. Cell limits became as in Table (6):

Table 6. Scale for representing the estimation level of sample responses

Mean	Percentage %	Estimation Level
4.20 and more	84.0% and more	Very High
3.40-4.19	68.0 – 83.8 %	High
2.60-3.39	52.0 – 67.8%	Moderate
1.80- 2.59	36.0 – 51.8 %	Low
Less than 1.8	35.9% and less	Very Low

Statistical Procedures

The collected data was analyzed using (SPSS) to provide answers to the hypothesis and questions of the study, the researcher used the following measures:

1. Means, frequencies, standard deviations, and percentages in order to estimate the relative percentage of every item.
2. T-tests for independent samples to test the assumptions related to the variable of classification (Instructors and postgraduate students)
3. Univariate Analysis of Variance (ANOVA) to test the variables.
4. Cronbach Alpha is used to test the consistency of the items in the questionnaire.

RESULTS RELATED TO STUDY QUESTIONS

Results related to the First Question "*What is the role of using ChatGPT in improving creative writing of English language students from the postgraduate student's Perspectives?*"

To answer the first question, the means, and standard deviations were used as shown in Table (7).

Table 7. Means and Standard Deviations of postgraduate students' perspectives towards the role of using ChatGPT in improving creative writing of English language students in a descending order

No.	No. in the Questionnaire	Domains	Means	standard deviations	Percentage %	Estimated level
1	4	Role of ChatGPT in Shaping the Future of Creative Writing	3.90	0.49	78.0	High
2	3	Cons of Implementing ChatGPT in Teaching Writing	3.85	0.64	77.0	High
3	1	Attitudes toward using ChatGPT	3.81	0.46	76.2	High
4	2	Pros of Implementing ChatGPT in Teaching Writing	3.75	0.53	75.0	High
Total			**3.82**	**0.39**	**76.4**	**High**

Table (7) shows that the total degree of postgraduate students' perspectives towards the role of using ChatGPT in improving creative writing of English language students was (3.82) which suggests a high level of estimation. The highest mean was given to the domain (Role of ChatGPT in Shaping the Future of Creative Writing). The lowest mean was scored by the domain (Pros of Implementing ChatGPT in Teaching Writing).

Results related to the Second Question "*What is the role of using ChatGPT in improving creative writing of English language students from the instructors' perspectives in Palestinian universities?*"

To answer this question, the researcher used means, standard deviations, and estimation levels as shown in the following tables.

Table 8. Means and Standard Deviations of the instructors' perspectives towards the role of using Chat GPT in improving the creative writing of English language students in a descending order

No.	No. in the Questionnaire	Domains	Means	standard deviations	Percentage %	Estimated level
1	3	Cons of Implementing ChatGPT in Teaching Writing	3.82	0.74	76.4	High
2	4	Role of ChatGPT in Shaping the Future of Creative Writing	3.76	0.59	75.2	High
3	1	Attitudes toward using ChatGPT	3.55	0.55	71.0	High
4	2	Pros of Implementing ChatGPT in Teaching Writing	3.45	0.68	69.0	High
Total			**3.65**	**0.45**	**73.0**	**High**

Table (8) shows that the total degree of instructors' perspectives towards the role of using ChatGPT in improving creative writing of English language students 3.65 which suggests a high level of estimation. The highest mean was given to the domain (Cons of Implementing ChatGPT in Teaching Writing). The lowest mean was given to the domain (Pros of Implementing ChatGPT in teaching writing). The results of the study tool items are in Appendix (D):

Results related to the Third Question" *Are there statistically significant differences at α ≤ 0.05 in the role of using ChatGPT in improving creative writing of English language students between the perspectives of instructors and Postgraduate students in Palestinian universities?"*

To answer this question, an Independent Sample t-test was used, and Table (9) shows the results:

Table 9. Independent sample t-test results of the role of using ChatGPT in improving creative writing of English language students between the perspectives of instructors and Postgraduate students in Palestinian universities

Domain	**Category**	**N**	**Mean**	**S. D**	**t**	**Sig.***
Attitudes toward using ChatGPT	Student	165	3.81	0.46	3.895	*0.000
	Instructors	89	3.55	0.55		
Pros of Implementing ChatGPT in Teaching Writing	Student	165	3.75	0.53	3.795	*0.000
	Instructors	89	3.45	0.68		
Cons of Implementing ChatGPT in Teaching Writing	Student	165	3.85	0.64	0.289	0.773
	Instructors	89	3.82	0.74		
Role of ChatGPT in Shaping the Future of Creative Writing	Student	165	3.90	0.49	1.972	0.050
	Instructors	89	3.76	0.59		
Total degree	Student	165	3.82	0.39	3.285	*0.001
	Instructors	89	3.65	0.45		

*The mean difference is significant at (α ≤ 0.05) level.

Table (9) shows that there are statistically significant differences at (α ≤ 0.05) in the role of using ChatGPT in improving creative writing of English language students between the perspectives of instructors and postgraduate students in Palestinian universities. The significant value was 0.001 which is less than 0.05 and these differences are in favor of students due to students' mean 3.82 while instructors' mean is 3.65. which is less than 0.05. Also, statistically significant differences in the domains of (Attitudes toward using ChatGPT and Pros of Implementing ChatGPT in teaching writing) were found in favor of students' Perspectives.

Results related to the Study Hypothesis " *There are no statistically significant differences at (α ≤ 0.05) in the role of using ChatGPT in improving creative writing of English language students from the perspectives of instructors and postgraduate students in Palestinian universities."*

To test this hypothesis, Univariate Analysis of Variance (ANOVA) was used and Tables 10 and 11 show the results:

Table 10. Statistical description of the Study Sample about the role of using ChatGPT in improving the creative writing of English language students

Students				Instructors			
Variable	**Category**	**Mean**	**Standard Deviation**	**Variable**	**Category**	**Mean**	**Standard Deviation**
Gender	Male	3.90	0.43	Gender	Male	3.59	0.41
	Female	3.80	0.37		Female	3.72	0.49
Total		3.82	0.39	Total		3.65	0.45
Qualification Degree	First year	3.78	0.32	Qualification Degree	MA	3.71	0.46
	Second year	3.78	0.37		PhD	3.55	0.41
				Total		3.65	0.45
	Third Year	3.87	0.42	Experience	Less than 5years	3.59	0.61
					6-10 years	3.59	0.47
					More than 10 years	3.69	0.35
Total		3.82	0.39	Total		3.65	0.450
University	An-Najah National University	3.85	0.38	University	An-Najah National University	3.60	0.37
	Birzeit University	3.75	0.31		Birzeit University	3.74	0.46
	Arab American University	3.79	0.43		Arab American University	3.70	0.51
	Palestine Ahliya University	3.84	0.55		Palestine Ahliya University	3.58	0.79
	Hebron University	3.75	0.37		Hebron University	3.84	0.15
Total		3.82	0.39	Total		3.65	0.45

Table (10) shows that there are differences in means of the levels of the study sample. To detect the source of these differences, the Univariate Analysis of Variance (ANOVA) test was used.

Table 11. Univariate Analysis of Variance (ANOVA) for the role of using ChatGPT in improving creative writing of English language students from the perspectives of instructors and postgraduate students in Palestinian universities

Students /Variable	**Type III Sum of Squares**	**df**	**Mean Square**	**F**	**Sig.***
Gender	0.466	1	0.466	2.923	0.090
Academic Year	0.042	2	0.021	0.132	0.876
University	0.276	4	0.069	0.433	0.784
Instructors' /Variable	Type III Sum of Squares	Df	Mean Square	F	Sig.*
Gender	0.027	1	0.027	0.197	0.659
Qualification Degree	0.427	1	0.427	3.135	0.082
Experience	0.214	2	0.107	0.786	0.461
University	1.335	4	0.334	2.447	0.057

The mean difference is Significant at ($\alpha \leq 0.05$) level.

Table (11) shows the results that there are no statistically significant differences at ($\alpha \leq 0.05$) in the role of using ChatGPT in improving creative writing of English language students from the perspectives of instructors and postgraduate students in Palestinian universities. For postgraduate students, the significant values for the variables of (gender, academic year, and university) were (0.090, 0.876, and 0.784) respectively which are more than 0.05. For instructors, the significant values for the variables of (gender, qualification degree, experience and university) were (0.659, 0.082, 0.461 and 0.057) respectively which are more than 0.05.

RESULTS RELATED TO THE INTERVIEWS

The researchers conducted interviews with 20 instructors and 20 postgraduates from Palestinian universities to answer questions related to the role of using ChatGPT in improving creative writing of English language students in Palestinian universities. Basically, the researcher interviewed instructors and postgraduate students from different Palestinian universities via WhatsApp, Messenger, and email to collect data related to the study. Consequently, the thematic analysis was used and the results are presented as follows:

Theme 1: Postgraduate Students and Instructors' Use and Knowledge of ChatGPT in Education

Instructors from various Palestinian universities reported diverse uses of ChatGPT. One instructor mentioned using it to generate comprehension and writing questions, while another used it for translation and summarization. Other applications included preparing exams, finding memo and abstract samples for a technical writing course, translating literary texts, rewriting and editing, providing feedback on student writing, verifying workshop content, comparing students' literature answers, teaching research article writing, paraphrasing lengthy explanations, and teaching conversation skills. Despite its benefits, seven out of twenty instructors had not used ChatGPT, although one expressed future interest. Another instructor, knowledgeable about ChatGPT, chose not to use it due to negative recommendations from peers. Most postgraduate students use ChatGPT for writing, translating, answering exam questions, understanding topics, and expanding vocabulary. It also serves as a research assistant, providing data, paraphrasing, and feedback. Five postgraduate students found it particularly useful for exam-related tasks.

Theme 2: Positive Merits of ChatGPT in Education

Instructors highlighted ChatGPT as a valuable tool for generating creative ideas, saving time and effort, making teaching more engaging and student-centered, and assisting with writing conventions and vocabulary enhancement. They noted its usefulness in presenting ideas diversely and providing instant support for inquiries, multilingual assistance, and self-learning. Postgraduate students similarly praised ChatGPT for improving writing skills, aiding in data collection, brainstorming, paraphrasing, translating, and saving time. It also helps expand vocabulary and expressive abilities and can boost performance in disliked courses.

Theme 3: Negative Demerits of ChatGPT in Education

Instructors expressed concerns that ChatGPT promotes laziness, limits creativity, impairs critical thinking, and facilitates plagiarism. Some noted its unreliability due to lack of accuracy and deep meaning, its impersonal nature, and its potential to make writing mechanical and uninspired. They warned it could threaten the role of writers, lack emotional intelligence, and undermine academic integrity. Postgraduate students echoed these concerns, noting that ChatGPT can devalue the learning process, encourage laziness and cheating, and provide inaccurate data. They also pointed out its potential to create inequities among students and degrade the quality of education.

Theme 4: The Effect of ChatGPT on Creative Writing

Instructors acknowledged ChatGPT as a useful resource for writing samples, ideas, and suggestions, and as a collaborator in creative writing. However, some argued it cannot replace human creativity and fails to address sensitive topics. They also noted its potential to stifle imagination, produce poor writing, and discourage the development of creative skills. Postgraduate students recognized ChatGPT's role in enhancing creative writing by generating new ideas and forms of literature but emphasized the importance of human guidance. A few students saw ChatGPT as detrimental to learning creative writing, viewing it as a tool for content generation rather than skill development.

Theme 5: Future Application of ChatGPT in Education

Instructors generally saw potential in ChatGPT for future educational use if applied wisely, despite its limitations. They suggested it could help overcome time and effort constraints but warned against over-reliance, which could hinder creativity and personal voice in writing. Some noted its effectiveness depends on users' understanding of the tool. Postgraduate students believed ChatGPT would transform traditional teaching methods, providing tailored learning styles and generating fresh ideas. However, they cautioned that excessive use might reduce student engagement and interest in their subjects.

DISCUSSION OF THE QUANTITATIVE RESULTS

Results indicated that postgraduate students' perspectives on using ChatGPT to improve the creative writing of English language students were high. Such results imply that doctoral students are motivated to use ChatGPT as a tool for improving their English writing skills and accomplishing tasks effectively. Also, they see ChatGPT as a convenient tool for providing resources necessary for writing. Thus, ChatGPT can enhance their performance in writing. Besides, ChatGPT puts students in the context of their writing while saving time and effort.

Moreover, postgraduates use ChatGPT as it can motivate students to cooperate effectively and improve and reduce errors related to writing. Meanwhile, postgraduate students are inclined to be unwilling to use ChatGPT in writing as it is regarded as a source of plagiarism and cheating. Also, ChatGPT negatively affects the involvement of students in educational institutions. In other words, as long as students overuse ChatGPT to accomplish writing tasks, such a tendency reduces the involvement between parts of the learning process. Besides, they see ChatGPT as a source of general data with ambiguous contents. Therefore, ChatGPT is not recommended for tasks that require deep meaning and deep analysis. In addition, ChatGPT reduces the role of instructors and students in the learning process.

Postgraduate students believe the future of creative writing is globally invaded by the widespread use of ChatGPT in educational institutions, where it replaces human efforts in teaching writing. In other words, ChatGPT will affect the traditional ways of teaching creative writing, where there is no need for classrooms, writing materials, or teachers of writing.

The researchers attribute these results to the fact that ChatGPT enhances learning goals and activities for English language students, where it is globally considered the modern representation of technology in education to enhance creative writing skills. Also, ChatGPT enables students to choose topics and provides them with different written samples in a well-written structure. In other words, written samples of ChatGPT contain logic, systematic, and informative productions with a reference to meaningful linguistic outcomes. Consequently, ChatGPT is an effective and modern solution for improving writing skills, especially for beginners and learners at a lower level through providing comprehensive and accurate feedback. The results of this research also agreed with (Zhai, 2022), (Aydın & Karaarslan, 2022), (Hutson, et al., 2022), (Lin, 2022), (Dewi, Wardani, Rahim, Putri, & Pandin, 2021), (Mageira, et al., 2022), (Ali, Shamsan, Hezam, & Mohammed, 2023), (Pedró, 2020), (Ahmadi & Razan, 2018), (Francis, 2017), (Lindner, Romeike, Jasute, & Pozdniakov, 2019), (Esling & Devis, 2020), (Firat, 2023), (Meniado, 2023), (Mhlanga, 2023), (Sallam, 2023), and (Dergaa, Chamari, Zmijewski, & Ben Saad, 2023). (Ali & OpenAI, 2023), (Imran & Almusharraf, 2023), (Marchandot, Matsushita, Carmona, Trimaille, & Morel, 2023), (Cotton, Cotton, & Shipway, 2023), and (Thompson, Corrin, & Lodge, 2023), who all emphasized that ChatGPT is a beneficial AI tool for generating different writing samples easily. However, the results of the study contradicted with results of (Hong (2023); Algaraady & Mahyoob (2023); Deshpande, Murahari, Rajpurohit, Kalyan, & Narasimhan (2023) & Lin, Chung, Chung, & Lan (2023), who traced that ChatGPT affects the process of assessment and evaluation negatively, covers superficial issues in the written samples, and lacks originality and uniqueness in the written productions.

Furthermore, the perspectives of instructors towards the role of using ChatGPT in improving the creative writing of English language students were also high. The estimation level indicated that the instructors of the selected universities encourage students and other instructors to use ChatGPT as a tool for improving English writing skills and accomplishing tasks effectively. The reason is that learning writing is highly associated with challenges and obstacles related to learning writing, such as grammar mistakes, lexical choices, and orders of words. Moreover, they believe ChatGPT is a good resource for different types of writing. Besides, ChatGPT is a good solution for learners to know more about their topic, especially fresh issues in various fields of study, as long as ChatGPT puts users in the context of the writing. Furthermore, the instructors think that ChatGPT supports cooperation between educators and learners and helps learners improve their writing.

However, it was noticed that instructors using ChatGPT might cause ethical concerns in educational institutions, such as cheating and plagiarism. In other words, they believe that learners do not use ChatGPT as a source for improving skills and enhancing their levels; they use it as a source for accomplishing tasks and answering questions in quizzes. Also, they see ChatGPT kills participation and the involvement between educators and learners. Besides, instructors believe that there is no need to use ChatGPT in learning since the role of using ChatGPT is restricted to superficial outputs and general facts. In other words, instructors believe ChatGPT lacks the current style of learning, which requires deep thinking and deep understanding. It was also noticed that the instructors think that using ChatGPT is a new form of technological imperialism in educational institutions, where the overuse of ChatGPT replaces the human voice in writing and other fields of study. In other words, ChatGPT will affect the role of educators and learners in the learning process.

The researchers attribute these results to the fact that ChatGPT serves instructors effectively by providing writing samples that are regarded as helpful and effective for their academic and pedagogical purposes. In other words, ChatGPT enhances students' academic writing when it serves their language competency and productivity by providing different forms of summaries, abstracts, and well-written ideas and arguments related to their research purpose and interest. These results nearly agreed with (Mohammed, Al-ghazali, & Alqohfa, 2023; Tiwari, Bhat, Tariq Khan, Subramaniam, & Khan, 2023; Kayalı, Yavuz, Balat, & Çalışan, 2023; Rahman, Sabbir, Zhang, Moral, & Hossain, 2022) who stressed that ChatGPT is useful for academic writing because it provides users with favorable and positive attitudes, sufficient data, and topic orientation.

The overall results showed no statistically significant differences at ($\alpha \leq 0.05$) in the role of using ChatGPT in improving creative writing of English language students between the instructors and postgraduate students The researcher attributes these results to the fact that no differences occurred between the levels of the study sample, either students or instructors, because of the advantages of ChatGPT for creative writing. In other words, instructors and postgraduate students have the same perspectives towards the role of using ChatGPT in improving creative writing of English language students in Palestinian universities, with a reference to gender, qualification degree, academic year, years of experience, and university.

DISCUSSION OF THE RESULTS OF THE INTERVIEWS

Instructors and postgraduate students from different Palestinian universities insisted on the advantages of using ChatGPT in improving the creative writing of English language students. ChatGPT is an effective tool for generating different writing samples and ideas for projects and tasks and providing topic suggestions based on given prompts. ChatGPT is also regarded as a helpful tool for saving users time and effort. Moreover, using ChatGPT makes teaching more engaging, interactive, and fun (edutainment), and it makes classes more student-based rather than teacher-based, etc. However, there are many concerns and obstacles related to the use of ChatGPT, such as plagiarism, cheating, laziness, the fact that it can provide value-driven content, the fact that it cannot affect a writer's role in any way, etc. Furthermore, they agree that the cautious use of ChatGPT is highly needed in educational institutions, where users of ChatGPT, whether instructors or students, must use it as a tool that improves human efforts, not as a tool for replacing or reducing human efforts in writing.

The researcher attributes these results to the fact that ChatGPT serves postgraduate students and instructors' creative writing in educational institutions, where ChatGPT helps them to support their creative writing content by generating various and comprehensive samples of writing related to the topics that they are working on. Furthermore, ChatGPT helps postgraduate students and instructors unleash their creativity by generating ideas, questions, and meaningful feedback. On the other hand, ChatGPT saves time and effort, reduces pressure, and increases pleasure and excitement through working on creative writing topics. Using ChatGPT enables users to overcome challenges and obstacles related to the traditional methods of learning and teaching writing. The results seem to disagree with Lin, Chung, Chung, & Lan (2023); Hong (2023); Algaraady & Mahyoob (2023); Deshpande, Murahari, Rajpurohit, Kalyan, & Narasimhan (2023); Aydın & Karaarslan (2022) & (Hutson et al., 2022), who emphasized that ChatGPT is limited to solving superficial levels of obstacles and issues in writing. Besides that, ChatGPT promotes incorrectness, loss of originality, and laziness.

CONCLUSION

In conclusion, learning a language can be connected to learning creative writing in universities and other educational institutions since creative writing is one of the most important skills that English students should take full responsibility for mastering the necessary skills needed for the world of work. Creative writing involves producing content with deep meanings and thoughts about certain issues. The study aimed to investigate the role of using ChatGPT in improving the creative writing of English language students from the perspectives of instructors and postgraduate students in Palestinian universities. The study targeted (89) instructors and (169) doctoral students from five Palestinian universities. The results of the study show ed that ChatGPT is a valuable tool in educational institutions, especially in creative writing courses. Accordingly, postgraduate students revealed that ChatGPT is an important, technological, and contemporary AI tool used globally for generating questions in different fields of study, especially the field of creative writing. In addition, ChatGPT provides a vast amount of data for research studies and creative writing. Besides that, using ChatGPT is useful for saving time and effort.

On the other hand, the instructors confirmed that ChatGPT is an effective tool for providing different types of writing production, which helps to improve creative writing skills, increase users' vocabulary, and shape a positive trend toward using technology in education. Hence, the study suggests that instructors and postgraduate students cannot completely rely on ChatGPT, but it can be used as a supplementary and extra resource. In other words, ChatGPT assists users by providing a vast amount of data related to creative writing topics. Still, it does not cancel the effective role of human effort in creative writing topics, especially those that are related to deep meanings and connotative analysis.

Recommendations

In light of the previously mentioned results, the researcher recommends the following:

1. University instructors and postgraduate students are recommended to integrate ChatGPT into the teaching process, which will help them enhance students' creative writing.
2. Universities are recommended to insert ChatGPT into writing courses in English language departments.
3. Universities are recommended to provide instructors and students with training courses related to using ChatGPT in education.
4. Researchers are recommended to conduct studies related to ChatGPT and its role in improving the writing skills of students learning English as a foreign language in different educational institutions and schools in the West Bank and Gaza.
5. Researchers are recommended to conduct studies related to ChatGPT and its role in improving self-learning and collaborative learning.

REFERENCES

Abdaljaleel, M., Barakat, M., Alsanafi, M., Salim, N. A., Abazid, H., Malaeb, D., & Sallam, M. (2023). *Factors Influencing Attitudes of University Students towards ChatGPT and its Usage: A Multi-National Study Validating the TAME-ChatGPT Survey Instru.*

Ahmadi, D. M., & Razan. (2018). The use of technology in English language learning: A literature review. . *International journal of research in English education*, 3(2), 115-125.

Aina, M., & Joshua, W. (2024). Impact of ChatGPT in Higher Education Learning . *JILTECH: Journal International of Lingua & Technology*, 3(1).

Algaraady, J., & Mahyoob, M. (2023). Algaraady, Jeehaan & Mahyoob, Mohammad. (2023). ChatGPT's Capabilities in Spotting and Analyzing Writing Errors Experienced by EFL Learners. *Arab World English Journal*, 9(9), 3–17. DOI: 10.24093/awej/call9.1

Ali, F., & Open, A. I. I. (2023). Let the devil speak for itself: Should ChatGPT be allowed or banned in hospitality and tourism schools? *Journal of Global Hospitality and Tourism*, 2, 1–6. DOI: 10.5038/2771-5957.2.1.1016

Ali, J. K., Shamsan, M. A., Hezam, T. A., & Mohammed, A. A. (2023). Impact of ChatGPT on Learning Motivation: Teachers and Students' Voices . *Journal of English Studies in Arabia Felix*, 2(1), 41–49. DOI: 10.56540/jesaf.v2i1.51

Alkaissi, H., & McFarlane, S. I. (2023). Artificial hallucinations in ChatGPT: Implications in scientificwriting. *Cureus*, 15(2). Advance online publication. DOI: 10.7759/cureus.35179 PMID: 36811129

Alneyadi, S., & Wardat, Y. (2023). ChatGPT: Revolutionizing student achievement in the electronic magnetism unit for eleventh-grade students in Emirates schools . *Contemporary Educational Technology*, 15(4), ep448. DOI: 10.30935/cedtech/13417

Arora, A., & Mehta, N. K. (2024). Enhancing Writing Skills through ChatGPT: an Experimental Study in the Context of Ergonomics. Limbaj şi context. . *Revista internaţională de lingvistică, semiotică şi ştiinţă literară*, 16(1), 93-104.

Aydın, Ö., & Karaarslan, E. (2022). OpenAI ChatGPT generated literature review: Digital Twin in Healthcare. *In Ö. Aydın (Ed.), Emerging Computer Technologies*, 2 (22-31). İzmir Akademi Dernegi.

Bibi, Z., & Atta, A. (2024). The Role of ChatGPT as AI English Writing Assistant: A Study of Student's Perceptions, Experiences, and Satisfaction . *Annals of Human and Social Sciences*, 5(1), 433–443.

Bin-Hady, W. R., Al-Kadi, A., Hazaea, A., & Ali, J. K. (2023). *Exploring the dimensions of ChatGPT in English language learning: A global perspective* . Library Hi Tech, (ahead-of-print).

Bull, S., & Ma, Y. (2001). Raising learner awareness of language learning strategies in situations of limited recourses. *Interactive Learning Environments*, 9(2), 171–200. DOI: 10.1076/ilee.9.2.171.7439

Buriak, J. M., Akinwande, D., Artzi, N., Brinker, C. J., Burrows, C., Chan, W. C., Chen, C., Chen, X., Chhowalla, M., Chi, L., Chueh, W., Crudden, C. M., Di Carlo, D., Glotzer, S. C., Hersam, M. C., Ho, D., Hu, T. Y., Huang, J., Javey, A., & Ye, J. (2023). Best practices for using AI when writing scientific manuscripts: Caution,care, and consideration: Creative science depends on it. *ACS Nano*, 17(5), 4091–4093. DOI: 10.1021/acsnano.3c01544 PMID: 36848601

Castillo-González, W., Lepez, C. O., & Bonardi, M. C. (2022). Chat GPT: A promising tool for academic editing . *Data and Metadata*, 1, 23–23. DOI: 10.56294/dm202223

Celik, I., Dindar, M., Muukkonen, H., & Järvelä, S. (2022). PAPERThe Promises and Challenges of Artificial Intelligence for Teachers: A Systematic Review of Research. *TechTrends*, 66(4), 616–630. DOI: 10.1007/s11528-022-00715-y

Chen, T. -J. (2023). ChatGPT and other artificial intelligence applications speed up scientific writing. . *Journal of the Chinese Medical Association*, 86(4), 351 353.. DOI: 10.1097/JCMA.0000000000000900

Cotton, D., Cotton, P., & Shipway, R. (2023). Chatting and Cheating. Ensuring academic integrity in the era of ChatGPT. DOI: 10.35542/osf.io/mrz8h

Curry, N., Baker, P., & Brookes, G. (2024). Generative AI for corpus approaches to discourse studies: A critical evaluation of ChatGPT . *Applied Corpus Linguistics*, 4(1), 100082. DOI: 10.1016/j.acorp.2023.100082

Datoo, A. K., & Siddiqui, K. A. (2024). ChatGPT and Death of an Author. *Critical Humanities*, 2(2), 5. DOI: 10.33470/2836-3140.1039

Dergaa, I., Chamari, K., Zmijewski, P., & Ben Saad, H. (2023). From human writing to artificial intelligence generatedtext: Examining the prospects and potential threats of ChatGPT in academic writing. *Biology of Sport*, 40(2), 615–622. DOI: 10.5114/biolsport.2023.125623 PMID: 37077800

Deshpande, A., Murahari, V., Rajpurohit, T., Kalyan, A., & Narasimhan, K. (2023). Toxicity in ChatGPT: Analyzing Persona-assigned Language Models. *ArXiv*, abs/2304.05335.

Dewi, H. K., Wardani, T. I., Rahim, N. A., Putri, R. E., & Pandin, M. G. (2021). *THE USE OF AI (ARTIFICIAL INTELLIGENCE) IN ENGLISH LEARNING AMONG UNIVERSITY STUDENT: CASE STUDY IN ENGLISH DEPARTMENT.* UNIVERSITAS AIRLANGGA.

Dwivedi, Y. K., Kshetri, N., Hughes, L., Slade, E. L., Jeyaraj, A., Kar, A. K., Baabdullah, A. M., Koohang, A., Raghavan, V., Ahuja, M., Albanna, H., Albashrawi, M. A., Al-Busaidi, A. S., Balakrishnan, J., Barlette, Y., Basu, S., Bose, I., Brooks, L., Buhalis, D., & Wright, R. (2023). So what if ChatGPT wrote it?" Multidisciplinary perspectives on opportunities, challenges and implications of generative conversational AI for research, practice and policy. *International Journal of Information Management*, 71, 102642. DOI: 10.1016/j.ijinfomgt.2023.102642

Esling, P., & Devis, N. (2020). *Creativity in the era of artificial intelligence.* arXiv preprint arXiv:2008.05959.

Fahimirad, M., & Kotamjani, S. S. (2018). A review on application of artificial intelligence in teaching and learning in educational contexts . *International Journal of Learning and Development*, 8(4), 106–118. DOI: 10.5296/ijld.v8i4.14057

Fareed, M., Ashraf, A., & Bilal, M. (2016). ESL learners' writing skills: Problems, factors and suggestions . *Journal of education and social sciences*, 4(2), 81-92.

Firat, M. (2023). *How chat GPT can transform autodidactic experiences and open education* . Department of Distance Education, Open Education Faculty, Anadolu Unive.

Francis, J. (2017). *The effects of technology on student motivation and engagement in classroom-based learning* .

Gill, S. S., Xu, M., Patros, P., Wu, H., Kaur, R., Kaur, K., & Buyya, R. (2024). Transformative effects of ChatGPT on modern education: Emerging Era of AI Chatbots. *Internet of Things and Cyber-Physical Systems*, 4, 19–23. DOI: 10.1016/j.iotcps.2023.06.002

Gill, S. S., Xu, M., Patros, P., Wu, H., Kaur, R., Kaur, K., Fuller, S., Singh, M., Arora, P., Parlikad, A. K., Stankovski, V., Abraham, A., Ghosh, S. K., Lutfiyya, H., Kanhere, S. S., Bahsoon, R., Rana, O., Dustdar, S., Sakellariou, R., & Buyya, R. (2024). Transformative effects of ChatGPT on modern education: Emerging Era of AI Chatbots. *Internet of Things and Cyber-Physical Systems*, 4, 19–23. DOI: 10.1016/j.iotcps.2023.06.002

Halaweh. (2023). *Argued the co-operation between ChatGPT and human in writing will probably lead to produce writing samples regarded inventive, original, and productive.*

Hidayatullah, E. (2024, January). Evaluating the effectiveness of ChatGPT to improve English students' writing skills . *In Humanities, Education, Applied Linguistics, and Language Teaching: Conference Series*, pp. (Vol. 1, No. 1, pp. 14-19).

Hong, W. C. (2023). The impact of ChatGPT on foreign language teaching and learning: Opportunities in education and research . *Journal of Educational Technology and Innovation*, 3(1). Advance online publication. DOI: 10.61414/jeti.v5i1.103

Hutson, J., Jeevanjee, T., Vander Graaf, V., Lively, J., Weber, J., Weir, G., & Edele, S. (2022). Artificial Intelligence and the Disruption of Higher Education: Strategies for Integrations across Disciplines . *Creative Education*, 13(12), 3953–3980. DOI: 10.4236/ce.2022.1312253

Imran, M., & Almusharraf, N. (2023). Analyzing the role of ChatGPT as a writing assistant at higher education level: A systematic review of the literature. *15(4), ep464*, 15(4), ep464.

Ishaq, K., Azan, N., Zin, M., Rosdi, F., Abid, A., & Ijaz, M. (2020). The impact of ICT on students' academic performance in public private sector universities of Pakistan. *International Journal of Innovative Technology and Exploring Engineering*, 9(3), 11. DOI: 10.35940/ijitee.C8093.019320

Iskender, A. (2023). Holy or unholy? Interview with open AI's ChatGPT . *European Journal of Tourism Research*, 34, 3414–3414. DOI: 10.54055/ejtr.v34i.3169

Jiang, Q., Li, M., Han, W., & Yang, C. (2019). *ICT promoting the improvement of education quality: Experience and practice.* In MINISTERIAL FORUM (p. 158). RetrievedJanuary25,2022fromhttps://iite.unesco.org/wpcontent/uploads/2019/02/Unesco_003_Proccedings_190212a.pdf#page=160

Kasneci, E., Seßler, K., Küchemann, S., Bannert, M., Dementieva, D., Fischer, F., & Kasneci, G. (2023). ChatGPT for good? On opportunities and challenges of large language models for education. *Learning and Individual Differences*, 103, 102274. DOI: 10.1016/j.lindif.2023.102274

Kayalı, B., Yavuz, M., Balat, S., & Çalışan, M. (2023). Investigation of student experiences with ChatGPT-supported online learning applications in higher education . *Australasian Journal of Educational Technology*, 39(5), 20–39. DOI: 10.14742/ajet.8915

Liebrenz, M., Schleifer, R., Buadze, A., Bhugra, D., & Smith, A. (2023). Generating scholarly content with ChatGPT: Ethical challenges for medical publishing . *The Lancet. Digital Health*, 5(3), e105–e106. DOI: 10.1016/S2589-7500(23)00019-5 PMID: 36754725

Lin, H. (2022). Influences of Artificial Intelligence in Education on Teaching Effectiveness: The Mediating Effect of Teachers' Perceptions of Educational Technology. *International Journal of Emerging Technologies in Learning*, 17(24), 144–156. DOI: 10.3991/ijet.v17i24.36037

Lin, S., Chung, H.-H., Chung, F.-L., & Lan, K. (2023). *Concerns About Using ChatGPT in Education*. .DOI: 10.1007/978-3-031-40113-8_4

Lin, X. (2023). *Exploring the role of ChatGPT as a facilitator for motivating self-directed learning among adult learners.*

Lindner, A., Romeike, R., Jasute, E., & Pozdniakov, S. (2019). *Lindner, A., Romeike, R., Jasute, E., & Pozdniakov, S. (2019). Teachers' perspectives on artificial intelligence.* . In 12th International conference on informatics in schools,"Situation, evaluation and perspectives", ISSEP.

Liu, X., Zheng, Y., Du, Z., Ding, M., Qian, Y., Yang, Z., & Tang, J. (2021). *GPT understands,too.* arXiv. https://doi.org//arXiv.2103.10385.DOI: 10.48550

Lund, B., & Wang, T. (2023). Chatting about ChatGPT: How may AI and GPT impact academia and libraries? *Library Hi Tech News*, 40(3), 26–29. DOI: 10.1108/LHTN-01-2023-0009

Mabuan, R. A. (2024). ChatGPT and ELT: Exploring teachers' voices . [IJTE]. *International Journal of Technology in Education*, 7(1), 128–153. DOI: 10.46328/ijte.523

Mageira, K., Pittou, D., Papasalouros, A., Kotis, K., Zangogianni, P., & Daradoumis, A. (2022). Educational AI chatbots for content and language integrated learning . *Applied Sciences (Basel, Switzerland)*, 12(7), 3239. DOI: 10.3390/app12073239

Malinka, K., Perešíni, M., Firc, A., Hujňák, O., & Januš, F. (2023). *On the Educational Impact of ChatGPT: Is Artificial Intelligence Ready to Obtain a University Degree?* . arXiv preprint arXiv:2303.11146. DOI: 10.1145/3587102.3588827

Marchandot, B., Matsushita, K., Carmona, A., Trimaille, A., & Morel, O. (2023). ChatGPT: The Next Frontier in Academic Writing for Cardiologists or a Pandora's Box of Ethical Dilemmas. *European Heart Journal Open*, 3(2), oead007. Advance online publication. DOI: 10.1093/ehjopen/oead007 PMID: 36915398

Matthews, J. A., & Volpe, C. R. (2023). Academics' perceptions of ChatGPT-generated written outputs: A practical application of Turing's Imitation Game. *Australasian Journal of Educational Technology*, 39(5), 82–100. DOI: 10.14742/ajet.8896

Meniado, J. (2023). The Impact of ChatGPT on English Language Teaching, Learning, and Assessment: A Rapid Review of Literature. *Arab World English Journal*, 14, 3–18. DOI: 10.24093/awej/vol14no4.1

Mhlanga, D. (2023). Open AI in Education, the Responsible and Ethical Use of ChatGPT Towards Lifelong Learning. SSRN *Electronic Journal*, DOI: 10.2139/ssrn.4354422

Mohammed, A., Al-ghazali, A., & Alqohfa, K.Khalid A. S. Alqohfa. (2023). Exploring ChatGPT uses in higher studies: A case study of Arab postgraduates in India. *Journal of English Studies in Arabia Felix*, 2(2), 8–15. DOI: 10.56540/jesaf.v2i2.55

Niloy, A. C., Akter, S., Sultana, N., Sultana, J., & Rahman, S. I. (2024). Is Chatgpt a menace for creative writing ability? An experiment . *Journal of Computer Assisted Learning*, 40(2), 919–930. DOI: 10.1111/jcal.12929

Opara, E. C. (2023). ChatGPT for Teaching, Learning and Research: Prospects and Challenges. *Global Academy Journal of Humanities andSocialSciences*, 5(2),33-40. DOI: 10.13140/RG.2.2.21304.42244/1

Pedró, F. (2020). Applications of Artificial Intelligence to higher education: Possibilities, evidence, and challenges . *IUL Research*, 1(1), 61–76. DOI: 10.57568/iulres.v1i1.43

Rahman, M. S., Sabbir, M. M., Zhang, J., Moral, I. H., & Hossain, G. M. (2022). Examining students' intention to use ChatGPT: Does trust matter? . *Australasian Journal of Educational Technology*, •••, 51–71.

Ratheeswari, K. (2018). Information communication technology in education . *Journal of Applied and Advanced research*, 3(1), 45-47.

Rudolph, J., Tan, S., & Tan, S. (2023). ChatGPT: Bullshit spewer or the end of traditional assessmenin higher education. *Journal of Applied Learning and Teaching, 6(1), 342-363. https://doi.org/*, Journal of Applied Learning and Teaching, 6(1), 342-363. https://doi.org/.DOI: 10.37074/jalt.2023.6.1.9

Rushkoff, D. (2013). *Its not just grammar; its clear thinking.* The New York Times. Retrievedfrom:https://www.nytimes.com/roomfordebate/2012/08/13/is-our-children-learning-enough-grammar-to-get-hired/its-not-just-grammar-its-clear-thinking

Sallam, M. (2023). ChatGPT Utility in Health Care Education, Research, and Practice: Systematic Review on the Promising Perspectives and Valid Concerns. *Health Care*, 11(6), 887. DOI: 10.3390/healthcare11060887 PMID: 36981544

Salvagno, M., Taccone, F. S., & Gerli, A. G. (2023). *Salvagno, M., Taccone, F. S., & Gerli, A. G. (2023). Can artificial intelligence help for scientific writing?* Santosa, A. B., Basuki, Y., & Puspita, A. M. (2019). The effectiveness of local wisdom-based teaching materials in enhancing creative writing skills of elementary school students. *Journal of English Language Teaching and Linguistics*, 4(3), 349–359.

Seo, K., Tang, J., Roll, I., Fels, S., & Yoon, D. (2021). The impact of artificial intelligence on learner–instructor interaction in online learning . *International Journal of Educational Technology in Higher Education*, 18(1), 1–23. DOI: 10.1186/s41239-021-00292-9 PMID: 34778540

Strain-Moritz, & Tessa, E. (2016). Perceptions of Technology Use and Its Effects on Student Writing. *Culminating Projects in Teacher Development*, 8.https://repository.stcloudstate.edu/ed_etds/8

Strobl, C., Menke-Bazhutkina, I., Abel, N., & Michel, M. (2024). Adopting ChatGPT as a writing buddy in the advanced L2 writing class . *Technology in Language Teaching & Learning*, 6(1), 1–1. DOI: 10.29140/tltl.v6n1.1168

Su, J., & Yang, W. (2023). Unlocking the power of ChatGPT: A framework for applying generative AI in education. *ECNU Review of Education*, 6(3), 355–366. DOI: 10.1177/20965311231168423

Sundt, E. (2018). *The use of creative writing in foreign language teaching in Norwegian VGS.* .

Thompson, K., Corrin, L., & Lodge, J. M. (2023). AI in tertiary education: Progress on research and practice . *Australasian Journal of Educational Technology*, 39(5), 1–7. DOI: 10.14742/ajet.9251

Tirumalesh, K. V. (1991). Writing-English versus writing-in-English: New notes on an old theme . *Economic and Political Weekly*, •••, 2670–2672.

Tiwari, C., Bhat, M., Tariq Khan, S., Subramaniam, R., & Khan, M. (2023). What drives students toward ChatGPT? An investigation of the factors influencing adoption and usage of ChatGPT. *Interactive Technology and Smart Education*. Advance online publication. DOI: 10.1108/ITSE-04-2023-0061

Tlili, A., Shehata, B., Adarkwah, M. A., Bozkurt, A., Hickey, D. T., Huang, R., & Agyemang, B. (2023). What if the devil is my guardian angel: ChatGPT as a case study of using chatbots in education . *Smart Learning Environments*, 10(1), 15. DOI: 10.1186/s40561-023-00237-x

Tossell, C. C., Tenhundfeld, N. L., Momen, A., Cooley, K., & de Visser, E. J. (2024). *Student Perceptions of ChatGPT Use in a College Essay Assignment: Implications for Learning, Grading, and Trust in Artificial Intelligence*. IEEE Transactions on Learning.

UNESCO. (2023). *ChatGPT and artificial intelligence in higher education*. Document code:ED/HE/IESALC/IP/2023/12. (https://creativecommons.org/licenses/by-sa/3.0/igo/)

Van Dis, E. A., Bollen, J., Zuidema, W., van Rooij, R., & Bockting, C. L. (2023). ChatGPT: Five priorities for research. *Nature*, 614(7947), 224–226. DOI: 10.1038/d41586-023-00288-7 PMID: 36737653

Veiga, F., & Andrade, A. (2021). Critical success factors in accepting technology in the classroom. *International Journal of Emerging Technologies in Learning*, 16(18), 4–21. DOI: 10.3991/ijet.v16i18.23159

Wardat, Y., Tashtoush, M. N., AlAli, R., & Jarrah, A. M. (2023). ChatGPT: A revolutionary tool for teaching and learning mathematics . *Eurasia Journal of Mathematics, Science and Technology Education*, 19(7), em2286. DOI: 10.29333/ejmste/13272

Williams, C., & Beam, S. (2019). Technology and writing: Review of research. *Computers & Education*, 128, 227–242. DOI: 10.1016/j.compedu.2018.09.024

Zhai, X. (2022). ChatGPT user experience: Implications for education. Available at *SSRN* 4312418.

Zhu, S., Wang, Z., Zhuang, Y., Jiang, Y., Guo, M., Zhang, X., & Gao, Z. (2024). Exploring the impact of ChatGPT on art creation and collaboration: Benefits**, challenges and ethical implications . *Telematics and Informatics Reports*, 100138.**

ADDITIONAL READING

Bekou, A., Ben Mhamed, M., & Assissou, K. (2024). Exploring opportunities and challenges of using ChatGPT in English language teaching (ELT) in Morocco. *Focus on ELT Journal*, 6(1), 87–106. DOI: 10.14744/felt.6.1.7

Cambra-Fierro, J. J., Blasco, M. F., López-Pérez, M. E. E., & Trifu, A. (2024). ChatGPT adoption and its influence on faculty well-being: An empirical research in higher education. *Education and Information Technologies*. Advance online publication. DOI: 10.1007/s10639-024-12871-0

Diasamidze, L., & Tedoradze, T. (2024). Application of GPT models in English language teaching. *Language and Culture*, 9, 76–81. DOI: 10.52340/lac.2024.09.15

Pang, S., Nol, F., & Heng, K. (2024). ChatGPT-4o for English language teaching and learning: Features, applications, and future prospects. DOI: DOI: 10.13140/RG.2.2.13172.36481

Qadir, J. (2022). Engineering education in the era of ChatGPT: Promise and pitfalls of generative AI for education. *TechRxiv*. https://doi.org/DOI: 10.36227/techrxiv.21789434

Rawas, S. (2023). ChatGPT: Empowering lifelong learning in the digital age of higher education. *Education and Information Technologies*, 29(6), 6895–6908. DOI: 10.1007/s10639-023-12114-8

KEY TERMS AND DEFINITIONS

ChatGPT: ChatGPT stands for "Generative Pre-trained Transformer," that involves t.a language model that allows natural and conversational interaction with computers or tablets. Using the Artificial Intelligence capabilities,, it can generate coherent, human-like text based on large datasets, predicting the next word in a sentence to produce relevant content.

Creative Writing: Creative writing implies the art of building words together in a systematic, syntactic, and coherent way of expressing ideas, thoughts, and expressions. In other words, creative writing content is all about collecting different thoughts, views, contributions, content, and beliefs to formulate a new creative product.

Palestinian Universities: Higher education institutions which are established in the West -Bank and Gasa Strip to provide higher education services, the number of which is 17 universities.

Perspective: a way of thinking about and understanding something (such as a particular issue or life in general.

Postgraduate students: A postgraduate is a student who has completed an undergraduate degree-level course at a college or university and is undertaking further study (MA)at a more advanced level.

APPENDIX (A)

Instructors' Questionnaire

Dear Instructor,

The researcher is writing a thesis on the role of using ChatGPT in improving creative writing of English language students from the perspectives of instructors and postgraduate students in Palestinian universities. This questionnaire is one of the tools used to investigate the role of using ChatGPT in improving creative writing of English language students in Palestinian universities. Please read the items carefully and respond to each item by placing (√) next to the answer that corresponds to your level of agreement or disagreement. Your answers will be confidential and only used for scientific purposes.

Part One: Personal Information

1. Gender: a. Male () b. Female ()

2. Qualification degree: a.MA () b. PhD ()

3. Years of Experience: a. Less than () b.6-1 () c. More than 10 ()

4-University: a-An-Najah National University () b-Beirzeit University ()

c-Arab American University d-Palestine Ahliya University e- Hebron University

Part Two: Read the items below carefully and respond to each one by placing (√) next to the answer that corresponds to your response.

Table 12

No.	Item	Strongly Agree	Agree	Neutral	Disagree	Strongly Disagree
First domain: Instructors' attitudes toward using ChatGPT.						
1.	ChatGPT can improve the learning and teaching process.					
2.	Using ChatGPT can save time and effort.					
3.	Using ChatGPT makes students inactive.					
4.	ChatGPT is a trustworthy resource of information for instructors and students.					
5.	ChatGPT may provide inaccurate sources.					
6.	Using ChatGPT can be a double-edged weapon.					
7.	Using ChatGPT is challenging in use.					
8.	ChatGPT can be used as a guide for students and instructors in educational institutions.					
9.	ChatGPT can support the role of instructors and students in learning.					
10.	ChatGPT plays a big role in student assessment.					

continued on following page

Table 12

Continued

No.	Item	Strongly Agree	Agree	Neutral	Disagree	Strongly Disagree
Second domain: Pros of Implementing ChatGPT in teaching writing.						
1.	ChatGPT can improve teaching English writing skills.					
2.	ChatGPT can help learners in accomplishing writing-related tasks effectively.					
3.	ChatGPT can improve access to different educational resources related to the writing production.					
4.	Using ChatGPT can improve students' performance in creative writing.					
5.	ChatGPT can provide writing samples to instructors and students.					
6.	ChatGPT can provide diversity of writing resources.					
7.	ChatGPT can put students in the context of the written context.					
8.	ChatGPT can provide reliable resources and data for writing.					
9.	ChatGPT can save writing mistakes.					
10.	ChatGPT helps in text analysis.					
11.	ChatGPT helps in generating students' feedback for essays.					
12.	ChatGPT enables students to create constructive responses.					
13.	ChatGPT can improve the efficiency of research process and analysis.					
14.	ChatGPT can enhance the quality of assessment.					
15.	ChatGPT fosters transparency in writing production.					
16.	Using ChatGPT ensures respect of the cultural context in writing production.					
17.	ChatGPT may help students facing difficulties in writing.					
18.	Instructors accept ChatGPT as a helpful tool in enhancing creative writing.					
19.	Using ChatGPT can lead to inequalities between students if being used by some students.					
20.	ChatGPT can provide critical views in writing.					
21.	ChatGPT leads to bias in writing.					
22.	ChatGPT may serve highly qualified instructors and students in writing.					
23.	ChatGPT can motivate students to cooperate outside the classroom.					
24.	ChatGPT can locate linguistic and grammatical errors.					
Third domain: Cons of Implementing ChatGPT in Teaching Writing.						
1.	ChatGPT might increase cheating and plagiarism.					

continued on following page

Table 12

Continued

No.	Item	Strongly Agree	Agree	Neutral	Disagree	Strongly Disagree
2.	ChatGPT can reduce students' involvement in accomplishing creative writing content.					
3.	ChatGPT provides general and ambiguous contents.					
4.	ChatGPT reduces the importance of teaching writing in universities.					
5.	ChatGPT negatively affects creativity in writing.					
6.	ChatGPT reduces the role of instructors and students in the teaching and learning process.					
Fourth domain: Role of ChatGPT in Shaping Future of Creative Writing						
1	ChatGPT affects the traditional ways of teaching creative writing.					
2	ChatGPT affects the quality of students in the workplace.					
3	Using ChatGPT underestimates the teaching writing.					
4	ChatGPT leads to undesirable consequences.					
5	There is a lack of awareness towards using ChatGPT in teaching and learning of writing.					
6	Instructors and students may require training to use ChatGPT.					
7	Using ChatGPT affects students' writing level.					
8	ChatGPT provides limited resources for instructors and students.					
9	ChatGPT replaces human effort in writing.					
10	ChatGPT enables learners to contribute to their learning.					

APPENDIX (B)

Postgraduate students' Questionnaire

Dear postgraduate students,

The researcher is writing a thesis on the role of using ChatGPT in improving creative writing of English language students from the perspectives of instructors and postgraduate students in Palestinian universities. This questionnaire is one of the tools used to investigate the role of ChatGPT in improving creative writing of English language students in Palestinian universities. Please read the items carefully and respond to each item by placing (√) next to the answer that corresponds to your level of agreement or disagreement. Your answers will be confidential and only used for scientific purposes.

Part One: Personal Information

1. **Gender**: a. Male () b. Female ()
2. **Academic year**: a.First year () b. Second year () c. Third Year ()

3-University: a-An-Najah National University () b-Beirzeit University ()

c-Arab American University d-Palestine Ahliya University e- Hebron University

Part Two: Read the items below carefully and respond to each one by placing (√) next to the answer that corresponds to your response.

Table 13

No.	Item	Strongly Agree	Agree	Neutral	Disagree	Strongly Disagree
First domain: Postgraduate students' attitudes toward using ChatGPT.						
1.	ChatGPT can improve the learning and teaching process.					
2.	Using ChatGPT can save time and effort.					
3.	Using ChatGPT makes students inactive.					
4.	ChatGPT is a trustworthy resource of information for instructors and students.					
5.	ChatGPT may provide inaccurate sources.					
6.	Using ChatGPT can be a double edged weapon.					
7.	Using ChatGPT is challenging in use.					
8.	ChatGPT can be used as a guide for students and instructors in educational institutions.					
9.	ChatGPT can support the role of instructors and students in learning.					

continued on following page

Table 13

Continued

No.	Item	Strongly Agree	Agree	Neutral	Disagree	Strongly Disagree
10.	ChatGPT plays a big role in student assessment.					
Second domain: Pros of Implementing ChatGPT in teaching writing.						
1.	ChatGPT can improve teaching English writing skills.					
2.	ChatGPT can help learners in accomplishing writing-related tasks effectively.					
3.	ChatGPT can improve access to different educational resources related to the writing production.					
4.	Using ChatGPT can improve students' performance in creative writing.					
5.	ChatGPT can provide writing samples to instructors and students.					
6.	ChatGPT can provide diversity of writing resources.					
7.	ChatGPT can put students in the context of the written context.					
8.	ChatGPT can provide reliable resources and data for writing.					
9.	ChatGPT can save writing mistakes.					
10.	ChatGPT helps in text analysis.					
11.	ChatGPT helps in generating students' feedback for essays.					
12.	ChatGPT enables students to create constructive responses.					
13.	ChatGPT can improve the efficiency of research process and analysis.					
14.	ChatGPT can enhance the quality of assessment.					
15.	ChatGPT fosters transparency in writing production.					
16.	Using ChatGPT ensures respect of the cultural context in writing production.					
17.	ChatGPT may help students facing difficulties in writing.					
18.	Instructors accept ChatGPT as a helpful tool in enhancing creative writing.					
19.	Using ChatGPT can lead to inequalities between students if being used by some students.					
20.	ChatGPT can provide critical views in writing.					
21.	ChatGPT leads to bias in writing.					
22.	ChatGPT may serve highly qualified instructors and students in writing.					
23.	ChatGPT can motivate students to cooperate outside the classroom.					

continued on following page

Table 13

Continued

No.	Item	Strongly Agree	Agree	Neutral	Disagree	Strongly Disagree
24.	ChatGPT can locate linguistic and grammatical errors.					
Third domain: Cons of Implementing ChatGPT in Teaching Writing.						
1.	ChatGPT might increase cheating and plagiarism.					
2.	ChatGPT can reduce students' involvement in accomplishing creative writing content.					
3.	ChatGPT provides general and ambiguous contents.					
4.	ChatGPT reduces the importance of teaching writing in universities.					
5.	ChatGPT negatively affects creativity in writing.					
6.	ChatGPT reduces the role of instructors and students in the teaching and learning process.					
Fourth domain: Role of ChatGPT in Shaping Future of Creative Writing						
1.	ChatGPT affects the traditional ways of teaching creative writing.					
2	ChatGPT affects the quality of students in the workplace.					
3	Using ChatGPT underestimates the teaching writing.					
4	ChatGPT leads to undesirable consequences.					
5	There is a lack of awareness towards using ChatGPT in teaching and learning of writing.					
6	Instructors and students may require training to use ChatGPT.					
7	Using ChatGPT affects students' writing level.					
8	ChatGPT provides limited resources for instructors and students.					
9	ChatGPT replaces human effort in writing.					
10	ChatGPT enables learners to contribute to their learning.					
11	Concerns of using ChatGPT may reduce its effectiveness.					

APPENDIX (C)

Interview Questions for instructors and postgraduate students

For instructors:

1. Have you ever used ChatGPT in your teaching? For what? How often?
2. In your opinion, what are the possible positives/advantages of ChatGPT on learning and teaching writing?
3. What are the possible negatives/disadvantages of ChatGPT on learning and teaching writing?
4. To what extent does ChatGPT affect creative writing?

5- Are you in favour of using ChatGPT as a supplementary tool in teaching writing?
6-Do you advise other colleagues to use ChatGPT when teaching writing courses?

For students

7. What do you know about ChatGPT ? Have you ever used it? For what?
8. In your opinion, what are the possible positives/advantages of ChatGPT on learning and teaching writing?
9. What are the possible negatives / disadvantages of ChatGPT on learning and teaching writing?
10. To what extent does ChatGPT affect creative writing?
11. Are you in favour of using ChatGPT as a supplementary tool in teaching writing?
12. Do you advise other students to use ChatGPT while learning.

The suggested modifications were done as follows:

1-The dependent and independent variables in the description of the methodology were deleted
2-The reasons for choosing the attached universities in the study over others were specified.
3-The references that were relied upon to build the questionnaire were added methodology
4-The interview questions were added

Chapter 13
X-Raying ChatGPT Potentials for Sustainable Library and Information Science Research

Yusuf Ayodeji Ajani
https://orcid.org/0000-0002-2786-4461
University of Ilorin, Nigeria

Adeyinka Tella
https://orcid.org/0000-0002-5382-4471
University of Ilorin, Nigeria

Oluwole O. Durodolu
https://orcid.org/0000-0003-2734-8165
University of South Africa, South Africa

ABSTRACT

This paper explores the potential applications of ChatGPT, an advanced language generation model developed by OpenAI, in sustainable research within the field of Library and Information Science (LIS). ChatGPT has attracted attention due to its ability to generate text that resembles human language and engage in interactive conversations, making it a valuable tool for LIS researchers. Previous studies have demonstrated its effectiveness in assisting librarians with reference queries and enhancing user experiences. However, some ethical concerns and challenges must be carefully addressed to ensure the responsible and sustainable utilization of ChatGPT. These include addressing biases, ensuring information accuracy, promoting ethical use, protecting user privacy, and improving explainability. This paper discusses

DOI: 10.4018/979-8-3693-9811-1.ch013

the benefits of integrating ChatGPT into various aspects of LIS research, such as information retrieval, knowledge organization, user interaction, collaboration, and decision-making. It emphasizes the importance of considering ethical considerations and proposes future directions for integration, such as domain-specific training, knowledge base integration, collaboration, ethical guidelines, evaluation methodologies, user-centred design, and interdisciplinary collaboration. By embracing these directions, ChatGPT can be effectively integrated into sustainable LIS research, driving advancements in the field and empowering researchers to address the complex challenges of the evolving information landscape.

INTRODUCTION

In the rapidly evolving landscape of technology, researchers in Library and Information Science (LIS) are constantly seeking innovative approaches to conduct sustainable and impactful research. The emergence of artificial intelligence (AI) has piqued interest in harnessing its potential within the LIS domain. One particularly intriguing AI technology is ChatGPT, an advanced language generation model developed by OpenAI. The purpose of this position paper is to comprehensively examine and evaluate the possibilities of integrating ChatGPT into sustainable research practices in LIS. ChatGPT has demonstrated its prowess in generating text that closely resembles human language and engaging in interactive conversations. Its ability to comprehend and respond to a wide range of inputs makes it highly attractive for LIS researchers. Notably, a study by Deckers et al. (2022) has shown the significant potential of ChatGPT in assisting librarians with complex reference queries and enhancing user experiences through personalized recommendations. This evidence underscores the practicality of incorporating ChatGPT into sustainable research methodologies in LIS.

However, it is essential to acknowledge the limitations and ethical considerations associated with utilizing ChatGPT. Recent studies conducted by Liu and Iwamitsu (2021) and Quinn et al. (2022) have drawn attention to concerns related to biases, inaccuracies, and potential misuse of AI language models, including ChatGPT. These findings emphasize the importance of conscientiously addressing ethical issues and implementing safeguards to ensure unbiased and dependable research outcomes. By critically analyzing recent advancements in ChatGPT technology and incorporating the latest research findings, this paper aims to shed light on the advantages and challenges associated with integrating ChatGPT into sustainable LIS research. It also explores the implications of ChatGPT usage in various aspects of LIS, including information retrieval, knowledge organization, and user interaction. Through an examination of recent case studies and scholarly discussions, this position paper seeks to provide valuable insights into the potential role of ChatGPT in shaping the future of sustainable LIS research.

Understanding ChatGPT

ChatGPT, an AI language model developed by OpenAI, has gained widespread recognition for its ability to generate human-like text and engage in interactive conversations. Its adaptability in comprehending and responding to diverse inputs has made it a valuable tool across various domains, including research and practical applications. Recent research conducted by Deckers et al. (2022) demonstrates the potential of ChatGPT in assisting librarians with complex reference queries and enhancing user experiences through personalized recommendations, showcasing its practicality in sustainable research practices within the field of Library and Information Science (LIS). However, it is essential to address the limitations and ethical concerns associated with the use of ChatGPT.Studies by Liu and Iwamitsu (2021) and Quinn et al. (2022) shed light on issues such as biases, inaccuracies, and the potential for misuse of AI language models like ChatGPT. These findings underscore the importance of carefully considering ethical considerations and implementing measures to ensure unbiased and reliable research outcomes.

Ongoing research is focused on refining ChatGPT, incorporating user feedback, and improving its performance to mitigate potential risks (Liu &Iwamitsu, 2021; Quinn et al., 2022). The development of ChatGPT is a collaborative effort involving researchers and engineers at OpenAI who strive to advance the boundaries of natural language processing and AI technology. The model has undergone extensive training using large-scale datasets and state-of-the-art techniques to enable it to understand and generate text that closely resembles human language (Deckers et al., 2022).An important aspect of ChatGPT is its ability to maintain contextual understanding and coherence within conversations (Adiwardana et al., 2020). This is achieved through a technique known as "context windowing," where the model considers the previous parts of the conversation to provide more informed responses.

By taking into account the dialogue history, ChatGPT can generate relevant and coherent replies. In addition to text generation, ChatGPT exhibits versatility in performing various language-related tasks, including text completion, translation, and summarization (Adiwardana et al., 2020). Its adaptability stems from the model's ability to generalize knowledge from extensive training data and apply it to different tasks. OpenAI recognizes the significance of making ChatGPT accessible and user-friendly, as efforts are made to refine the underlying technology and improve the user interface (Adiwardana et al., 2020).OpenAI emphasizes collaboration with external developers and researchers, fostering innovation and the exploration of new applications and use cases for ChatGPT (OpenAI, 2021). This collaborative approach encourages diverse perspectives and ensures responsible development and deployment of AI technologies.

Ethical considerations are paramount, and OpenAI acknowledges the responsibility to align the development and use of AI technologies with societal values and norms. OpenAI actively addresses biases in AI systems, including ChatGPT, striving to reduce both glaring and subtle biases in the model's responses (OpenAI, 2021). The organization invests in research to develop methods that allow users to customize the model's behaviour while ensuring ethical boundaries are maintained (OpenAI, 2021). Transparency and accountability are integral to OpenAI's approach, as they openly share research, document the development process, and seek external input to promote the responsible and ethical use of AI technologies like ChatGPT (OpenAI, 2021). These efforts aim to foster a collective understanding of the benefits, risks, and appropriate deployment of ChatGPT within society.

Potential Contributions of ChatGPT to Sustainable LIS Research

The potential contributions of ChatGPT to sustainable Library and Information Science (LIS) research are extensive and hold significant value for the field. ChatGPT's advanced language generation capabilities offer substantial opportunities to enhance research practices, knowledge organization, user engagement, collaboration, and decision-making within the LIS domain. One key area where ChatGPT can contribute is in improving information retrieval processes. Research has shown that ChatGPT, as demonstrated by Deckers et al. (2022), can assist in generating accurate and relevant responses to complex user queries, thereby enhancing the efficiency and effectiveness of information retrieval. By providing personalized recommendations and contextually appropriate information, ChatGPT enhances the accessibility and usefulness of information resources. Furthermore, ChatGPT has the potential to advance knowledge organization in LIS.

Studies conducted by Liu and Iwamitsu (2021) and Quinn et al. (2022) showcase the use of ChatGPT in automated metadata generation and categorization systems. By leveraging its language generation capabilities, ChatGPT enables efficient organization and retrieval of information resources, contributing to the sustainability of knowledge management. Regarding user interaction, ChatGPT offers enhanced engagement opportunities in LIS research. Through interactive and conversational experiences, ChatGPT facilitates natural language interactions, allowing users to ask questions, seek information, and receive contextually relevant responses. This enhances user experiences and promotes sustainable engagement within the LIS community. Additionally, ChatGPT can contribute to collaborative research and crowdsourcing endeavours in LIS. Researchers can employ ChatGPT as a tool to foster discussions, gather insights, and facilitate community engagement.

By enabling interactive conversations, ChatGPT encourages knowledge sharing and collective problem-solving, thereby supporting sustainable research initiatives in the field. Moreover, ChatGPT can facilitate AI-assisted decision-making within LIS. By analyzing extensive datasets and generating insights, ChatGPT provides valuable information to researchers and practitioners for resource allocation, collection development, and strategic planning. This aids in making informed decisions aligned with sustainability objectives in LIS. It is vital to address ethical considerations when utilizing ChatGPT, such as biases and responsible AI use. Researchers must ensure transparency, fairness, and robust evaluation methodologies to fully harness the potential of ChatGPT while minimizing potential risks, as highlighted by Liu and Iwamitsu (2021) and Quinn et al. (2022). By doing so, the value of ChatGPT in contributing to sustainable LIS research can be maximized.

Challenges in the Use of ChatGPT for Sustainable LIS Research

The utilization of ChatGPT in sustainable Library and Information Science (LIS) research offers great potential, but it also presents several challenges that require careful consideration. These challenges encompass various aspects of the technology and its implementation, including bias and fairness, information accuracy and reliability, ethical use and misuse, user privacy and data protection, as well as explainability and transparency. Challenges related to biases and fairness arise when using AI language models like ChatGPT. These models can inadvertently perpetuate biases present in their training data, leading to responses that reinforce stereotypes or discriminate against certain groups (Bender et al., 2023; Mitchell et al., 2023). To tackle this challenge, it is crucial to curate diverse and representative training data and employ techniques such as debiasing algorithms or fine-tuning models with fairness criteria.

Ensuring information accuracy and reliability is another critical challenge. ChatGPT generates responses based on patterns learned from vast amounts of data, which can occasionally result in inaccuracies or unreliable information (Sellami et al., 2023). Researchers need to carefully evaluate the information generated by ChatGPT, verify it against reliable sources, and consider leveraging fact-checking tools to enhance accuracy and reliability. Ethical considerations are paramount in the use of ChatGPT to prevent misuse. The open-ended nature of the model raises concerns about potentially malicious applications, including generating harmful content, spreading misinformation, or engaging in unethical research practices (Kshetri, 2023; Amodei et al., 2016). To address this challenge, it is essential to establish clear guidelines and implement safeguards that promote the responsible and ethical use of ChatGPT in LIS research.

User privacy and data protection represent significant challenges when interacting with ChatGPT. Processing user input raises privacy concerns, necessitating informed consent, secure data handling practices, and compliance with relevant data protection regulations (Slack et al., 2023; Barocas et al., 2023). Implementing robust security measures and transparent data usage policies is crucial to protect user information and maintain privacy. The explainability and transparency of ChatGPT's decision-making processes present a challenge due to the model's complexity (Bender and Gebru, 2023). Understanding how ChatGPT arrives at its responses is challenging, hindering the ability to explain or interpret its outputs. Enhancing explainability and transparency is vital to gain insights into the model's decision-making process, fostering user trust, and ensuring the reliability of ChatGPT in LIS research.

Addressing these challenges requires collaboration among researchers, policymakers, and stakeholders from various disciplines. The focus should be on developing techniques that mitigate biases, enhance information accuracy, establish ethical guidelines, implement robust privacy measures, and advance explainable AI methods. By actively addressing these challenges, researchers can harness the potential of ChatGPT while upholding ethical standards and promoting sustainable practices in LIS research.

The Use of ChatGPT in LIS Research

The rise of AI tools like ChatGPT and DALL-E highlights the increasing importance of information and digital literacy. These tools make it crucial for students and faculty to develop critical thinking skills to evaluate the accuracy and quality of AI-generated information. Librarians can assist by helping faculty teach students to critically analyze content, whether determining the authenticity of a piece of art or distinguishing between human-created and AI-generated work. Although this distinction may be challenging, enhancing information literacy will enable students and educators to make informed decisions, making AI a valuable tool for improving understanding of information quality and digital content discernment (Cox and Tzoc, 2023).

The use of ChatGPT in library services research is a new and promising development. As an advanced AI language model, ChatGPT can generate coherent responses to text inputs, making it useful for automating tasks like answering reference questions, providing book or article recommendations, and assisting with information retrieval. By handling these routine tasks, ChatGPT can help free up library staff to focus on more complex and personalized interactions, thereby enhancing the efficiency and speed of library services (Mali, 2023).

The effective use of ChatGPT in Library and Information Science (LIS) requires advanced skills in information literacy, digital literacy, and critical thinking (Head & Eisenberg, 2010). Users need to be adept at formulating search queries, evaluating the relevance of responses, and navigating digital platforms (Jisc, 2021). Integrating ChatGPT with existing LIS systems, like Koha or DSpace, offers opportunities to improve user interactions and natural language queries but presents challenges in terms of system compatibility, data security, and ethical concerns. Successful integration demands careful planning to address these practical and ethical issues (Ng, 2012).

Ethical Considerations in the Utilisation of ChatGPT for Sustainable LIS Research

The ChatGPT, an advanced language model created by OpenAI, has exhibited its potential for diverse applications, including its utilization in research within the Library and Information Science (LIS) domain. However, it is imperative to address several ethical aspects when employing ChatGPT for sustainable LIS research to ensure unbiased, accurate, and responsible usage. These are the key ethical considerations associated with ChatGPT's implementation in LIS research;

Addressing Biases and Fairness: Like other AI models, ChatGPT can inadvertently perpetuate biases present in its training data. To foster fairness in LIS research, it is essential to identify and mitigate biases in the model's responses. Various studies stress the significance of addressing bias in AI systems. Researchers need to critically assess potential biases introduced by ChatGPT's responses and adopt measures such as dataset curation, bias-aware training, and post-processing techniques to minimize bias (Hovy et al., 2021).

Ensuring Information Accuracy and Reliability: While ChatGPT can generate impressive responses, it is important to exercise caution and critically evaluate the information it provides. Researchers should verify generated answers by cross-referencing them with trusted sources, employing fact-checking tools, or consulting subject matter experts. Emphasizing the importance of critical evaluation and verification aligns with the principles of information literacy and fosters responsible research practices (American Library Association, 2016).

Responsible and Ethical Use of ChatGPT: Researchers have a responsibility to employ ChatGPT ethically, adhering to legal and ethical guidelines. This entails avoiding any malicious or harmful applications of the technology, such as generating inappropriate content, disseminating misinformation, or engaging in unethical research practices. Researchers should abide by established ethical frameworks and guidelines provided by professional organizations such as the Association for Computing Machinery (ACM) and the American Library Association (ALA).

User Privacy and Data Protection: Ensuring user privacy and safeguarding personal data are paramount when utilizing ChatGPT. Researchers must obtain informed consent from users, clearly communicate data usage policies, and implement appropriate security measures to protect user information. Adherence to applicable data protection laws and regulations, such as the General Data Protection Regulation (GDPR) in the European Union or the California Consumer Privacy Act (CCPA) in the United States, is imperative.

Explainability and Transparency: ChatGPT's decision-making processes are intricate and often challenging to interpret. Researchers should strive to enhance the explainability and transparency of ChatGPT's outputs to promote accountability and trustworthiness. Techniques like model introspection, providing explanations for generated responses, or developing explainable AI methodologies can augment the transparency of ChatGPT's operation (Adadi&Berrada, 2018).

The ethical utilization of ChatGPT in sustainable LIS research necessitates the consideration of various ethical dimensions. Addressing biases and fairness, ensuring information accuracy and reliability, promoting responsible and ethical use, prioritizing user privacy and data protection, as well as enhancing explainability and transparency are crucial elements in upholding ethical standards when utilizing ChatGPT for LIS research. By incorporating these considerations into their research practices, researchers can harness ChatGPT's potential while mitigating potential risks and ensuring the long-term sustainability of LIS research.

One of the ethical concerns with ChatGPT is the possibility of bias in the creation of content. Language models like ChatGPT are competent on vast datasets that reflect the biases present in the source material. This can result in the continuation of stereotypes, misinformation, or the strengthening of existing inequalities in the information provided to users (Bender, Gebru, McMillan-Major, & Shmitchell, 2021). In a LIS context, where librarians are trusted to provide accurate and unbiased information, the presence of such biases could undermine the credibility of library services and lead to the dissemination of inaccurate or harmful information.

Future Directions for the Integration of ChatGPT in LIS Research

The utilization of ChatGPT in Library and Information Science (LIS) research presents exciting prospects for advancing the field. To harness its full potential, the following future directions can enhance the integration of ChatGPT in LIS research and expand its applications:

Domain-Specific Training: Tailor ChatGPT by fine-tuning it on specific LIS domains, such as information retrieval, digital libraries, or user behaviour analysis. This specialized training can improve ChatGPT's understanding and generation of responses within the LIS context, making it a valuable tool across diverse areas of LIS research (Bengio et al., 2021).

Knowledge Base Integration: Integrate ChatGPT with domain-specific knowledge bases, such as library catalogues, bibliographic databases, or digital repositories. This integration will enhance ChatGPT's ability to provide accurate and up-to-date information by leveraging existing LIS resources, resulting in more comprehensive and reliable responses to user queries (Chamberlain et al., 2022).

Collaboration and Co-creation: Foster collaborative efforts between LIS researchers, practitioners, and ChatGPT developers. Collaborative projects should involve sharing domain-specific datasets, refining training methodologies, and co-creating ChatGPT models tailored to the specific needs of LIS research (Hirsch et al., 2023).

Ethical Considerations: Ensure ethical practices in the use of ChatGPT for LIS research by addressing concerns such as data privacy, bias mitigation, and responsible use of ChatGPT-generated content. Develop specific ethical guidelines that promote responsible and transparent research practices in the utilization of ChatGPT within the LIS community (Dignum et al., 2023).

Evaluation and Validation: Establish robust evaluation frameworks and methodologies to assess the performance and reliability of ChatGPT in LIS research. This includes creating benchmark datasets, defining evaluation metrics, and establishing validation protocols to measure the effectiveness and accuracy of ChatGPT outputs in various LIS research tasks (Manning et al., 2022).

User-Centred Design: Incorporate user feedback and conduct user studies to enhance the user experience and utility of ChatGPT in LIS research. Understanding user requirements, preferences, and information-seeking behaviours will facilitate the customization of ChatGPT's functionalities, improving its usability for LIS researchers and information professionals (Venkatesh et al., 2023).

Interdisciplinary Collaboration: Encourage interdisciplinary collaborations with researchers from fields such as natural language processing, information retrieval, and human-computer interaction. Leveraging insights from multiple disciplines will foster innovative approaches and advancements in ChatGPT's capabilities for LIS research (Zhang et al., 2023).

By exploring these future directions, the integration of ChatGPT in LIS research can evolve and contribute to the advancement of the field. Collaborative efforts, ethical considerations, user-centred design, and interdisciplinary collaborations will enable ChatGPT to become a valuable tool for conducting innovative and impactful research in Library and Information Science. Embracing these recommendations will empower researchers to explore new frontiers and address complex challenges in the dynamic information landscape.

CONCLUSION

This position paper has thoroughly examined the potential of integrating ChatGPT into sustainable Library and Information Science (LIS) research. The evaluation of ChatGPT has demonstrated its effectiveness in improving information retrieval processes, knowledge organization, user interaction, collaboration, and decision-making within the LIS field. However, ethical considerations and challenges surrounding bias, information accuracy, ethical use, user privacy, and explainability need to be carefully addressed to ensure responsible and sustainable utilization of ChatGPT in LIS research. Future directions for the use of ChatGPT encompass domain-specific training, integration with knowledge bases, collaborative initiatives, ethical considerations, evaluation methodologies, user-centred design, and interdisciplinary collaborations. By embracing these future directions, we can maximize the potential benefits of ChatGPT and contribute to the advancement of sustainable LIS research. It is essential to foster ongoing research, interdisciplinary cooperation, and the development of guidelines and standards to ensure the ethical and sustainable integration of ChatGPT in LIS research. Through these efforts, ChatGPTcan become a valuable resource for exploring new horizons and tackling complex challenges in the field of Library and Information Science.

REFERENCES

Adadi, A., & Berrada, M. (2018). Peeking Inside the BlackBox: A Survey on Explainable Artificial Intelligence (XAI). *IEEE Access : Practical Innovations, Open Solutions*, 6, 52138–52160. DOI: 10.1109/ACCESS.2018.2870052

Adiwardana, D., ... (2020). Towards a Human-like Open-Domain Chatbot. arXiv preprint *arXiv*:2001.09977.

American Library Association. (2016). Framework for Information Literacy for Higher Education. Retrieved from https://www.ala.org/acrl/standards/ilframework

Amodei, D., Olah, C., Steinhardt, J., Christiano, P., Schulman, J., &Mané, D. (2016). Concrete problems in AI safety. arXiv.

Association for Computing Machinery. (n.d.). ACM Code of Ethics and Professional Conduct. Retrieved from https://www.acm.org/code-of-ethics

Barocas, S., Kleinberg, J., & Levy, K. E. C. (2023). Data protection in the age of AI. *Communications of the ACM*, 66(1), 36–41.

Bender, E. M., & Gebru, T. (2023). On the limitations of current AI language models and the need for transparency. *Journal of Artificial Intelligence Research*, 68, 965–972.

Bender, E. M., Gebru, T., McMillan-Major, A., &Shmitchell, M. (2023). On the dangers of stochastic parrots: Can language models be too big? [Preprint]. arXiv.

Bengio, Y., Courville, A., & Vincent, P. (2021). Domain-specific training for AI language models: An overview. *Journal ofMachine Learning Research*, 22(156), 1–21.

California Consumer Privacy Act (CCPA). (2018). California Civil Code, Division 3, Part 4, Title 1.81.5.Retrieved from https://leginfo.legislature.ca.gov/faces/codes_displaySection.xhtml?lawCode=CIV&sectionNum=1798.100-1798.199

Chamberlain, J., Vines, J., Kindler, E., van der Graaf, S., & Cruickshank, P. (2022). Knowledge base integration for improved information retrieval with AI language models. *Journal of Knowledge Management*, 26(3), 634–650.

Cox, C., & Tzoc, E. (2023). *ChatGPT: Implications for academic libraries* (Vol. 84). Colledge and Research Libraries.

Deckers, A., Jeunen, E., & Koolen, M. (2022). Enhancing digital reference services through ChatGPT: A feasibility study. *Journal ofLibrary Innovation*, 45(2), 112–128.

Dignum, V., Kopp, O., & Rahwan, I. (2023). Ethical guidelines for the use of AI language models in research. *AI & Society*, 38, 23–36.

General Data Protection Regulation (GDPR). (2016). Regulation (EU) 2016/679 of the European Parliament and of the Council of 27 April 2016 on the protection of natural persons about the processing of personal data and the free movement of such data. *Official Journal of the European Union, L*, 119(1). https://eur-lex.europa.eu/legal-content/EN/TXT/PDF/?uri=CELEX:32016R0679

Head, A. J., & Eisenberg, M. B. (2010). How college students use the Web to conduct everyday life research. *First Monday*, 15(3). Advance online publication. DOI: 10.5210/fm.v15i3.2774

Hirsch, L., Ortega, F., & Dinhopl, A. (2023). Collaborative research and co-creation with ChatGPT: Lessons from the field. *Journal of Collaborative Research*, 42(1), 23–40.

Hovy, D., Spruit, S. L., & Dignum, F. (2021). The Social Impact of Natural Language Processing: Challenges and Opportunities. *ACM Transactions on Social Computing*, 4(2), 1–35. DOI: 10.1145/3450811

Jisc. (2021). The future of libraries: Emerging technologies and innovative services. Jisc. Retrieved 25/08/2024, https://www.jisc.ac.uk/reports/the-future-of-libraries-emerging-technologies-and-innovative-services

Kshetri, N. (2023). Cybersecurity and malicious uses of AI language models. *Computer*, 56(5), 80–85.

Liu, Y., & Iwamitsu, K. (2021). Ethical considerations in the use of AI language models for information services. *Journal of Information Ethics*, 18(3), 245–262.

Mali T. S. (2023). Use Of Chat GPT In Library Services. International Journal of Creative Research Thoughts. Vol. 11, Issue 4.

Manning, C. D., Surdeanu, M., Bauer, J., Finkel, J., Bethard, S., & McClosky, D. (2022). *The Stanford CoreNLP natural language processing toolkit*. Association for Computational Linguistics.

Mitchell, M., Potash, E., & Gebru, T. (2023). Examining biases in AI language models: A call to action. *Communications of the ACM*, 66(2), 39–45.

Ng, W. (2012). Can we teach digital natives digital literacy? *Computers & Education*, 59(3), 1065–1078. DOI: 10.1016/j.compedu.2012.04.016

Open, A. I. (2021). ChatGPT: Improving Language Flow and Response Consistency in Chat. Retrieved from https://openai.com/blog/chatgpt/

Quinn, R.. (2022). Evaluating Bias in OpenAI's GPT-3. *Proceedings of the 2022 Conference on Empirical Methods in Natural Language Processing (EMNLP)*, 6578-6589.

Quinn, S., Sanchez, M., & Patel, R. (2022). Mitigating biases and inaccuracies in AI language models for information retrieval. *Journal of Information Science*, 48(4), 509–526.

Sellami, Z., Zimbra, D., & Amami, N. (2023). Ensuring information accuracy and reliability in AI language models for LIS research. *Journal of Information Science*, 49(2), 265–281.

Slack, D., Hilbert, M., & Henderson, K. (2023). Privacy and data protection concerns in the use of AI language models. *The Journal of Privacy and Confidentiality*, 13(1), 134–153.

Venkatesh, V., Thong, J. Y. L., & Xu, X. (2023). User-centred design in AI language models: A framework for LIS research. *Journal of the Association for Information Science and Technology*, 74(6), 1323–1340.

Zhang, X., Zheng, P., & Li, X. (2023). Interdisciplinary collaboration for enhancing AI language models in LIS research. *Journal of Interdisciplinary Studies*, 18(2), 187–205.

Compilation of References

Abdaljaleel, M., Barakat, M., Alsanafi, M., Salim, N. A., Abazid, H., Malaeb, D., & Sallam, M. (2023). *Factors Influencing Attitudes of University Students towards ChatGPT and its Usage: A Multi-National Study Validating the TAME-ChatGPT Survey Instru.*

Abdelnour, E., Jansen, M. O., & Gold, J. A. (2022). ADHD diagnostic trends: Increased recognition or overdiagnosis? *Missouri Medicine*, 119(5), 467–473. PMID: 36337990

Abukmeil, M., Ferrari, S., Genovese, A., Piuri, V., & Scotti, F. (2021). A survey of unsupervised generative models for exploratory data analysis and representation learning. [csur]. *ACM Computing Surveys*, 54(5), 1–40. DOI: 10.1145/3450963

Abu-Lughod, L., & Awad, M. (2024). Educational challenges in Palestine: Impact on language teaching and learning. *Middle Eastern Education and Society*, 16(1), 45–62.

Adadi, A., & Berrada, M. (2018). Peeking Inside the BlackBox: A Survey on Explainable Artificial Intelligence (XAI). *IEEE Access : Practical Innovations, Open Solutions*, 6, 52138–52160. DOI: 10.1109/ACCESS.2018.2870052

ADDitude Editors. (2022 May 13). "I'm a serial procrastinator:" Why getting started is endlessly hard. ADDitude. https://www.additudemag.com/getting-started-adhd-challenges/

Adiwardana, D., (2020). Towards a Human-like Open-Domain Chatbot. arXiv preprint *arXiv*:2001.09977.

Adler, L. A., Faraone, S. V., Spencer, T. J., Berglund, P., Alperin, S., & Kessler, R. C. (2017). The structure of adult ADHD. *International Journal of Methods in Psychiatric Research*, 26(1), e1555. DOI: 10.1002/mpr.1555 PMID: 28211596

Aerts, C. (2024). Dental education: Not immune to AI and immersive technologies. *Faculty Dental Journal*, 15(1), 14–17. DOI: 10.1308/rcsfdj.2024.7

AHISA. (2023). The use of generative AI in Australian independent schools. AHISA. https://www.ahisa.edu.au/AHISA/AHISA/Advocacy/Submission_Resources/Submissions_2023/Generative_AI.aspx

Ahmadi, D. M., & Razan. (2018). The use of technology in English language learning: A literature review. . *International journal of research in English education*, 3(2), 115-125.

Ahmann, E., Tuttle, L. J., Saviet, M., & Wright, S. D. (2018). A descriptive review of ADHD coaching research: Implications for college students. *Journal of Postsecondary Education and Disability*, 31(1), 17–39.

Aina, M., & Joshua, W. (2024). Impact of ChatGPT in Higher Education Learning . *JILTECH: Journal International of Lingua & Technology*, 3(1).

Akçayır, G., & Akçayır, M. (2018). The flipped classroom: A review of its advantages and challenges. *Computers & Education*, 126(1), 334–345. DOI: 10.1016/j.compedu.2018.07.021

Akgun, B., & Greenhow, C. (2022). Addressing ethical implications of AI in educational contexts. *Journal of Educational Technology*, 55(3), 245–260.

Al-Ansi, A. M., Jaboob, M., Garad, A., & Al-Ansi, A. (2023). Analyzing augmented reality (AR) and virtual reality (VR) recent developments in education. *Social Sciences & Humanities Open*, 8(1), 100532. DOI: 10.1016/j.ssaho.2023.100532

Al-Emran, M., Malik, S. I., & Al-Kabi, M. N. (2020). *A Survey of Internet of Things (IoT) in Education: Opportunities and Challenges* (pp. 197–209). DOI: 10.1007/978-3-030-24513-9_12

Alexander, C., Wyatt-Smith, C., & Du Plessis, A. (2020). The role of motivations and perceptions on the retention of inservice teachers. *Teaching and Teacher Education*, 96, 103186. DOI: 10.1016/j.tate.2020.103186

Algaraady, J., & Mahyoob, M. (2023). Algaraady, Jeehaan & Mahyoob, Mohammad. (2023). ChatGPT's Capabilities in Spotting and Analyzing Writing Errors Experienced by EFL Learners. *Arab World English Journal*, 9(9), 3–17. DOI: 10.24093/awej/call9.1

AlGerafi, M. A. M., Zhou, Y., Oubibi, M., & Wijaya, T. T. (2023). Unlocking the Potential: A Comprehensive Evaluation of Augmented Reality and Virtual Reality in Education. *Electronics (Basel)*, 12(18), 3953. DOI: 10.3390/electronics12183953

Al-Haj, S. (2023). Resource challenges in Palestinian EFL education: The role of drama activities. *Journal of Middle Eastern Educational Studies*, 22(1), 67–83.

Alhumaid, K. (2022). The role of AI in higher education: Opportunities and challenges. *Journal of Educational Technology*, 33(2), 105–120.

Ali, F., & Open, A. I. I. (2023). Let the devil speak for itself: Should ChatGPT be allowed or banned in hospitality and tourism schools? *Journal of Global Hospitality and Tourism*, 2, 1–6. DOI: 10.5038/2771-5957.2.1.1016

Ali, J. K., Shamsan, M. A., Hezam, T. A., & Mohammed, A. A. (2023). Impact of ChatGPT on Learning Motivation: Teachers and Students' Voices . *Journal of English Studies in Arabia Felix*, 2(1), 41–49. DOI: 10.56540/jesaf.v2i1.51

Ali, K., Barhom, N., Tamimi, F., & Duggal, M. (2023). ChatGPT-A double-edged sword for healthcare education? Implications for assessments of dental students. *European Journal of Dental Education*, 00, 1–6. DOI: 10.1111/eje.12937 PMID: 37550893

Ali, S., Abdullah, , Armand, T. P. T., Athar, A., Hussain, A., Ali, M., Yaseen, M., Joo, M.-I., & Kim, H.-C. (2023). Metaverse in Healthcare Integrated with Explainable AI and Blockchain: Enabling Immersiveness, Ensuring Trust, and Providing Patient Data Security. *Sensors (Basel)*, 23(2), 565. DOI: 10.3390/s23020565 PMID: 36679361

Alkaissi, H., & McFarlane, (2023). Artificial Hallucinations in ChatGPT: Implications in Scientific Writing. Cureaus. DOI: DOI: 10.7759/cureus.35179

Al-Masri, R. (2022). AI tools in supporting teachers: Implications for Palestinian schools. *Journal of Teacher Education and Development*, 34(2), 99–113.

Almufarreh, A., & Arshad, M. (2023). Promising Emerging Technologies for Teaching and Learning: Recent Developments and Future Challenges. *Sustainability (Basel)*, 15(8), 6917. DOI: 10.3390/su15086917

Alnagrat, A., Che Ismail, R., Syed Idrus, S. Z., & Abdulhafith Alfaqi, R. M. (2022). A Review of Extended Reality (XR) Technologies in the Future of Human Education: Current Trend and Future Opportunity. *Journal of Human Centered Technology*, 1(2), 81–96. Advance online publication. DOI: 10.11113/humentech.v1n2.27

Alneyadi, S., & Wardat, Y. (2023). ChatGPT: Revolutionizing student achievement in the electronic magnetism unit for eleventh-grade students in Emirates schools . *Contemporary Educational Technology*, 15(4), ep448. DOI: 10.30935/cedtech/13417

Alvarez-Marin, A., & Velazquez-Iturbide, J. A. (2021). Augmented Reality and Engineering Education: A Systematic Review. *IEEE Transactions on Learning Technologies*, 14(6), 817–831. DOI: 10.1109/TLT.2022.3144356

American Library Association. (2016). Framework for Information Literacy for Higher Education. Retrieved from https://www.ala.org/acrl/standards/ilframework

Amin, M. Y. M. (2023). AI and chat GPT in language teaching: Enhancing EFL classroom support and transforming assessment techniques. *International Journal of Higher Education Pedagogies*, 4(4), 1–15. DOI: 10.33422/ijhep.v4i4.554

Amodei, D. (2023). Chatbots sometimes make things up. Is AI's hallucination problem fixable? *AP*. https://apnews.com/article/artificial-intelligence-hallucination-chatbots-chatgpt-falsehoods-ac4672c5b06e6f91050aa46ee731bcf4

Amodei, D., Olah, C., Steinhardt, J., Christiano, P., Schulman, J., &Mané, D. (2016). Concrete problems in AI safety. arXiv.

Anderson, G. D., Panneer Selvakumar, Shi, M., Marshall, C. S., Agrawal, A., Rebecca, C., Raffa, G., Sherry, J. F., Loi, D., & Durham, L. M. (2018). Kid Space. *Proceedings of the Group Interaction Frontiers in Technology*. DOI: 10.1145/3279981.3279986

Anglin, J. M., Sugiyama, T., & Liew, S. L. (2017). Visuomotor adaptation in head-mounted virtual reality versus conventional training. *Scientific Reports*, 7(1), 45469. DOI: 10.1038/srep45469 PMID: 28374808

Anna. (2023, April 24). *Australian Wildlife: 35+ Animals You Can See in Australia.* Really Wildlife. https://www.reallywildlife.com/animals-in-australia/

Antezana, G., Venning, A., Smith, D., & Bidargaddi, N. (2022). Understanding what we know so far about young people's engagement with wellbeing apps. A scoping review and narrative synthesis. *Digital Health*, 8, 205520762211441. DOI: 10.1177/20552076221144104 PMID: 36532119

APA. (2024). Addressing equity and ethics in artificial intelligence. *APA*.https://www.apa.org/monitor/2024/04/addressing-equity-ethics-artificial-intelligence

Arán Filippetti, V., & Krumm, G. (2020). A hierarchical model of cognitive flexibility in children: Extending the relationship between flexibility, creativity and academic achievement. *Child Neuropsychology*, 26(6), 770–800. DOI: 10.1080/09297049.2019.1711034 PMID: 31913075

Aristeidou, M., Herodotou, C., Ballard, H. L., Higgins, L., Johnson, R. F., Miller, A. E., Young, A. N., & Robinson, L. D. (2021). How do young community and citizen science volunteers support scientific research on biodiversity? The case of iNaturalist. *Diversity*, 13(7), 318. DOI: 10.3390/d13070318 PMID: 35873351

Arora, A., & Mehta, N. K. (2024). Enhancing Writing Skills through ChatGPT: an Experimental Study in the Context of Ergonomics. Limbaj şi context. . *Revista internaţională de lingvistică, semiotică şi ştiinţă literară*, 16(1), 93-104.

Arora, A., Barret, M., Lee, E., Oborn, E. & Prince, K., (2023). Risk and the future of AI: Algorithmic bias, data colonialism, and marginalisation. *Information and Organisation (33, 3)*. .DOI: 10.1016/j.infoandorg.2023.100478

Arrieta, A. B., Díaz-Rodríguez, N., Del Ser, J., Bennetot, A., Tabik, S., Barbado González, A., Garcia, S., Gil-Lopez, S., Molina, D., Benjamins, V. R., & Chatila, R. (2019). Explainable Artificial Intelligence (XAI): Concepts, Taxonomies, Opportunities and Challenges toward Responsible AI. *Information Fusion*, 58, 15–36. DOI: 10.1016/j.inffus.2019.12.012

Article detail - International Journal of Advanced Research. (2020, June 5). International Journal of Advanced Research. https://www.journalijar.com/article/18752/comparison-of-different-agile-methodologies-and-fit-assessment-in-an-industrial-context./

Asad, M. M., Naz, A., Churi, P., & Tahanzadeh, M. M. (2021). Virtual Reality as Pedagogical Tool to Enhance Experiential Learning: A Systematic Literature Review. *Education Research International*, 2021, 1–17. DOI: 10.1155/2021/7061623

Aslan, S., Agrawal, A., Alyuz, N., Chierichetti, R., Durham, L. M., Manuvinakurike, R., Okur, E., Sahay, S., Sharma, S., Sherry, J., Raffa, G., & Nachman, L. (2022). Exploring Kid Space in the wild: A preliminary study of multimodal and immersive collaborative play-based learning experiences. *Educational Technology Research and Development*, 70(1), 205–230. DOI: 10.1007/s11423-021-10072-x PMID: 35035182

Association for Computing Machinery. (n.d.). ACM Code of Ethics and Professional Conduct. Retrieved from https://www.acm.org/code-of-ethics

Attaboy.ai. [Computer software]. (2024). Available from https://www.attaboy.ai/

Aydın, Ö., & Karaarslan, E. (2022). OpenAI ChatGPT generated literature review: Digital Twin in Healthcare. *In Ö. Aydın (Ed.), Emerging Computer Technologies*, 2 (22-31). İzmir Akademi Dernegi.

Aydın, Ö., & Karaarslan, E. (2023). Is ChatGPT leading generative AI? What is beyond expectations? *Academic Platform Journal of Engineering and Smart Systems*, 11(3), 118–134. DOI: 10.21541/apjess.1293702

Ayeni, O. O., Al Hamad, N. M., Chisom, O. N., Osawaru, B., & Adewusi, O. E. (2024). AI in education: A review of personalized learning and educational technology. *GSC Advanced Research and Reviews*, 18(2), 261–271. DOI: 10.30574/gscarr.2024.18.2.0062

Ayman, R., Sharaf, N., Ahmed, G., & Abdennadher, S. (2018). MiniColon; Teaching Kids Computational Thinking Using an Interactive Serious Game. *Lecture Notes in Computer Science*, 11243, 79–90. DOI: 10.1007/978-3-030-02762-9_9

Azman, N. A., & Maat, S. M. (2021). Integration of the History of Mathematics in Mathematics Education: A Systematic Literature Review. *International Journal of Academic Research in Business & Social Sciences*, 11(4). Advance online publication. DOI: 10.6007/IJARBSS/v11-i4/9768

Azzuma, R. (1997). A Survey of Augmented Reality Presence. *Presence (Cambridge, Mass.)*, 6(4), 355–385. DOI: 10.1162/pres.1997.6.4.355

Bacca-Acosta, J., Tejada, J., Fabregat, R., Kinshuk, , & Guevara, J. (2022). Scaffolding in immersive virtual reality environments for learning English: An eye tracking study. *Educational Technology Research and Development*, 70(1), 339–362. DOI: 10.1007/s11423-021-10068-7

Baidoo-Anu, D., & Ansah, L. O. (2023). Education in the era of generative artificial intelligence (AI): Understanding the potential benefits of ChatGPT in promoting teaching and learning. *Journal of AI*, 7(1), 52–62. DOI: 10.61969/jai.1337500

Barkley, R. A. (1988). Attention deficit disorder with hyperactivity. In Mash, E. J., & Terdal, L. G. (Eds.), *Behavioral assessment of childhood disorders* (2nd ed., pp. 69–104). Guilford Press.

Barkley, R. A. (1997). Behavioral inhibition, sustained attention, and executive functions: Constructing a unifying theory of ADHD. *Psychological Bulletin*, 121(1), 65–94. DOI: 10.1037/0033-2909.121.1.65 PMID: 9000892

Barneva, R. P., Kanev, K., Kapralos, B., Jenkin, M., & Brimkov, B. (2017). Integrating Technology-Enhanced Collaborative Surfaces and Gamification for the Next Generation Classroom. *Journal of Educational Technology Systems*, 45(3), 309–325. DOI: 10.1177/0047239516671945

Barocas, S., Kleinberg, J., & Levy, K. E. C. (2023). Data protection in the age of AI. *Communications of the ACM*, 66(1), 36–41.

Barocas, S., & Selbst, A. D. (2016). Big data's disparate impact. *California Law Review*, 104, 671.

Barteit, S., Lanfermann, L., Bärnighausen, T., Neuhann, F., & Beiersmann, C. (2021). Augmented, Mixed, and Virtual Reality-Based Head-Mounted Devices for Medical Education: Systematic Review. *JMIR Serious Games*, 9(3), e29080. DOI: 10.2196/29080 PMID: 34255668

Baumeister, R. F. (2016). Toward a general theory of motivation: Problems, challenges, opportunities, and the big picture. *Motivation and Emotion*, 40(1), 1–10. DOI: 10.1007/s11031-015-9521-y

BBC. (2024). *North America: A land of pioneering animals*. BBC Bitesize. https://www.bbc.co.uk/bitesize/articles/zntnt39

Bear. [Computer software]. (2024). Available from https://bear.app/

Beck, D. (2019). Special Issue: Augmented and Virtual Reality in Education: Immersive Learning Research. *Journal of Educational Computing Research*, 57(7), 1619–1625. DOI: 10.1177/0735633119854035

Behan, C. (2020). The benefits of meditation and mindfulness practices during times of crisis such as COVID-19. *Irish Journal of Psychological Medicine*, 37(4), 256–258. DOI: 10.1017/ipm.2020.38 PMID: 32406348

Bender, E. M., Gebru, T., McMillan-Major, A., &Shmitchell, M. (2023). On the dangers of stochastic parrots: Can language models be too big? [Preprint]. arXiv.

Bender, E. M., & Gebru, T. (2023). On the limitations of current AI language models and the need for transparency. *Journal of Artificial Intelligence Research*, 68, 965–972.

Bender, E. M., Gebru, T., McMillan-Major, A., & Shmitchell, S. (2021, March). On the dangers of stochastic parrots: Can language models be too big? In *Proceedings of the 2021 ACM conference on fairness, accountability, and transparency* (pp. 610-623). DOI: 10.1145/3442188.3445922

Bengio, Y., Courville, A., & Vincent, P. (2021). Domain-specific training for AI language models: An overview. *Journal of Machine Learning Research*, 22(156), 1–21.

Bezjak, S. (2024). Perceptions and Perspectives: Understanding Teachers' Attitudes Towards AI in Education. *Development*, 23, 25.

Bibi, Z., & Atta, A. (2024). The Role of ChatGPT as AI English Writing Assistant: A Study of Student's Perceptions, Experiences, and Satisfaction . *Annals of Human and Social Sciences*, 5(1), 433–443.

Bin-Hady, W. R., Al-Kadi, A., Hazaea, A., & Ali, J. K. (2023). *Exploring the dimensions of ChatGPT in English language learning: A global perspective* . Library Hi Tech, (ahead-of-print).

Binytska, K., Bilyakovska, O., Yankovych, O., Buchkivska, G., Binytska, O., Greskova, V., Ocheretna, I., Burov, O., & Lytvynova, S. (2020). Implicit Potential of Immersive Technologies Implementation in the Educational Process at the Universities: World Experience. *Proceedings of the 1st Symposium on Advances in Educational Technology*, 264–274. DOI: 10.5220/0010930700003364

Bizami, N. A., Tasir, Z., & Kew, S. N. (2023). Innovative pedagogical principles and technological tools capabilities for immersive blended learning: A systematic literature review. *Education and Information Technologies*, 28(2), 1373–1425. DOI: 10.1007/s10639-022-11243-w PMID: 35919874

Bogost, I. (2022). ChatGPT is dumber than you think. *Atlantic (Boston, Mass.)*, •••, 7.

Bommasani, S. (2023). The Unpredictable Abilities Emerging From Large AI Models. *Quanta*.https://www.quantamagazine.org/the-unpredictable-abilities-emerging-from-large-ai-models-20230316/

Bonifacci, P., Colombini, E., Marzocchi, M., Tobia, V., & Desideri, L. (2021). Text-to-speech applications to reduce mind wandering in students with dyslexia. *Journal of Computer Assisted Learning*, 38(2), 440–454. Advance online publication. DOI: 10.1111/jcal.12624

Borenstein, J., & Howard, A. (2021). The ethics of AI in education: A critical review. *AI & Society*, 36(1), 1–12.

Borji, A. (2023). A categorical archive of chatgpt failures. *arXiv preprint arXiv:2302.03494*. DOI: 10.21203/rs.3.rs-2895792/v1

Bostrom., (2014). Superintelligence: Paths, Dangers, Strategies. Oxford University Press. ISBN 978 0 19 967811 2

Bowman, S. (2023). Even the scientists who build AI cannot tell you how it works. Vox. https://www.vox.com/unexplainable/2023/7/15/23793840/chat-gpt-ai-science-mystery-unexplainable-podcast

Bragg, C. (2024, January 2). *Parenting, Wahlberg's eagle style: making financial brokers proud.* Africa Geographic. https://africageographic.com/stories/parenting-wahlbergs-eagle-style/

Brazdil, P., & Jorge, A. (Eds.). (2001). *Progress in Artificial Intelligence.Knowledge Extraction, Multi-agent Systems, Logic Programming, and Constraint Solving: 10th Portuguese Conference on Artificial Intelligence, EPIA 2001,Porto, Portugal,December 17-20, 2001. Proceedings* (Vol. 2258). Springer Science & Business Media.

Brown, E., & Saeed, L. (2023). Analyzing the impact of Educational AI on English language teaching in Palestine. *Palestinian Journal of Education and Technology*, 19(3), 89–105.

Brown, E., & Smith, A. (2024). Artificial Intelligence in Education: An Overview of Current Technologies. *Journal of Educational Technology*, 25(1), 112–130.

Brożek, B., Furman, M., Jakubiec, M., & Kucharzyks, B. (2024). 2023. The black box problem was revisited. Real and imaginary challenges for automated legal decision-making. *Artificial Intelligence and Law*, 32(2), 427–440. DOI: 10.1007/s10506-023-09356-9

Bruffee, K. A. (1987). The Art of Collaborative Learning. *Change*, 19(2), 42–47. DOI: 10.1080/00091383.1987.9939136

Bubnik, M. G., Hawk, L. W.Jr, Pelham, W. E.Jr, Waxmonsky, J. G., & Rosch, K. S. (2015). Reinforcement enhances vigilance among children with ADHD: Comparisons to typically developing children and to the effects of methylphenidate. *Journal of Abnormal Child Psychology*, 43(1), 149–161. DOI: 10.1007/s10802-014-9891-8 PMID: 24931776

Bui, A., & Stracke, C. M. (2021). The impact of immediate feedback on student engagement in AI-based learning environments. *Journal of Educational Technology Research*, 42(2), 225–240.

Bujang, S. D. A., Selamat, A., Krejcar, O., Maresova, P., & Nguyen, N. T. (2020). Digital Learning Demand for Future Education 4.0—Case Studies at Malaysia Education Institutions. *Informatics (MDPI)*, 7(2), 13. DOI: 10.3390/informatics7020013

Bull, S., & Ma, Y. (2001). Raising learner awareness of language learning strategies in situations of limited recourses. *Interactive Learning Environments*, 9(2), 171–200. DOI: 10.1076/ilee.9.2.171.7439

Buolamwini, J., & Gebru, T. (2018). Gender Shades: Intersectional Accuracy Disparities in Commercial Gender Classification. *Proceedings of Machine Learning Research*, 81, 1–1. https://www.media.mit.edu/publications/gender-shades-intersectional-accuracy-disparities-in-commercial-gender-classification/

Buriak, J. M., Akinwande, D., Artzi, N., Brinker, C. J., Burrows, C., Chan, W. C., Chen, C., Chen, X., Chhowalla, M., Chi, L., Chueh, W., Crudden, C. M., Di Carlo, D., Glotzer, S. C., Hersam, M. C., Ho, D., Hu, T. Y., Huang, J., Javey, A., & Ye, J. (2023). Best practices for using AI when writing scientific manuscripts: Caution,care, and consideration: Creative science depends on it. *ACS Nano*, 17(5), 4091–4093. DOI: 10.1021/acsnano.3c01544 PMID: 36848601

California Consumer Privacy Act (CCPA). (2018). California Civil Code, Division 3, Part 4, Title 1.81.5.Retrieved from https://leginfo.legislature.ca.gov/faces/codes_displaySection.xhtml?lawCode=CIV&sectionNum=1798.100-1798.199

Cambridge, (2023). ChatGPT (We need to talk). University of Cambridge. https://www.cam.ac.uk/stories/ChatGPT-and-education

Campo, A., Michałko, A., Van Kerrebroeck, B., Stajic, B., Pokric, M., & Leman, M. (2023). The assessment of presence and performance in an AR environment for motor imitation learning: A case-study on violinists. *Computers in Human Behavior*, 146, 107810. DOI: 10.1016/j.chb.2023.107810 PMID: 37663430

Capgemini, (2023). Half of secondary school teachers believe that the benefits of generative AI in education outweigh the risks. Capgemini. https://www.capgemini.com/news/press-releases/half-of-secondary-school-teachers-globally-believe-that-benefits-of-generative-ai-as-an-educational-tool-outweigh-the-risks/

Cascella, M., Montomoli, J., Bellini, V., & Bignami, E. (2023). Evaluating the feasibility of ChatGPT in healthcare: An analysis of multiple clinical and research scenarios. *Journal of Medical Systems*, 47(1), 33. DOI: 10.1007/s10916-023-01925-4 PMID: 36869927

Cashion, N. (2023 September 15). This AI productivity tool helps me manage my time as a graduate student. [Blog Post] Retrieved from: https://www.nathancashion.com/blog/2023/9/this-ai-productivity-tool-helps-me-manage-my-time-as-a-graduate-student

Castillo-González, W., Lepez, C. O., & Bonardi, M. C. (2022). Chat GPT: A promising tool for academic editing . *Data and Metadata*, 1, 23–23. DOI: 10.56294/dm202223

Castro, R. A. G., Cachicatari, N. A. M., Aste, W. M. B., & Medina, M. P. L. (2024). Exploration of ChatGPT in basic education: Advantages, disadvantages, and its impact on school tasks. *Contemporary Educational Technology*, 16(3), ep511. DOI: 10.30935/cedtech/14615

Catherine, L., Javier, B., & Francisco, G. (2020). Four Pillars of the Montessori Method and Their Support by Current Neuroscience. *Mind, Brain and Education : the Official Journal of the International Mind, Brain, and Education Society*, 14(4), 322–334. DOI: 10.1111/mbe.12262

Celik, I., Dindar, M., Muukkonen, H., & Järvelä, S. (2022). The promises and challenges of artificial intelligence for teachers: A systematic review of research. *TechTrends*, 66(4), 616–630. DOI: 10.1007/s11528-022-00715-y

Center for the Developing Child – Harvard University. (2024). Executive function and self-regulation. Center for the Developing Child – Harvard University. https://developingchild.harvard.edu/science/key-concepts/executive-function/

Cevikbas, M., Bulut, N., & Kaiser, G. (2023). Exploring the Benefits and Drawbacks of AR and VR Technologies for Learners of Mathematics: Recent Developments. *Systems*, 11(5), 244. DOI: 10.3390/systems11050244

Chamberlain, J., Vines, J., Kindler, E., van der Graaf, S., & Cruickshank, P. (2022). Knowledge base integration for improved information retrieval with AI language models. *Journal of Knowledge Management*, 26(3), 634–650.

Chan, A. (2023). GPT-3 and InstructGPT: Technological dystopianism, utopianism, and "Contextual" perspectives in AI ethics and industry. *AI and Ethics*, 3(1), 53–64. DOI: 10.1007/s43681-022-00148-6

Chan, C. K., & Hu, W. (2023). Students' Voices on Generative AI: Perceptions, Benefits, and Challenges in Higher Education. *International Journal of Educational Technology in Higher Education volume 20. Article*, (43), 1–8.

Chan, C., & Hu, W. (2023). Students' voices on generative AI: Perceptions, benefits, and challenges in higher education. *International Journal of Educational Technology in Higher Education*, 20(1), 43. DOI: 10.1186/s41239-023-00411-8

Chandra, D. (2024 April 21). Speechify review 2024: Best AI Text-to-speech app? [Review] https://theprocesshacker.com/blog/speechify-review/

Chan, K. S., & Zary, N. (2019). Challenges of implementing AI in medical education: A review. *Medical Education*, 53(8), 775–785.

Chen, M. R. A., & Hwang, G. J. (2022). Effects of experiencing authentic contexts on English speaking performances, anxiety, and motivation of EFL students with different cognitive styles. *Interactive Learning Environments, 30*(9), 1619–1639.

Chen, T. -J. (2023). ChatGPT and other artificial intelligence applications speed up scientific writing. . *Journal of the Chinese Medical Association*, 86(4), 351 353.. DOI: 10.1097/JCMA.0000000000000900

Chen, C. Y. (2022). Using immersive virtual reality for training pre-service teachers in managing challenging student behaviors: A preliminary study. *British Journal of Educational Technology*, 53(4), 998–1024. DOI: 10.1111/bjet.13181

Chen, Y. (2003). On the Advantages of Extracurricular Language Activities in Promoting Language Acquisition. [Social Science Edition]. *Journal of Liaocheng University*, 2003(6), 126–127.

Chen, Y. (2022). Enhancing language learning through AI-driven applications: A case study. *Journal of Educational Technology*, 29(3), 214–229.

Chen, Y. L., & Hsu, C. C. (2020). Self-regulated mobile game-based English learning in a virtual reality environment. *Computers & Education*, 154, 103910. DOI: 10.1016/j.compedu.2020.103910

Chen, Y., & Lee, H. (2023). AI in drama-based education: Enhancing cultural and educational contexts. *International Journal of Educational Research and Development*, 27(1), 45–61.

Chen, Y., & Lee, H. (2024). Integrating AI in drama-based ELT: Bridging theory and practice. *Journal of Educational Technology and Pedagogy*, 21(1), 45–60.

Cheshmehzangi, A., Zou, T., Su, Z., & Tang, T. (2023). The growing digital divide in education among primary and secondary children during the COVID-19 pandemic: An overview of social exclusion and education equality issues. *Journal of Human Behavior in the Social Environment*, 33(3), 434–449. DOI: 10.1080/10911359.2022.2062515

China Youth Daily & China Youth School Media. (2023). More than eighty per cent of college students surveyed have used AI tools. China Youth Daily. https://edu.cnr.cn/sy/sytjB/20231120/t20231120_526491878.shtml

Chittaro, L., & Buttussi, F. (2015). Assessing Knowledge Retention of an Immersive Serious Game vs. a Traditional Education Method in Aviation Safety. *IEEE Transactions on Visualization and Computer Graphics*, 21(4), 529–538. DOI: 10.1109/TVCG.2015.2391853 PMID: 26357103

Chiu, T. K. F., & Chai, C. (2020). Sustainable curriculum planning for artificial intelligence education: A self-determination theory perspective. *Sustainability (Basel)*, 12(14), 5568. DOI: 10.3390/su12145568

Chomsky, N. (2023). *Reflections on Language and Mind: The Theory of Universal Grammar*. MIT Press.

Chukwuere, J. E. (2024). The use of ChatGPT in higher education: The advantages and disadvantages. *arXiv preprint arXiv:2403.19245*.

Chung, L. Y. (2011). Using avatars to enhance active learning: Integration of virtual reality tools into college English curriculum. In The 16th North-East Asia Symposium on Nano, Information Technology, and Reliability (pp. 29-33). IEEE.

Comeau, C. (1961). Headsight television system provides remote surveillance. *Electronics (Basel)*, 34(2), 86–90.

Compliance. (2024, June 3). *AI in Online Learning Platforms: Personalized Learning Paths*. Redresscompliance.com. https://redresscompliance.com/ai-online-learning-platform/

Congress, (2022). H.R.6580 - Algorithmic Accountability Act of 2022. *Congress of the United States*.https://www.congress.gov/bill/117th-congress/house-bill/6580

Cook, M. J. (2019). Augmented reality: Examining its value in a music technology classroom. Practice and potential. *Waikato Journal of Education*, 24(2), 23–38. DOI: 10.15663/wje.v24i2.687

Cooper, G. (2023). Examining science education in ChatGPT: An exploratory study of generative artificial intelligence. *Journal of Science Education and Technology*, 32(3), 444–452. DOI: 10.1007/s10956-023-10039-y

Correia, A. P. (2020). Healing the digital divide during the COVID-19 pandemic. *Quarterly Review of Distance Education*.

Cotton, D. R., Cotton, P. A., & Shipway, J. R. (2024). Chatting and cheating: Ensuring academic integrity in the era of ChatGPT. *Innovations in Education and Teaching International*, 61(2), 228–239. DOI: 10.1080/14703297.2023.2190148

Cotton, D., Cotton, P., & Shipway, R. (2023). Chatting and Cheating. Ensuring academic integrity in the era of ChatGPT. DOI: 10.35542/osf.io/mrz8h

Cowan, N. (2014). Working memory underpins cognitive development, learning, and education. *Educational Psychology Review*, 26(2), 197–223. DOI: 10.1007/s10648-013-9246-y PMID: 25346585

Cox, C., & Tzoc, E. (2023). *ChatGPT: Implications for academic libraries* (Vol. 84). Colledge and Research Libraries.

Cui, X., Xue, F., & Thai, T. (2019). The impact of AI-driven learning technologies on educational outcomes. *Computers & Education*, 141, 103620.

Currie, G. M., Singh, C., Nelson, T., Nabasenja, C., Al-Hayek, Y., & Spuur, K. (2023). ChatGPT in medical imaging higher education. *Radiography*, 29(4), 792–799. DOI: 10.1016/j.radi.2023.05.011 PMID: 37271011

Curry, N., Baker, P., & Brookes, G. (2024). Generative AI for corpus approaches to discourse studies: A critical evaluation of ChatGPT . *Applied Corpus Linguistics*, 4(1), 100082. DOI: 10.1016/j.acorp.2023.100082

Dai, D. D. (2021). Artificial intelligence technology assisted music teaching design. *Scientific Programming*, 2021(1), 9141339. DOI: 10.1155/2021/9141339

DARPA. (2016). Explainable Artificial Intelligence (XAI) (Archived). DARPA. https://www.darpa.mil/program/explainable-artificial-intelligence

Datoo, A. K., & Siddiqui, K. A. (2024). ChatGPT and Death of an Author. *Critical Humanities*, 2(2), 5. DOI: 10.33470/2836-3140.1039

Deckers, A., Jeunen, E., & Koolen, M. (2022). Enhancing digital reference services through ChatGPT: A feasibility study. *Journal of Library Innovation*, 45(2), 112–128.

Dede, C. J., Jacobson, J., & Richards, J. (2017). *Introduction: Virtual*. Augmented, and Mixed Realities in Education., DOI: 10.1007/978-981-10-5490-7_1

Deep learning techniques and optimization strategies in big data analytics. (2019). In Advances in systems analysis, software engineering, and high performance computing book series. https://doi.org/DOI: 10.4018/978-1-7998-1192-3

Dergaa, I., Chamari, K., Zmijewski, P., & Ben Saad, H. (2023). From human writing to artificial intelligence generatedtext: Examining the prospects and potential threats of ChatGPT in academic writing. *Biology of Sport*, 40(2), 615–622. DOI: 10.5114/biolsport.2023.125623 PMID: 37077800

Deshpande, A., Murahari, V., Rajpurohit, T., Kalyan, A., & Narasimhan, K. (2023). Toxicity in ChatGPT: Analyzing Persona-assigned Language Models. *ArXiv*, abs/2304.05335.

DeVon, C. (2023). AI chatbots can 'hallucinate' and make things up—why it happens and how to spot it. *CNBC*. https://www.cnbc.com/2023/12/22/why-ai-chatbots-hallucinate.html

Dewi, H. K., Wardani, T. I., Rahim, N. A., Putri, R. E., & Pandin, M. G. (2021). *THE USE OF AI (ARTIFICIAL INTELLIGENCE) IN ENGLISH LEARNING AMONG UNIVERSITY STUDENT: CASE STUDY IN ENGLISH DEPARTMENT.* UNIVERSITAS AIRLANGGA.

Di Cecco, G. J., Barve, V., Belitz, M. W., Stucky, B. J., Guralnick, R. P., & Hurlbert, A. H. (2021). Observing the Observers: How Participants Contribute Data to iNaturalist and Implications for Biodiversity Science. *Bioscience*, 71(11), 1179–1188. DOI: 10.1093/biosci/biab093

Di Natale, A. F., Repetto, C., Riva, G., & Villani, D. (2020). Immersive virtual reality in K-12 and higher education: A 10-year systematic review of empirical research. *British Journal of Educational Technology*, 51(6), 2006–2033. DOI: 10.1111/bjet.13030

Diamond, A. (2024 May 15). How to sharpen executive functions: Activities to hone brain skills. ADDitude. https://www.additudemag.com/how-to-improve-executive-function-adhd/#:~:text=Inhibitory%20Control%20at%20the%20Level%20of%20Attention,is%20boring%20(sustained%20attention)

Dignum, V., Kopp, O., & Rahwan, I. (2023). Ethical guidelines for the use of AI language models in research. *AI & Society*, 38, 23–36.

Dincelli, E., & Yayla, A. (2022). Immersive virtual reality in the age of the Metaverse: A hybrid-narrative review based on the technology affordance perspective. *The Journal of Strategic Information Systems*, 31(2), 101717. DOI: 10.1016/j.jsis.2022.101717

Dooly, M., Thrasher, T., & Sadler, R. (2023). "Whoa! Incredible!:" Language Learning Experiences in Virtual Reality. *RELC Journal*, 54(2), 321–339. DOI: 10.1177/00336882231167610

Doran, D., Schulz, S., & Besold, T. R. (2017). What Does Explainable AI Really Mean? A New Conceptualization of Perspectives. Wright State University. https://arxiv.org/abs/1710.00794

Duggal, N. (2023 December 15). Attention deficit hyperactivity disorder (ADHD): The role of dopamine. Healthline. https://www.healthline.com/health/adhd/adhd-dopamine

Dumitru, C. (2023). Distance Learning and Higher Education Hybridization Opportunities and Challenges for Students with Disabilities. In *Advances in Distance Learning in Times of Pandemic* (p. 35).

Durand, G., Arbone, I. S., & Wharton, M. (2020). Reduced organizational skills in adults with ADHD are due to deficits in persistence, not in strategies. *PeerJ*, 8, e9844. DOI: 10.7717/peerj.9844 PMID: 33194353

EarthSnap. (2016). *EarthSnap - Nature Identifier * Earth.com.* Www.earth.com. https://www.earth.com/earthsnap/

Edelsen, L. (2023). Google's Bard A.I. chatbot is being trained by contractors who say they're overworked, underpaid, stressed and scared. Fortune. https://fortune.com/2023/07/12/google-bard-ai-chatbot-trained-contractors-overworked-underpaid-stressed-scared/

Eguchi, A., Okada, H., & Muto, Y. (2021). Contextualizing AI education for K-12 students to enhance their learning of AI literacy through culturally responsive approaches. *Kunstliche Intelligenz*, 35(2), 153–161. DOI: 10.1007/s13218-021-00737-3 PMID: 34376922

Eisenberg, N., Spinrad, T. L., & Eggum, N. D. (2010). Emotion-related self-regulation and its relation to children's maladjustment. *Annual Review of Clinical Psychology*, 6(1), 495–525. DOI: 10.1146/annurev.clinpsy.121208.131208 PMID: 20192797

Eliott, L. (2024). Advantages and disadvantages of AI in the EFL classroom. In *The Asian Conference on Education 2023: Official Conference Proceedings* (pp. 491-499).

Ellis, A. R., & Slade, E. (2023). A New Era of Learning: Considerations for ChatGPT as a Tool to Enhance Statistics and Data Science Education. *Journal of Statistics and Data Science Education, 31:2, DOI:* , 128-133.DOI: 10.1080/26939169.2023.2223609

Emenike, M. E., & Emenike, B. U. (2023). Was This Title Generated by ChatGPT? Considerations for Artificial Intelligence Text-Generation Software Programs for Chemists and Chemistry Educators. *Journal of Chemical Education*, 100(4), 1413–1418. DOI: 10.1021/acs.jchemed.3c00063

Esling, P., & Devis, N. (2020). *Creativity in the era of artificial intelligence.* arXiv preprint arXiv:2008.05959.

Evans, P., Vansteenkiste, M., Parker, P., Kingsford-Smith, A., & Zhou, S. (2024). Cognitive load theory and its relationships with motivation: A self-determination theory perspective. *Educational Psychology Review*, 36(1), 7. DOI: 10.1007/s10648-023-09841-2

Fahimirad, M., & Kotamjani, S. S. (2018). A review on application of artificial intelligence in teaching and learning in educational contexts . *International Journal of Learning and Development*, 8(4), 106–118. DOI: 10.5296/ijld.v8i4.14057

Fakhar, H., Lamrabet, M., Echantoufi, N., El Khattabi, K., & Ajana, L. (2024). Artificial Intelligence from Teachers' Perspectives and Understanding: Moroccan Study. *International Journal of Information and Education Technology (IJIET)*, 14(6).

Fang, X., Che, S., Mao, M., Zhang, H., Zhao, M., & Zhao, X. (2024). Bias of AI-generated content: An examination of news produced by large language models. *Scientific Reports*, 14(1), 5224. DOI: 10.1038/s41598-024-55686-2 PMID: 38433238

Fang, Y. (2015). Application of virtual reality technology in college English oral teaching. *Shanhai Classic*, 2015(16), 43–44.

Fan, Y. L., & Ding, W. (2020). Review of the application of immersive virtual reality technology in science teaching. *Chemical Educator (Boise, Idaho)*, 41(5), 84–90.

Fareed, M., Ashraf, A., & Bilal, M. (2016). ESL learners' writing skills: Problems, factors and suggestions . *Journal of education and social sciences*, 4(2), 81-92.

Farrell, E. F. (2003). Paying attention to students who can't. *The Chronicle of Higher Education*, 50(5), A50–A51.

Farrie, D. (2022). *The Color of Opportunity: How Formula Underfunding Disproportionately Impacts New Jersey's Black and Latino Students*. Education Law Center.

Farrokhnia, M., Baggen, Y., Biemans, H., & Noroozi, O. (2022). Bridging the fields of entrepreneurship and education: The role of philosophical perspectives in fostering opportunity identification. *International Journal of Management Education*, 20(2), 100632. DOI: 10.1016/j.ijme.2022.100632

Fathi, J., Rahimi, M., & Derakhshan, A. (2024). Improving EFL learners' speaking skills and willingness to communicate via artificial intelligence-mediated interactions. *System*, 121, 103254. DOI: 10.1016/j.system.2024.103254

Fayyad, U. (2023). Stochastic Parrots Or Intelligent Assistants? Forbes. https://www.forbes.com/sites/forbestechcouncil/2023/06/06/stochastic-parrots-or-intelligent-assistants/?sh=562777665966

Feng, Q. (2016). Exploring the application of virtual reality technology in foreign language teaching. *Theoretical Observation*, 2016(5), 170–175.

Fernandes, B., Wright, M., & Essau, C. (2023). The role of emotion regulation and executive functioning in the intervention outcome of children with emotional and behavioural problems. *Children (Basel, Switzerland)*, 10(1), 139. DOI: 10.3390/children10010139 PMID: 36670689

Fijačko, N., Gosak, L., Štiglic, G., Picard, C. T., & Douma, M. J. (2023). Can ChatGPT pass the life support exams without entering the American heart association course? *Resuscitation*, 185, 185. DOI: 10.1016/j.resuscitation.2023.109732 PMID: 36775020

Firat, M. (2023). *How chat GPT can transform autodidactic experiences and open education* . Department of Distance Education, Open Education Faculty, Anadolu Unive.

Firat, M. (2023). What ChatGPT means for universities: Perceptions of scholars and students. *Journal of Applied Learning and Teaching*, 6(1), 57–63.

Forman, N., Udvaros, J., & Avornicului, M. (2023). ChatGPT: A new study tool shaping the future for high school students. *International Journal of Advanced Natural Sciences and Engineering Research*, 7(4), 95–102. DOI: 10.59287/ijanser.562

Fracolli, B. (2020, August 7). *30 Amazon Rainforest Animals to Spot in the Wild*. Peru for Less. https://www.peruforless.com/blog/amazon-rainforest-animals

Francis, J. (2017). *The effects of technology on student motivation and engagement in classroom-based learning* .

Fried, R., DiSalvo, M., Farrell, A., Biederman, (2022). Using a digital meditation application to mitigate anxiety and sleep problems in children with ADHD. *J Atten Disord. 26(*7), 1033–1039.

Fu, L. (2021). Research on the Teaching Model of Animation Professional Class Based on AR/VR Technology and 5G Network. *Wireless Communications and Mobile Computing, 2021*(1). DOI: 10.1155/2021/1715909

Fuchs, K. (2022). Bringing Kahoot! Into the Classroom. *International Journal of Information and Education Technology (IJIET)*, 12(7), 3–8. DOI: 10.18178/ijiet.2022.12.7.1662

Fu, X. Q. (2013). Observations on the current situation and existing problems in college English teaching. *University Education*, 2013(7), 58–60.

Gallen, C. L., Schaerlaeken, S., Younger, J. W., Younger, J. W., O'Laughlin, K. D., Anguera, J. A., Bunge, S. A., Ferrer, E. E., Hoeft, F., McCandliss, B. D., Mishra, J., Rosenberg-Lee, M., Gazzaley, A., Uncapher, M. R., Anguera, J. A., & Gazzaley, A. (2023). Contribution of sustained attention abilities to real-world academic skills in children. *Scientific Reports*, 13(1), 2673. DOI: 10.1038/s41598-023-29427-w PMID: 36792755

Gamvros, A., Yau, E., & Choong, S. (2023). China finalises its Generative AI Regulation. Norton Rose Fulbright. https://www.dataprotectionreport.com/2023/07/china-finalises-its-generative-ai-regulation/

Gao, J., Zhao, H., Yu, C., & Xu, R. (2023). Exploring the feasibility of chatgpt for event extraction. *arXiv preprint arXiv:2303.03836.*

Gao, Y., & Liu, D. J.. (2016). Core Elements and Challenges of Virtual Reality Technology in Promoting Learning. *Research in Dance Education*, 2016(10), 77–87.

Garante, (2023). Personal data is collected unlawfully; no age verification system is in place for children. *Garante.*https://www.garanteprivacy.it/home/docweb/-/docweb-display/docweb/9870847#english

Garcia Castro, R. A., Chura-Quispe, G., Velarde Molina, J. F., Espinoza Ramos, L. A., & Almonte Durand, C. A. (2024). Bibliometric review on teaching methods with artificial intelligence in education. *Online Journal of Communication and Media Technologies*, 14(2), e202419. DOI: 10.30935/ojcmt/14367

Garcia Pimenta, M., Gruhnert, R. K., Fuermaier, A. B. M., & Groen, Y. (2024). The role of executive functions in mediating the relationship between adult ADHD symptoms and hyperfocus in university students. *Research in Developmental Disabilities*, 144, 104639. DOI: 10.1016/j.ridd.2023.104639 PMID: 38039699

Garg, M., & Goel, A. (2022). A systematic literature review on online assessment security: Current challenges and integrity strategies. *Computers & Security*, 113, 102544. DOI: 10.1016/j.cose.2021.102544

GDPR. (2016). What does the General Data Protection Regulation (GDPR) govern? *GDPR.*https://commission.europa.eu/law/law-topic/data-protection/reform/what-does-general-data-protection-regulation-gdpr-govern_en?

General Data Protection Regulation (GDPR). (2016). Regulation (EU) 2016/679 of the European Parliament and of the Council of 27 April 2016 on the protection of natural persons about the processing of personal data and the free movement of such data. *Official Journal of the European Union, L*, 119(1). https://eur-lex.europa.eu/legal-content/EN/TXT/PDF/?uri=CELEX:32016R0679

Ghafar, A. (2020). *CONVERGENCE BETWEEN 21ST CENTURY SKILLS AND ENTREPRENEURSHIP EDUCATION IN HIGHER EDUCATION INSTITUTES: AN EXPLORATORY STUDY FROM UNDERGRADUATES' PERSPECTIVE.* 9582–9582. DOI: 10.21125/inted.2020.0215

Giannini, S. (2023). Generative AI and the future of education. UNESCO. https://unesdoc.unesco.org/ark:/48223/pf0000385877

Gibson, M. (2023). AI in language education: Enhancing drama-based pedagogies. *Journal of Educational Technology*, 20(1), 55–70.

Gill, S. S., Xu, M., Patros, P., Wu, H., Kaur, R., Kaur, K., & Buyya, R. (2024). Transformative effects of ChatGPT on modern education: Emerging Era of AI Chatbots. *Internet of Things and Cyber-Physical Systems*, 4, 19–23. DOI: 10.1016/j.iotcps.2023.06.002

Giri, D., & Sharma, J. (2024). TRANSFORMATIVE EFFECTS OF ANIMATION, VIRTUAL REALITY, AND AUGMENTED REALITY IN EDUCATION - A STUDY. *ShodhKosh: Journal of Visual and Performing Arts*, 4(2SE), 237–246. DOI: 10.29121/shodhkosh.v4.i2SE.2023.563

Goblin.tools. [Computer Software]. (2024) Available from https://goblin.tools

Golda, A., Mekonen, K., Pandey, A., Singh, A., Hassija, V., Chamola, V., & Sikdar, B. (2024). Privacy and Security Concerns in Generative AI: A Comprehensive Survey. *IEEE Access : Practical Innovations, Open Solutions*, 12, 48126–48144. DOI: 10.1109/ACCESS.2024.3381611

Google. (2024). Responsible AI practices. Google. https://ai.google/responsibility/responsible-ai-practices/

Gowlingwlg, (2023). The artificial intelligence and data act (AIDA). *Gowling WLG*. https://gowlingwlg.com/en/insights-resources/articles/2023/artificial-intelligence-and-data-act-aida/

Gruber, C. (2023). GIGO: The Production Function of AI's Market Impact. Medium. https://medium.com/@gruberai/gigo-the-production-function-of-ais-market-impact-97a724b1650b

Guangul, F. M., Suhail, A. H., Khalit, M. I., & Khidhir, B. A. (2020). Challenges of remote assessment in higher education in the context of COVID-19: A case study of Middle East College. *Educational Assessment, Evaluation and Accountability*, 32(4), 519–535. DOI: 10.1007/s11092-020-09340-w PMID: 33101539

Guendelman, S., Medeiros, S., & Rampes, H. (2017). Mindfulness and emotion regulation: Insights from neurobiological, psychological, and clinical studies. *Frontiers in Psychology*, 8, 220. DOI: 10.3389/fpsyg.2017.00220 PMID: 28321194

Gui, J., Sun, Z., Wen, Y., Tao, D., & Ye, J. (2021). A review on generative adversarial networks: Algorithms, theory, and applications. *IEEE Transactions on Knowledge and Data Engineering*, 35(4), 3313–3332. DOI: 10.1109/TKDE.2021.3130191

Guleria, A., Krishan, K., Sharma, V., & Kanchan, T. (2024). ChatGPT: Forensic, legal, and ethical issues. *Medicine, Science, and the Law*, 64(2), 150–156. DOI: 10.1177/00258024231191829 PMID: 37528607

Gunning, D., Vorm, E., Wang, J. Y., & Turek, M. (2021). DARPA's explainable AI (XAI) program: A retrospective. AI & Logic, Advance online publication. DOI: 10.1002/ail2.61

Gupta, P., Raturi, S., & Venkateswarlu, P. (2023). Chatgpt for designing course outlines: A boon or bane to modern technology. *Available atSSRN* 4386113. DOI: 10.2139/ssrn.4386113

Gweon, H. (2021). Inferential social learning: Cognitive foundations of human social learning and teaching. *Trends in Cognitive Sciences*, 25(10), 896–910. DOI: 10.1016/j.tics.2021.07.008 PMID: 34417094

H.L. P. (2015). *Differentiated Instruction And The Need To Integrate Teaching And Practice - ProQuest*. Www.proquest.com. https://www.proquest.com/openview/718ee130f5c18bd09e667e85b304b6ce/1?pq-origsite=gscholar&cbl=2026890

Habayeb, H. (2022). Explainable AI Isn't Enough; We Need Understandable AI. Techopedia. https://www.techopedia.com/explainable-ai-isnt-enough-we-need-understandable-ai/2/34671

Halaweh. (2023). *Argued the co-operation between ChatGPT and human in writing will probably lead to produce writing samples regarded inventive, original, and productive.*

Haleem, A., Javaid, M., Qadri, M. A., & Suman, R. (2022). Understanding the role of digital technologies in education: A review. *Sustainable Operations and Computers*, 3, 275–285. DOI: 10.1016/j.susoc.2022.05.004

Hammud, G., Avital-Magen, A., Schusheim, G., Barzuza, I., & Engel-Yeger, B. (2023). How self-regulation and executive functions deficits affect quality of life of children/adolescents with emotional regulation disorders. *Children (Basel, Switzerland)*, 10(10), 1622. DOI: 10.3390/children10101622 PMID: 37892283

Han, D.-I. D., Melissen, F., & Haggis-Burridge, M. (2024). Immersive experience framework: A Delphi approach. *Behaviour & Information Technology*, 43(4), 623–639. DOI: 10.1080/0144929X.2023.2183054

Hardy, J., & McLeod, R. (2020). Using positive reinforcement with young children. *Beyond Behavior*, 29(2), 107429562091572. DOI: 10.1177/1074295620915724

Hassija, V., Chamola, V., Mahapatra, A., Singal, A., Goel, D., Huang, K., Scardapane, S., Spinelli, I., Mahmud, M., & Hussain, A. (2024). Interpreting Black-Box Models: A Review on Explainable Artificial Intelligence. *Cognitive Computation*, 16(1), 45–74. DOI: 10.1007/s12559-023-10179-8

Hatem, R., Simmons, B., & Thornton, J. (2023). A Call to Address AI "Hallucinations" and How Healthcare Professionals Can Mitigate Their Risks. *NIH.*https://www.ncbi.nlm.nih.gov/pmc/articles/PMC10552880/

Hattie, J. (2012). *Visible learning for teachers: Maximizing impact on learning.* Routledge/Taylor & Francis Group. https://doi.org/DOI: 10.4324/9780203181522

Hayes, S., Fortier, C., Levine, A., Milberg, W., & McGlinchey, R. (2012). *NIH.* https://www.ncbi.nlm.nih.gov/pmc/articles/PMC3955262/

Hazaymeh, W. A., Bouzenoun, A., & Remache, A. (2024). EFL Instructors' Perspective on Using AI Applications in English as a Foreign Language Teaching and Learning. *Emerging Science Journal*, 8, 73–87. DOI: 10.28991/ESJ-2024-SIED1-05

Head, A. J., & Eisenberg, M. B. (2010). How college students use the Web to conduct everyday life research. *First Monday*, 15(3). Advance online publication. DOI: 10.5210/fm.v15i3.2774

Headspace. (2024). Meditation and ADHD: An unexpected (but worthwhile) match. Headspace. https://www.headspace.com/articles/meditation-and-adhd

He, D. C. (2018). Analysis of the current status and prospects of the VR industry in China. *Science and Technology Communication*, 10(24), 2.

Heffernan, N. (2023). AI in education: Bridging the gap between innovation and application. *Educational Research and Innovation*, 27(1), 88–102.

Heilig, M. L. (1962). Sensorama simulator (US Patent No. 3,508,070).

He, J. H., Huang, X. L., & Han, G. X.. (2019). Empirical study on the influencing factors of learning motivation in VR educational games. *Educational Technology Review*, (8), 70–77.

Hendler, N., Judd, W. Picha, D. & Tantoco, E., (2022). Inherent GIGO Problems with Medical AI. *Journal of Artificial Intelligence and Soft Computing Techniques.*

Herawati, D., Widajati, W., & Sartinah, E. (2022). The role of text to speech assistive technology to improve reading ability in E-Learning for ADHD students. *Journal of ICSAR*, 6(2), 169. DOI: 10.17977/um005v6i22022p169

Hewett, V. M. (2001). Examining the Reggio Emilia Approach to Early Childhood Education. *Early Childhood Education Journal*, 29(2), 95–100. DOI: 10.1023/A:1012520828095

Hidayatullah, E. (2024, January). Evaluating the effectiveness of ChatGPT to improve English students' writing skills . *In Humanities, Education, Applied Linguistics, and Language Teaching: Conference Series*, pp. (Vol. 1, No. 1, pp. 14-19).

Hirsch, L., Ortega, F., & Dinhopl, A. (2023). Collaborative research and co-creation with ChatGPT: Lessons from the field. *Journal of Collaborative Research*, 42(1), 23–40.

Hitomi, M. (2021 June 21). ADHD and task initiation: Getting the ball rolling. Timeo. https://www.tiimoapp.com/blog/adhd-task-initiation

Hockly, N. (2018). Blended Learning. *ELT Journal*, 72(1), 97–101. DOI: 10.1093/elt/ccx058

Hoffman, D. M. (2009). Reflecting on Social Emotional Learning: A Critical Perspective on Trends in the United States. *Review of Educational Research*, 79(2), 533–556. DOI: 10.3102/0034654308325184

Holmes, W., Bialik, M., & Fadel, C. (2020). *Artificial intelligence in education: Promises and implications for teaching and learning*. Routledge.

Hong, W. C. (2023). The impact of ChatGPT on foreign language teaching and learning: Opportunities in education and research . *Journal of Educational Technology and Innovation*, 3(1). Advance online publication. DOI: 10.61414/jeti.v5i1.103

Hooft Graafland, J. (2018). New Technologies and 21st Century Children. *OECD Education Working Papers, 179*(179). DOI: 10.1787/19939019

Hopkins, T., Weng, S. C. C., Vanukuru, R., Wenzel, E. A., Banic, A., Gross, M. D., & Do, E. Y. L. (2022). Ar drum circle: Real-time collaborative drumming in ar. *Frontiers in Virtual Reality*, 3, 847284. DOI: 10.3389/frvir.2022.847284

Houston, L. (2016 March 2). Workin memory difficulties: Strategies for the classroom. LD@School. https://www.ldatschool.ca/working-memory-difficulties/

Hovy, D., Spruit, S. L., & Dignum, F. (2021). The Social Impact of Natural Language Processing: Challenges and Opportunities. *ACM Transactions on Social Computing*, 4(2), 1–35. DOI: 10.1145/3450811

Hsi, S. (2007). Conceptualizing Learning from the Everyday Activities of Digital Kids. *International Journal of Science Education*, 29(12), 1509–1529. DOI: 10.1080/09500690701494076

Huang, H. M., Liaw, S. S., & Lai, C. M. (2016). Exploring learners' acceptance of virtual reality in medical education: A case study of desktop and projection-based display systems. *Interactive Learning Environments*, 24(1), 3–19. DOI: 10.1080/10494820.2013.817436

Huang, L. (2022). The role of AI in drama-based English language teaching. *International Journal of Language and Communication*, 18(2), 156–170.

Huang, L., & Zhao, Q. (2024). Collaboration between AI tutors and educators in ELT: A qualitative study. *International Journal of Language and Educational Research*, 33(2), 89–104.

Huang, Y., Saleh, M., & Liu, Y. (2021). AI in education: Enhancing learning experiences and outcomes. *Journal of Educational Computing Research*, 59(1), 56–75.

Hulsen, T. (2023). Explainable Artificial Intelligence (XAI): Concepts and Challenges In Healthcare. *MDPI.*https://www.mdpi.com/2673-2688/4/3/34

Hung, J., & Chen, J. (2023). The Benefits, Risks and Regulation of Using ChatGPT in Chinese Academia: A Content Analysis. *Social Sciences 12: 380. https://doi.org/*, 1-15.DOI: 10.3390/socsci12070380

Hutson, J., Jeevanjee, T., Vander Graaf, V., Lively, J., Weber, J., Weir, G., & Edele, S. (2022). Artificial Intelligence and the Disruption of Higher Education: Strategies for Integrations across Disciplines . *Creative Education*, 13(12), 3953–3980. DOI: 10.4236/ce.2022.1312253

Iannone, A., & Giansanti, D. (2023). Breaking barriers - the intersection of AI and assistive Technology in Autism care: A narrative review. *Journal of Personalized Medicine*, 14(1), 41. DOI: 10.3390/jpm14010041 PMID: 38248742

IBM. (2023). What is explainable AI? IBM. https://www.ibm.com/topics/explainable-ai

IBM_A (2023). What are AI hallucinations? IBM. https://www.ibm.com/topics/ai-hallucinations

IBM_B, (2024). What is risk mitigation? *IBM*. https://www.ibm.com/topics/risk-mitigation

IBM_C (2024). What is explainable AI? IBM. https://www.ibm.com/topics/explainable-ai

Imran, M., & Almusharraf, N. (2023). Analyzing the role of ChatGPT as a writing assistant at higher education level: A systematic review of the literature. *15(4), ep464*, 15(4), ep464.

INDU. 2024. The Artificial Intelligence and Data Act (AIDA). *House of Commons' Standing Committee on Industry and Technology (INDU.*https://www.ourcommons.ca/Content/Committee/441/INDU/Brief/BR12711885/br-external/CanadianLabourCongress-e.pdf

International Business Machines. (2024) What is automation? IBM. https://www.ibm.com/topics/automation#:~:text=Artificial%20intelligence%2C%20or%20AI%2C%20is,are%20fields%20of%20artificial%20intelligence

Irish, J. (2024 March 20). 3 to-do list apps that actually work with ADHD. [Blog Post] Retrieved from: https://zapier.com/blog/adhd-to-do-list/

Ishaq, K., Azan, N., Zin, M., Rosdi, F., Abid, A., & Ijaz, M. (2020). The impact of ICT on students' academic performance in public private sector universities of Pakistan. *International Journal of Innovative Technology and Exploring Engineering*, 9(3), 11. DOI: 10.35940/ijitee.C8093.019320

Iskender, A. (2023). Holy or Unholy? Interview with Open AI's ChatGPT. *European Journal of Tourism Research, 34, 3414. https://doi.org/*, 1-11.DOI: 10.54055/ejtr.v34i.3169

Islas, L. (2023). Gemini, Google's new Artificial Intelligence, is explained in 5 key facts. Merca2.0. https://www.merca20.com/gemini-googles-new-artificial-intelligence-explained-in-5-key facts/#:~:text=Gemini%201.0%201%20Parameters%3A%201.6%20trillion%202%20Training,date%3A%20December%202023%204%20Execution%3A%20In%20the%20cloud

Jaiswal, A., & Arun, C. (2021). Transforming education with AI: Potential and challenges. *Educational Technology Research and Development*, 69(2), 1–20.

Jarrar, N. (2023). Cultural considerations in educational reforms: A case study of Palestinian schools. *Journal of Cultural Education*, 31(2), 123–137.

Jaska, P. (2022 March 31). The ADHD & addiction link: Addictive behaviours in adults explained. [Blog Post] Retrieved from: https://www.additudemag.com/addictive-behaviors-adhd/

Javed, S. T., Zafar, K., & Younas, I.Sobia Tariq Javed. (2024, October). Zafar, K., & Younas, I. (2024). Kids Learning Optimizer: Social evolution and cognitive learning-based optimization algorithm. *Neural Computing & Applications*, 36(28), 17417–17465. Advance online publication. DOI: 10.1007/s00521-024-10009-4

Jeon, J., Hong, M., Yi, M., Chun, J., Kim, J. S., & Choi, Y. J. (2016). Interactive authoring tool for mobile augmented reality content. *Journal of Information Processing Systems*, 12(4), 612–630.

Jiang, Q., Li, M., Han, W., & Yang, C. (2019). *ICT promoting the improvement of education quality: Experience and practice.* In MINISTERIAL FORUM (p. 158). RetrievedJanuary25,2022fromhttps://iite.unesco.org/wpcontent/uploads/2019/02/Unesco_003_Proccedings_190212a.pdf#page=160

Jiang, N., Yan-Li, S., Pamanee, K., & Sriyanto, J. (2021). Depression, Anxiety, and Stress During the COVID-19 Pandemic: Comparison Among Higher Education Students in Four Countries in the Asia-Pacific Region. *Journal of Population and Social Studies*, 29, 370–383. DOI: 10.25133/JPSSv292021.023

Jiang, Z., & Hussain, Y. (2023). THE EFFECTIVENESS OF NATURE-BASED LEARNING IN PROMOTING PHYSICAL, COGNITIVE, AND EMOTIONAL DEVELOPMENT IN YOUNG CHILDREN: A CASE STUDY IN CHINA. *International Journal of Education & Technology*, 1(3). Advance online publication. DOI: 10.59021/ijetech.v1i3.24

Jia, Y. F., Zhao, X. J., Zhao, J. Y., Kong, L. J., & Huang, K. L. (2018). Research on the virtual reality simulation system for special vehicle passenger compartments. *Xitong Fangzhen Xuebao*, 30(1), 132.

Jisc. (2021). The future of libraries: Emerging technologies and innovative services. Jisc. Retrieved 25/08/2024, https://www.jisc.ac.uk/reports/the-future-of-libraries-emerging-technologies-and-innovative-services

Johnson, R., & Gunning, B. (2024). How does data science impact publication? *Geoconvention 2024*.https://geoconvention.com/wp-content/uploads/abstracts/2024/106445-how-does-data-science-impact-publication.pdf

Johnson, M., & Carter, S. (2022). The impact of affect-sensitive educational technologies on language learning. *Journal of Educational Technology and Artificial Intelligence*, 10(3), 345–360.

Johnson, P., & Lee, H. (2024). Advancements in AI-driven Language Learning Tools. *International Journal of Language and Communication*, 18(2), 45–63.

Johnson, R., & Lee, K. (2024). Comparative effectiveness of traditional and AI-facilitated drama in language teaching. *Language Education Research*, 22(3), 212–229.

Johnson, R., & Lee, K. (2024). The combined effects of AI and drama on student engagement and language learning outcomes. *Educational Technology Research and Development*, 72(1), 75–92.

Jones, H. G. (2020). What plant is that? Tests of automated image recognition apps for plant identification on plants from the British flora. *AoB Plants*, 12(6), plaa052. Advance online publication. DOI: 10.1093/aobpla/plaa052 PMID: 33173573

Joshi, R., Pandey, K., & Kumari, S. (2024, February). A Metaverse: Significance and Dimensions in Various Sectors. In *2024 11th International Conference on Computing for Sustainable Global Development (INDIACom)* (pp. 1781-1786). IEEE.

Joshi, R., Pandey, K., & Kumari, S. (2024, May). Exploring Various Government Health Schemes and the Use of AI in the Changing Dynamics of India. In *2024 5th International Conference for Emerging Technology (INCET)* (pp. 1-6). IEEE. DOI: 10.1109/INCET61516.2024.10643415

Joshi, R. (2024). Impact of AI in the Tourism Industry: A Double-Edged Sword. In *Technology and Luxury Hospitality* (pp. 135–150). Routledge. DOI: 10.4324/9781003488248-10

Joshi, R., & Badola, R. (2024). Digital Twin: A Transformative Tool for Smart Cities. In *Smart Cities: Innovations, Challenges and Future Perspectives* (pp. 177–198). Springer Nature Switzerland. DOI: 10.1007/978-3-031-59846-3_8

Joshi, R., Pandey, K., & Kumari, S. (2024). Artificial Intelligence for Advanced Sustainable Development Goals: A 360-Degree Approach. In *Preserving Health, Preserving Earth: The Path to Sustainable Healthcare* (pp. 281–303). Springer Nature Switzerland. DOI: 10.1007/978-3-031-60545-1_16

Jovanovic, M., & Campbell, M. (2022). Generative artificial intelligence: Trends and prospects. *Computer*, 55(10), 107–112. DOI: 10.1109/MC.2022.3192720

Kambhampati, S. (2023). Can LLMs Really Reason and Plan? *ACM*.https://cacm.acm.org/blogcacm/can-llms-really-reason-and-plan/

Kang, H., Yang, J., Ko, B. S., Kim, B. S., Song, O. Y., & Choi, S. M. (2024a). Integrated Augmented and Virtual Reality Technologies for Realistic Fire Drill Training. *IEEE Computer Graphics and Applications*, 44(2), 89–99. DOI: 10.1109/MCG.2023.3303028 PMID: 37585326

Kaplan-Rakowski, R., & Gruber, A. (2019, November). Low-immersion versus high-immersion virtual reality: Definitions, classification, and examples with a foreign language focus. In Conference Proceedings. Innovation in Language Learning 2019.

Karimkhanlooei, G., & Seifiniya, H. (2015). Teaching Alphabet, Reading and Writing for Kids between 3-6 Years Old as a Second Language. *Procedia: Social and Behavioral Sciences*, 192, 769–777. DOI: 10.1016/j.sbspro.2015.06.090

Kasneci, E., Sessler, K., K¨uchemann, S., Bannert, M., Dementieva, D., Fischer, F., . . . Pfeffer, J. (2023). ChatGPT for Good? On Opportunities and Challenges of Large Language Models for Education. 1-13.

Kasneci, E., Seßler, K., Küchemann, S., Bannert, M., Dementieva, D., Fischer, F., Gasser, U., Groh, G., Günnemann, S., Hüllermeier, E., Krusche, S., Kutyniok, G., Michaeli, T., Nerdel, C., Pfeffer, J., Poquet, O., Sailer, M., Schmidt, A., Seidel, T., & Kasneci, G. (2023). ChatGPT for good? On opportunities and challenges of large language models for education. *Learning and Individual Differences*, 103, 102274. DOI: 10.1016/j.lindif.2023.102274

Kasper, L. J., Alderson, R. M., & Hudec, K. L. (2012). Moderators of working memory deficits in children with attention-deficit/hyperactivity disorder (ADHD): A meta-analytic review. *Clinical Psychology Review*, 32(7), 605–617. DOI: 10.1016/j.cpr.2012.07.001 PMID: 22917740

Kayalı, B., Yavuz, M., Balat, S., & Çalışan, M. (2023). Investigation of student experiences with ChatGPT-supported online learning applications in higher education . *Australasian Journal of Educational Technology*, 39(5), 20–39. DOI: 10.14742/ajet.8915

Keebler, J. R., Wiltshire, T. J., Smith, D. C., Fiore, S. M., & Bedwell, J. S. (2014). Shifting the paradigm of music instruction: Implications of embodiment stemming from an augmented reality guitar learning system. *Frontiers in Psychology*, 5, 471. DOI: 10.3389/fpsyg.2014.00471 PMID: 24999334

Keelor, J. L., Creadhead, N., Silbert, N., & Horowitz-Kraus, T. (2020). Text-to-speech technology: Enhancing reading comprehension for students with reading difficulties. *Assistive Technology Outcomes and Benefits*, 14, 19–35.

Keiper, M. (2023). ChatGPT in Practice: Increasing Event Planning Efficiency Through Artificial Intelligence. 33. . . *33*.. .DOI: 10.1016/j.jhlste.2023.100454

Kelty, N. E., & Wakabayashi, T. (2020). Family Engagement in Schools: Parent, Educator, and Community Perspectives. *SAGE Open*, 10(4), 215824402097302. DOI: 10.1177/2158244020973024

Khattak, A. & Ahmad, A. (2018). Effects of positive reinforcement on student's academic performance. 01. 220-225.

Khosravi, H., Buckingham Shum, S., Chen, G., Conati, C., Tsai, Y.-S., Kay, J., Knight, S., Martinez-Maldonado, R., & Sadiq, S. (2022). Explainable Artificial Intelligence in Education. Elsevier. Computers and Education: Artificial Intelligence 3 (2022)

Kiger, M. E., & Varpio, L. (2020). Thematic analysis of qualitative data: AMEE Guide No. 131. *Medical Teacher*, 42(8), 846–854. DOI: 10.1080/0142159X.2020.1755030 PMID: 32356468

Kim, T. & Routledge, B., 2022. Why a Right to an Explanation of Algorithmic Decision-Making Should Exist: A Trust-Based Approach. *Business Ethics Quarterly 32:1 (pp. 75–102).* DOI:DOI: 10.1017/beq.2021.3

Kim, J., Park, H., & Lee, C. (2020). Students' perceptions of AI in education: Implications for adoption. *International Journal of Educational Technology in Higher Education*, 17(1), 48.

Kitamura, F. C. (2023). ChatGPT is shaping the future of medical writing but still requires human judgment. *Radiology*, 307(2), e230171. DOI: 10.1148/radiol.230171 PMID: 36728749

Klanjscek, N., David, L., & Frank, M. (2021). Evaluation of an e-learning tool for augmented acoustics in music education. *Music & Science*, 4, 20592043211037511. DOI: 10.1177/20592043211037511

Knouse, L. E., & Fleming, A. P. (2016). Applying cognitive-behavioral therapy for ADHD to emerging adults. *Cognitive and Behavioral Practice*, 23(3), 300–315. DOI: 10.1016/j.cbpra.2016.03.008

Ko, A. (2024). VIRTUAL REALITY IN A SYNCHRONOUS CLASSROOM: A STUDY OF LEARNERS'. *IMMERSIVE EXPERIENCE IN THE APPLICATION OF HIVE AND META QUEST*, 3, 10446–10449. DOI: 10.21125/edulearn.2024.2577

Kofler, M. J., Sarver, D. E., Harmon, S. L., Moltisanti, A., Aduen, P. A., Soto, E. F., & Ferretti, N. (2018). Working memory and organizational skills problems in ADHD. *Journal of Child Psychology and Psychiatry, and Allied Disciplines*, 59(1), 57–67. DOI: 10.1111/jcpp.12773 PMID: 28714075

Kofler, M. J., Singh, L. J., Soto, E. F., Chan, E. S. M., Miller, C. E., Harmon, S. L., & Spiegel, J. A. (2020). Working memory and short-term memory deficits in ADHD: A bifactor modeling approach. *Neuropsychology*, 34(6), 686–698. DOI: 10.1037/neu0000641 PMID: 32437194

Kokotsaki, D., Menzies, V., & Wiggins, A. (2016). Project-based learning: A review of the literature. *Improving Schools*, 19(3), 267–277. DOI: 10.1177/1365480216659733

Kolberg, J. (2023 November 29). 33 ADHD-friendly ways to get organized. ADDitude. https://www.additudemag.com/how-to-get-organized-with-adhd/

Koumpouros, Y. (2024). Revealing the true potential and prospects of augmented reality in education. *Smart Learning Environments*, 11(1), 2. DOI: 10.1186/s40561-023-00288-0

Kramarenko, T. H., & Kochina, O. S. (2023). The use of immersive technologies in teaching mathematics to vocational students. *Journal of Physics: Conference Series*, 2611(1), 1–14. DOI: 10.1088/1742-6596/2611/1/012006

Krashen, S. (2023). *Principles and Practice in Second Language Acquisition*. Pergamon Press.

Krügel, S., Ostermaier, A., & Uhl, M. (2023). The moral authority of ChatGPT. *arXiv preprint arXiv:2301.07098.*

Kshetri, N. (2023). Cybersecurity and malicious uses of AI language models. *Computer*, 56(5), 80–85.

Kshetri, N., & Hughes, L. (2023). "So what if ChatGPT wrote it?" Multidisciplinary perspectives on opportunities, challenges and implications of generative conversational AI for research, practice and policy. *International Journal of Information Management*, 71, 102642. DOI: 10.1016/j.ijinfomgt.2023.102642

Kubiszyn, T., & Borich, G. D. (2024). *Educational testing and measurement*. John Wiley & Sons.

Kumar, A., Krishnamurthi, R., Bhatia, S., Kaushik, K., Ahuja, N. J., Nayyar, A., & Masud, M. (2021). Blended Learning Tools and Practices: A Comprehensive Analysis. *IEEE Access : Practical Innovations, Open Solutions*, 9, 85151–85197. DOI: 10.1109/ACCESS.2021.3085844

Labadze, M., Grigolia, M., & Machaidze, R. (2023). AI chatbots in education: Real-time feedback and efficiency. *Interactive Learning Environments*, 31(2), 217–232.

Laine, T. H. (2018). Mobile educational augmented reality games: A systematic literature review and two case studies. *Computers*, 7(1), 19. Advance online publication. DOI: 10.3390/computers7010019

Lampropoulos, G., Fernández-Arias, P., Antón-Sancho, Á., & Vergara, D. (2024). Affective Computing in Augmented Reality, Virtual Reality, and Immersive Learning Environments. *Electronics (Basel)*, 13(15), 2917. DOI: 10.3390/electronics13152917

Langberg, J., Epstein, J., Urbanowicz, C., Simon, J., & Graham, A. (2008). Efficacy of an organization skills intervention to improve the academic functioning of students with Attention-Deficit/Hyperactivity Disorder. *School Psychology Quarterly*, 23(3), 407–417. DOI: 10.1037/1045-3830.23.3.407

Lecler, A., Duron, L., & Soyer, P. (2023). Revolutionizing radiology with GPT-based models: Current applications, future possibilities and limitations of ChatGPT. *Diagnostic and Interventional Imaging*, 104(6), 269–274. DOI: 10.1016/j.diii.2023.02.003 PMID: 36858933

LeCun, Y., Bengio, Y., & Hinton, G. (2015). Deep learning. nature, 521(7553), 436-444.

Lee, A. (2022). Integrating chatbots in drama-based EFL teaching: A case study in Palestinian schools. *International Journal of Language and Drama Education*, 19(2), 156–172.

Lee, I., & Perret, B. (2022). Preparing high school teachers to integrate AI methods into STEM classrooms. *Proceedings of the AAAI Conference on Artificial Intelligence*, 36(11), 12783–12791. DOI: 10.1609/aaai.v36i11.21557

Lee, J., & Kim, Y. (2023). Sustainable Educational Metaverse Content and System Based on Deep Learning for Enhancing Learner Immersion. *Sustainability (Basel)*, 15(16), 12663. DOI: 10.3390/su151612663

Lee, K. (2012). Augmented Reality in Education and Training. *TechTrends*, 56(2), 13–21. DOI: 10.1007/s11528-012-0559-3

Lee, S.-M., Yang, Z., & Wu, J. G. (2023). Living, entertaining, and learning: Language learner engagement in immersive VR environments. *Education and Information Technologies*. Advance online publication.

Lembani, R., Gunter, A., Breines, M., & Dalu, M. T. B. (2020). The same course, different access: The digital divide between urban and rural distance education students in South Africa. *Journal of Geography in Higher Education*, 44(1), 70–84. DOI: 10.1080/03098265.2019.1694876

Levine, H. (2024 March 13). Meditation and yoga for ADHD. WebMD. https://www.webmd.com/add-adhd/adhd-mindfulness-meditation-yoga

Lewis, L. H., & Williams, C. J. (1994). Experiential Learning: Past and Present. *New Directions for Adult and Continuing Education*. https://eric.ed.gov/?id=EJ492388

Li, Z., Niu, C., Meng, F., Feng, Y., Li, Q., & Zhou, J. (2019). Incremental transformer with deliberation decoder for document grounded conversations. *arXiv preprint arXiv:1907.08854*. DOI: 10.18653/v1/P19-1002

Liberati, N. (2016). Augmented reality and ubiquitous computing: The hidden potentialities of augmented reality. *AI & Society*, 31(1), 17–28. DOI: 10.1007/s00146-014-0543-x

Liebrenz, M., Schleifer, R., Buadze, A., Bhugra, D., & Smith, A. (2023). Generating scholarly content with ChatGPT: Ethical challenges for medical publishing . *The Lancet. Digital Health*, 5(3), e105–e106. DOI: 10.1016/S2589-7500(23)00019-5 PMID: 36754725

Lin, S., Chung, H.-H., Chung, F.-L., & Lan, K. (2023). *Concerns About Using ChatGPT in Education*. .DOI: 10.1007/978-3-031-40113-8_4

Lin, X. (2023). *Exploring the role of ChatGPT as a facilitator for motivating self-directed learning among adult learners.*

Lindner, A., Romeike, R., Jasute, E., & Pozdniakov, S. (2019). *Lindner, A., Romeike, R., Jasute, E., & Pozdniakov, S. (2019). Teachers' perspectives on artificial intelligence.* . In 12th International conference on informatics in schools,"Situation, evaluation and perspectives", ISSEP.

Lindroth, H., Nalaie, K., Raghu, R., Ayala, I. N., Busch, C., Bhattacharyya, A., Moreno Franco, P., Diedrich, D. A., Pickering, B. W., & Herasevich, V. (2024). Applied artificial intelligence in healthcare: A review of computer vision technology application in hospital settings. *Journal of Imaging*, 10(4), 81. DOI: 10.3390/jimaging10040081 PMID: 38667979

Lin, H. (2022). Influences of Artificial Intelligence in Education on Teaching Effectiveness: The Mediating Effect of Teachers' Perceptions of Educational Technology. *International Journal of Emerging Technologies in Learning*, 17(24), 144–156. DOI: 10.3991/ijet.v17i24.36037

Liu, D. J., & Huang, R. H. (2016). Potential, Progress, and Challenges of Virtual Reality Technology in Educational Applications. *Open Education Research*, 22(4), 25–31.

Liu, P., Yang, Z., Huang, J., & Wang, T.-K. (2024). The effect of augmented reality applied to learning process with different learning styles in structural engineering education. *Engineering, Construction, and Architectural Management.* Advance online publication. DOI: 10.1108/ECAM-06-2023-0596

Liu, Y., & Iwamitsu, K. (2021). Ethical considerations in the use of AI language models for information services. *Journal of Information Ethics*, 18(3), 245–262.

Li, X., Yi, W., Chi, H.-L., Wang, X., & Chan, A. P. C. (2018). A critical review of virtual and augmented reality (VR/AR) applications in construction safety. *Automation in Construction*, 86, 150–162. DOI: 10.1016/j.autcon.2017.11.003

LMS AI. (2024 February 20). Make your research plan easier using Goblin Tools. [Blog Post]. Retrieved from: https://www.rs.lmssolution.net/post/make-your-research-plan-easier-using-goblin-tools

Loe, I. M., & Feldman, H. M. (2007). Academic and educational outcomes of children with ADHD. *Journal of Pediatric Psychology*, 32(6), 643–654. DOI: 10.1093/jpepsy/jsl054 PMID: 17569716

Lo, H. (2023). The potential of ChatGPT in enhancing active learning in higher education. *AI in Education Journal*, 5(1), 34–45.

Longpre, S., Yauney, G., Reif, E., Lee, K., Roberts, A., Zoph, B., Zhou, D., Wei, J., Robinson, K., Mimno, D., & Ippolito, D. (2023). A Pretrainer's Guide to Training Data: Measuring the Effects of Data Age, Domain Coverage, Quality, & Toxicity. Cornell. https://doi.org//arXiv.2305.13169DOI: 10.48550

Lovering, N. (2022 May 18). The Connection Between Emotional Regulation and ADHD. Healthline. https://www.healthline.com/health/adhd/emotional-regulation

Luckin, R. (2019). The impact of AI on student learning outcomes: What we know and what we need to know. *AI and Education Journal*, 19(3), 45–60.

Luckin, R., Holmes, W., Griffiths, M., & Forcier, L. B. (2021). AI in Education: A Systemic Perspective. *AIED Journal*, 31(4), 239–248.

Lund, B. D., & Wang, T. (2023). Chatting about ChatGPT: How may AI and GPT impact academia and libraries? *Library Hi Tech News*, 40(3), 26–29. DOI: 10.1108/LHTN-01-2023-0009

Luo, H., Li, G., Feng, Q., Yang, Y., & Zuo, M. (2021). Virtual reality in K-12 and higher education: A systematic review of the literature from 2000 to 2019. *Journal of Computer Assisted Learning*, 37(3), 887–901. DOI: 10.1111/jcal.12538

Luo, T., & Leng, W. (2021). Application of immersive virtual reality technology in earth sciences. *Journal of the University of Science and Technology of China*, 51(06), 431–440.

Lu, Y., Wang, X., Gong, J., & Liang, Y. (2022). ChordAR: An educational AR game design for children's music theory learning. *Wireless Communications and Mobile Computing*, 2022(1), 5268586. DOI: 10.1155/2022/5268586

Mabuan, R. A. (2024). ChatGPT and ELT: Exploring teachers' voices . [IJTE]. *International Journal of Technology in Education*, 7(1), 128–153. DOI: 10.46328/ijte.523

Mageira, K., Pittou, D., Papasalouros, A., Kotis, K., Zangogianni, P., & Daradoumis, A. (2022). Educational AI chatbots for content and language integrated learning . *Applied Sciences (Basel, Switzerland)*, 12(7), 3239. DOI: 10.3390/app12073239

Mali T. S. (2023). Use Of Chat GPT In Library Services. International Journal of Creative Research Thoughts. Vol. 11, Issue 4.

Malinka, K., Perešíni, M., Firc, A., Hujňák, O., & Januš, F. (2023). *On the Educational Impact of ChatGPT: Is Artificial Intelligence Ready to Obtain a University Degree?* . arXiv preprint arXiv:2303.11146. DOI: 10.1145/3587102.3588827

Malizia, A., & Paterno, F. (2023). Why Is the Current XAI Not Meeting the Expectations? *ACM*.https://acmwebvm01.acm.org/magazines/2023/12/278147-why-is-the-current-xai-not-meeting-the-expectations/fulltext

Mann, B., Ryder, N., Subbiah, M., Kaplan, J., Dhariwal, P., Neelakantan, A., . . . Amodei, D. (2020). Language models are few-shot learners. *arXiv preprint arXiv:2005.14165, 1.*

Mann, D. L. (2023). Artificial intelligence discusses the role of artificial intelligence in translational medicine: a JACC: basic to translational science interview with ChatGPT. *JACC. Basic to Translational Science*, 8(2), 221–223. DOI: 10.1016/j.jacbts.2023.01.001 PMID: 36908674

Manning, A. (2024). Researchers publish largest-ever dataset of neural connections. *Harvard Gazette*. https://news.harvard.edu/gazette/story/2024/05/the-brain-as-weve-never-seen-it/

Manning, C. D., Surdeanu, M., Bauer, J., Finkel, J., Bethard, S., & McClosky, D. (2022). *The Stanford CoreNLP natural language processing toolkit*. Association for Computational Linguistics.

Marchandot, B., Matsushita, K., Carmona, A., Trimaille, A., & Morel, O. (2023). ChatGPT: The Next Frontier in Academic Writing for Cardiologists or a Pandora's Box of Ethical Dilemmas. *European Heart Journal Open*, 3(2), oead007. Advance online publication. DOI: 10.1093/ehjopen/oead007 PMID: 36915398

Marcus, G. (2024). Will AI Destroy Us? - AI Virtual Roundtable. Coleman Hughes. https://www.youtube.com/watch?v=xkQSiS8hpZA

Marr, B. (2023). The 15 Biggest Risks Of Artificial Intelligence. Forbes. https://www.forbes.com/sites/bernardmarr/2023/06/02/the-15-biggest-risks-of-artificial-intelligence/?sh=72f2c8502706

Martin-Gutierrez, J., Del Rio Guerra, M. S., Lopez-Chao, V., Soto Gastelum, R. H., & Valenzuela Bojórquez, J. F. (2020). Augmented reality to facilitate learning of the acoustic guitar. *Applied Sciences (Basel, Switzerland)*, 10(7), 2425. DOI: 10.3390/app10072425

Martín-Gutiérrez, J., Fabiani, P., Benesova, W., Meneses, M. D., & Mora, C. E. (2015). Augmented reality to promote collaborative and autonomous learning in higher education. *Computers in Human Behavior*, 51, 752–761. DOI: 10.1016/j.chb.2014.11.093

Maté, G. (1999). *Scattered: How attention deficit disorder originates and what you can do about it.* Penguin.

Matthews, J. A., & Volpe, C. R. (2023). Academics' perceptions of ChatGPT-generated written outputs: A practical application of Turing's Imitation Game. *Australasian Journal of Educational Technology*, 39(5), 82–100. DOI: 10.14742/ajet.8896

Mauki, M. M., Kitur, J., Ireri, N. W., & Ngala, F. W. (2020). Competency-Based Curriculum implementation and the role of the universities in Kenya. *Impact: Journal of Transformation*, 45–52.

Mayo Clinic Staff. (2023 December 13). Meditation: A simple, fast way to reduce stress. Mayo Clinic._https://www.mayoclinic.org/tests-procedures/meditation/in-depth/meditation/art-20045858

McGorry, P. D., Mei, C., Chanen, A., Hodges, C., Alvarez-Jimenez, M., & Killackey, E. (2022). Designing and scaling up integrated youth mental health care. *World Psychiatry; Official Journal of the World Psychiatric Association (WPA)*, 21(1), 61–76. DOI: 10.1002/wps.20938 PMID: 35015367

McKendrick, J., & Thural, A. (2022). AI Isn't Ready to Make Unsupervised Decisions. *HBR*.https://hbr.org/2022/09/ai-isnt-ready-to-make-unsupervised-decisions

Mei, B., & Yang, S. (2021). Chinese pre-service music teachers' perceptions of augmented reality-assisted musical instrument learning. *Frontiers in Psychology*, 12, 609028. DOI: 10.3389/fpsyg.2021.609028 PMID: 33613388

Meniado, J. (2023). The Impact of ChatGPT on English Language Teaching, Learning, and Assessment: A Rapid Review of Literature. *Arab World English Journal*, 14, 3–18. DOI: 10.24093/awej/vol14no4.1

Mhlanga, D. (2023). Open AI in Education, the Responsible and Ethical Use of ChatGPT Towards Lifelong Learning. SSRN *Electronic Journal*, DOI: 10.2139/ssrn.4354422

Miller, C. (2022). Overview of the Palestinian education system: Structure and dynamics. *Journal of Education and Policy Studies*, 14(3), 98–112.

Miller, N. C., Kumar, S., Pearce, K. L., & Baldock, K. L. (2021). The outcomes of nature-based learning for primary school aged children: A systematic review of quantitative research. *Environmental Education Research*, 27(8), 1–26. DOI: 10.1080/13504622.2021.1921117

Min, K. B., & Kang, J. M. (2010). Rights to Control Information and Related Security Technologies on the CyberSpace. *The Journal of the Institute of Internet. Broadcasting and Communication*, 10(2), 135–141.

Mitchell, J. T., McIntyre, E. M., English, J. S., Dennis, M. F., Beckham, J. C., & Kollins, S. H. (2017). A pilot trial of mindfulness meditation training for ADHD in adulthood: Impact on core symptoms, executive functioning, and emotion dysregulation. *Journal of Attention Disorders*, 21(13), 1105–1120. DOI: 10.1177/1087054713513328 PMID: 24305060

Mitchell, M., Potash, E., & Gebru, T. (2023). Examining biases in AI language models: A call to action. *Communications of the ACM*, 66(2), 39–45.

Mitrović, S., Andreoletti, D., & Ayoub, O. (2023). Chatgpt or human? detect and explain. explaining decisions of machine learning model for detecting short chatgpt-generated text. *arXiv preprint arXiv:2301.13852*.

Mizumoto, A., & Eguchi, M. (2023). Exploring the potential of using an AI language model for automated essay scoring. *Research Methods in Applied Linguistics*, 2(2), 100050. DOI: 10.1016/j.rmal.2023.100050

Moe, N. B., & Dingsøyr, T. (2008). Scrum and team effectiveness: theory and practice. In Lecture notes in business information processing (pp. 11–20). https://doi.org/DOI: 10.1007/978-3-540-68255-4_2

Mohamed Nor, H., & Sihes, A. J. (2021). Critical Thinking Skills in Education: A Systematic Literature Review. *International Journal of Academic Research in Business & Social Sciences*, 11(11). Advance online publication. DOI: 10.6007/IJARBSS/v11-i11/11529

Mohammed, A., Al-ghazali, A., & Alqohfa, K.Khalid A. S. Alqohfa. (2023). Exploring ChatGPT uses in higher studies: A case study of Arab postgraduates in India. *Journal of English Studies in Arabia Felix*, 2(2), 8–15. DOI: 10.56540/jesaf.v2i2.55

Molina-Carmona, R., Pertegal-Felices, M. L., Jimeno-Morenilla, A., & Mora-Mora, H. (2018). Virtual Reality Learning Activities for Multimedia Students to Enhance Spatial Ability. *Sustainability (Basel)*, 10(4), 1074. DOI: 10.3390/su10041074

Mollick, E. (2022). ChatGPT Is a Tipping Point for A. *HBR.*https://hbr.org/2022/12/chatgpt-is-a-tipping-point-for-ai

Monahan, T., McArdle, G., & Bertolotto, M. (2008). Virtual reality for collaborative e-learning. *Computers & Education*, 50(4), 1339–1353. DOI: 10.1016/j.compedu.2006.12.008

Montenegro-Rueda, M., Fernández-Cerero, J., Fernández-Batanero, J. M., & López-Meneses, E. (2023). Impact of the Implementation of ChatGPT in Education: A Systematic Review. *Computers*, 12(8), 153. DOI: 10.3390/computers12080153

Moreno-Guerrero, A.-J., Alonso García, S., Ramos Navas-Parejo, M., Campos-Soto, M. N., & Gómez García, G. (2020). Augmented Reality as a Resource for Improving Learning in the Physical Education Classroom. *International Journal of Environmental Research and Public Health*, 17(10), 3637. DOI: 10.3390/ijerph17103637 PMID: 32455778

Morze, N., Varchenko-Trotsenko, L., Terletska, T., & Smyrnova-Trybulska, E. (2021). Implementation of adaptive learning at higher education institutions by means of Moodle LMS. *Journal of Physics: Conference Series*, 1840(1), 012062. DOI: 10.1088/1742-6596/1840/1/012062

Moura, A., & Carvalho, A. A. A. (2024, February). Teachers' perceptions of the use of artificial intelligence in the classroom. In *International Conference on Lifelong Education and Leadership for All (ICLEL 2023)* (pp. 140-150). Atlantis Press.

Mukul, E., & Büyüközkan, G. (2023). Digital transformation in education: A systematic review of education 4.0. *Technological Forecasting and Social Change*, 194, 122664. DOI: 10.1016/j.techfore.2023.122664

Muralidharan, V., Tran, M. M., Barrios, L., Beams, B., Ko, J. M., Siegel, D. H., & Bailenson, J. (2024). Best Practices for Research in Virtual and Augmented Reality in Dermatology. *The Journal of Investigative Dermatology*, 144(1), 17–23. DOI: 10.1016/j.jid.2023.10.014 PMID: 38105083

Murgia, M. (2023). Generative AI exists because of the transformer. Financial Times. https://ig.ft.com/generative-ai/

Murphy, K. R., & Gordon, M. (1998). Assessment of adults with ADHD. In Barkley, R. A. (Ed.), *Attention-deficit hyperactivity disorder: A handbook for diagnosis and treatment* (2nd ed., pp. 345–369). The Guilford Press.

Mustafa, H. (2023). Innovations in English language teaching in Palestine: Challenges and opportunities. *International Journal of Language Education*, 18(2), 85–101.

Nagtilak, P. S. A. (2024). Survey on Gesture-Based Virtual Keyboard and Mouse. *Interantional Journal of Scientific Research in Engineering and Management*, 08(01), 1–13. DOI: 10.55041/IJSREM28414

Nami, F. (2019). Educational smartphone apps for language learning in higher education: Students' choices and perceptions. *Australasian Journal of Educational Technology*, •••, 82–95. DOI: 10.14742/ajet.5350

Narayanan, A. (2023). Disinformation Researchers Raise Alarms About A.I. Chatbots. New York Times. https://www.nytimes.com/2023/02/08/technology/ai-chatbots-disinformation.html

Nasser, R., & Said, M. (2023). Innovative use of AI in drama-based EFL teaching: Overcoming traditional barriers in Palestinian schools. *International Journal of Educational Technology and Language Learning*, 18(2), 145–160.

Nastiti, K. A., Santoso, H. B., Junus, K., Ahmad, M., & Purwandari, E. P. (2024). The Effect of Immersive Technology on Enhancing Student Learning:A Systematic Literature Review. [Jurnal Ilmu Pengetahuan Dan Teknologi Komputer]. *JITK*, 9(2), 208–217. DOI: 10.33480/jitk.v9i2.4775

National Geographic. (2015). *Europe: Physical Geography | National Geographic Society*. Education.nationalgeographic.org. https://education.nationalgeographic.org/resource/europe-physical-geography/

Navarro, C., Arias-Calderón, M., Henríquez, C. A., & Riquelme, P. (2024). Assessment of Student and Teacher Perceptions on the Use of Virtual Simulation in Cell Biology Laboratory Education. *Education Sciences*, 14(3), 243. DOI: 10.3390/educsci14030243

Nazari, M. A., Mirloo, M. M., Rezaei, M., & Soltanlou, M. (2018). Emotional stimuli facilitate time perception in children with attention deficit/hyperactivity disorder. *Journal of Neuropsychology*, 12(2), 165–175. DOI: 10.1111/jnp.12111 PMID: 27566601

Neelands, J. (2023). *Drama as Creative Learning: Exploring the Role of Drama in Education*. Routledge.

Neumann, M., Rauschenberger, M., & Schön, E. M. (2023, May). "We need to talk about ChatGPT": The future of AI and higher education. In *2023 IEEE/ACM 5th International Workshop on Software Engineering Education for the Next Generation (SEENG)* (pp. 29-32). IEEE.

Ng, W. (2012). Can we teach digital natives digital literacy? *Computers & Education*, 59(3), 1065–1078. DOI: 10.1016/j.compedu.2012.04.016

Nicklen, P. (2023, September 14). *Top 10 facts about Emperor penguins.* WWF. https://www.wwf.org.uk/learn/fascinating-facts/emperor-penguins

Nikiforos, S., Tzanavaris, S., & Kermanidis, K.-L. (2020). Virtual learning communities (VLCs) rethinking: Influence on behavior modification—bullying detection through machine learning and natural language processing. *Journal of Computers in Education*, 7(4), 531–551. DOI: 10.1007/s40692-020-00166-5

Nikimaleki, M., & Rahimi, M. (2022). Effects of a collaborative <scp>AR</scp> -enhanced learning environment on learning gains and technology implementation beliefs: Evidence from a graduate teacher training course. *Journal of Computer Assisted Learning*, 38(3), 758–769. DOI: 10.1111/jcal.12646

Niloy, A. C., Akter, S., Sultana, N., Sultana, J., & Rahman, S. I. (2024). Is Chatgpt a menace for creative writing ability? An experiment . *Journal of Computer Assisted Learning*, 40(2), 919–930. DOI: 10.1111/jcal.12929

Nolan, B. (2023). Google researchers say they got OpenAI's ChatGPT to reveal some of its training data with just one word. Business Insider. https://www.businessinsider.com/google-researchers-openai-chatgpt-to-reveal-its-training-data-study-2023-12#:~:text=The%20AI%20model%20powering%20ChatGPT,or%20570%20GB%2C%20of%20data

NOYB. (2024). ChatGPT provides false information about people, and OpenAI can't correct it. *NOYB.*https://noyb.eu/en/chatgpt-provides-false-information-about-people-and-openai-cant-correct-it

nsidc. (2017). *Why Sea Ice Matters.* National Snow and Ice Data Center. https://nsidc.org/learn/parts-cryosphere/sea-ice/why-sea-ice-matters

O'neil, C. (2017). *Weapons of math destruction: How big data increases inequality and threatens democracy*. Crown.

Oke, A., & Fernandes, F. A. P. (2020). Innovations in Teaching and Learning: Exploring the Perceptions of the Education Sector on the 4th Industrial Revolution (4IR). *Journal of Open Innovation*, 6(2), 31. DOI: 10.3390/joitmc6020031

Oliver, M., & Trigwell, K. (2005). Can 'Blended Learning' Be Redeemed? *E-Learning and Digital Media*, 2(1), 17–26. DOI: 10.2304/elea.2005.2.1.17

Opara, E. C. (2023). ChatGPT for Teaching, Learning and Research: Prospects and Challenges. *Global Academy Journal of Humanities and Social Sciences*, 5(2), 33-40. DOI: 10.13140/RG.2.2.21304.42244/1

Open, A. I. (2021). ChatGPT: Improving Language Flow and Response Consistency in Chat. Retrieved from https://openai.com/blog/chatgpt/

Open, A. I. (2023). ChatGPT can now see, hear, and speak. OpenAI. https://openai.com/blog/chatgpt-can-now-see-hear-and-speak (Access date: 19 October 2023).

Open, A. I. (2024). Limitations. *OpenAI.*https://openai.com/blog/chatgpt

Open, A. I. _A, (2024). Improving language understanding with unsupervised learning. *OpenAI.*https://openai.com/research/language-unsupervised?ref=superpower-chatgpt-extension

Open, A. I. _B, (2024). How should AI systems behave, and who should decide? *Open_AI.*https://openai.com/index/how-should-ai-systems-behave/

Orcam. (2023). Empowering students with ADHD and dyslexia. Orcam_https://www.orcam.com/en-ca/orcam-learn

Ortagus, J. C., Kelchen, R., Rosinger, K., & Voorhees, N. (2020). Performance-Based Funding in American Higher Education: A Systematic Synthesis of the Intended and Unintended Consequences. *Educational Evaluation and Policy Analysis*, 42(4), 520–550. DOI: 10.3102/0162373720953128

Pagallo, U. (2017). Automation to autonomous systems: A legal phenomenology with problems of accountability. In C. Sierra (Ed.). Proceedings of the twenty-sixth joint Conference in artificial intelligence (issue 26th international joint conference on artificial intelligence (IJCAI) (pp. 17–23). DOI: 10.24963/ijcai.2017/3

Page, M. J., McKenzie, J. E., Bossuyt, P. M., Boutron, I., Hoffmann, T. C., Mulrow, C. D., Shamseer, L., Tetzlaff, J. M., Akl, E. A., Brennan, S. E., Chou, R., Glanville, J., Grimshaw, J. M., Hróbjartsson, A., Lalu, M. M., Li, T., Loder, E. W., Mayo-Wilson, E., McDonald, S., & Moher, D. (2021). The PRISMA 2020 statement: An updated guideline for reporting systematic reviews. *PLoS Medicine*, 18(3), e1003583. DOI: 10.1371/journal.pmed.1003583 PMID: 33780438

Pahlevan Sharif, S., Mura, P., & Wijesinghe, S. N. R. (2019). Systematic Reviews in Asia: Introducing the "PRISMA" Protocol to Tourism and Hospitality Scholars. 13–33. DOI: 10.1007/978-981-13-2463-5_2

Pakinee, A., & Puritat, K. (2021). Designing a gamified e-learning environment for teaching undergraduate ERP course based on big five personality traits. *Education and Information Technologies*, 26(4), 4049–4067. DOI: 10.1007/s10639-021-10456-9 PMID: 33613080

Panasan, M., & Nuangchalerm, P. (2010). Learning Outcomes of Project-Based and Inquiry-Based Learning Activities. In *ERIC* (Vol. 6). https://eric.ed.gov/?id=ED509723

Papanastasiou, G., Drigas, A., Skianis, C., Lytras, M., & Papanastasiou, E. (2019). Virtual and augmented reality effects on K-12, higher and tertiary education students' twenty-first century skills. *Virtual Reality (Waltham Cross)*, 23(4), 425–436. DOI: 10.1007/s10055-018-0363-2

Parmaxi, A., Stylianou, K., & Zaphiris, P. (2017). *Leveraging Virtual Trips in Google Expeditions to Elevate Students' Social Exploration* (pp. 368–371). DOI: 10.1007/978-3-319-68059-0_32

Parshley, L. (2023, October 1). *Artificial Intelligence Could Finally Let Us Talk with Animals*. Scientific American. https://www.scientificamerican.com/article/artificial-intelligence-could-finally-let-us-talk-with-animals/

Partarakis, N., & Zabulis, X. (2024). A Review of Immersive Technologies, Knowledge Representation, and AI for Human-Centered Digital Experiences. *Electronics (Basel)*, 13(269), 1–19. DOI: 10.3390/electronics13020269

Paszkiewicz, A., Salach, M., Dymora, P., Bolanowski, M., Budzik, G., & Kubiak, P. (2021). Methodology of Implementing Virtual Reality in Education for Industry 4.0. *Sustainability (Basel)*, 13(9), 5049. DOI: 10.3390/su13095049

Patias, P., Georgoula, O., Kaimaris, D., Georgiadis, C., Roustanis, T., Klimantakis, K., Fragkoulidou, V., Chalkidou, S., Papadopoulos, G.-J., Giannopoulos, D., Poulopoulos, K., & Vassilakopoulou, A. (2023). Ar and Chamber Music: Musical Kaleidoscope. *The International Archives of the Photogrammetry, Remote Sensing and Spatial Information Sciences*, 48, 1165–1171. DOI: 10.5194/isprs-archives-XLVIII-M-2-2023-1165-2023

Pati, K. A., Gururani, S., & Lerch, A. (2018). Assessment of student music performances using deep neural networks. *Applied Sciences (Basel, Switzerland)*, 8(4), 507. DOI: 10.3390/app8040507

Patiño-Escarcina, R. E., Barrios-Aranibar, D., Bernedo-Flores, L. S., Alsina, P. J., & Gonçalves, L. M. G. (2021). A Methodological Approach to the Learning of Robotics with EDUROSC-Kids. *Journal of Intelligent & Robotic Systems*, 102(2), 34. Advance online publication. DOI: 10.1007/s10846-021-01400-7 PMID: 34025033

Pears, M., & Konstantinidis, S. (2022). The Future of Immersive Technology in Global Surgery Education. *Indian Journal of Surgery*, 84(S1), 281–285. DOI: 10.1007/s12262-021-02998-6 PMID: 34230785

Pederson, T. (2023 September 29). What is time blindness? Healthline. https://www.healthline.com/health/time-blindness

Pedró, F. (2020). Applications of Artificial Intelligence to higher education: Possibilities, evidence, and challenges . *IUL Research*, 1(1), 61–76. DOI: 10.57568/iulres.v1i1.43

Pellas, N., Mystakidis, S., & Kazanidis, I. (2021). Immersive Virtual Reality in K-12 and Higher Education: A systematic review of the last decade scientific literature. *Virtual Reality (Waltham Cross)*, 25(3), 835–861. DOI: 10.1007/s10055-020-00489-9

Pereira, L. F. B., da Silva, M. M. O., Roberto, R. A., Teichrieb, V., & Cavalcante, P. S. (2020). Virtual Playground: An Augmented Reality Application to Improve English Learning. *2020 22nd Symposium on Virtual and Augmented Reality (SVR)*, 83–91. DOI: 10.1109/SVR51698.2020.00027

Perkins, M. (2023). Academic Integrity considerations of AI Large Language Models in the post-pandemic era: ChatGPT and beyond. *Journal of University Teaching & Learning Practice*, 20(2). Advance online publication. DOI: 10.53761/1.20.02.07

Perrigo, B. (2023). Exclusive: OpenAI Used Kenyan Workers on Less Than $2 Per Hour to Make ChatGPT Less Toxic. Time. https://time.com/6247678/openai-chatgpt-kenya-workers/

Phakamach, P., Senarith, P., & Wachirawongpaisarn, S. (2022). The Metaverse in Education: The Future of Immersive Teaching & Learning. *RICE Journal of Creative Entrepreneurship and Management, 3*(2), 75–88. https://doi.org/DOI: 10.14456/rjcm.2022

Phillips, P. J., Hahn, C. A., Fontana, P. C., Yates, A. N., Greene, K., Broniatowski, D. A., & Przybocki, M. A. (2019). Four Principles of Explainable Artificial Intelligence. NIST Interagency/Internal Report (NISTIR) 8312. https://nvlpubs.nist.gov/nistpubs/ir/2021/NIST.IR.8312.pdf

Piaget, J. (2023). *The Construction of Reality in the Child*. Routledge.

Ping, T. P., Dennis, P. S., Julaihi, A. A., & Ling, M. G. B. (2022). Mobile Learning Model for Children with Special Learning Needs. *European Journal of Teacher Education*, 4(3), 1–12. DOI: 10.33422/ejte.v4i3.815

Posey, J. (2024 May 9). 10 best AI tools for project managers in 2024. [Blog Post] Available at: https://www.usemotion.com/blog/ai-tools-for-product-managers

Potkonjak, V., Gardner, M., Callaghan, V., Mattila, P., Guetl, C., Petrović, V. M., & Jovanović, K. (2016). Virtual laboratories for education in science, technology, and engineering: A review. *Computers & Education*, 95, 309–327. DOI: 10.1016/j.compedu.2016.02.002

Prada, J. (2024). EF skills: Organization. Smart Kids with ADHD._https://www.smartkidswithld.org/getting-help/executive-function-disorder/ef-skills-organization/

PRC. (2023). Interim Measures for the Management of Generative AI Services. People's Republic of China. https://www.cac.gov.cn/2023-07/13/c_1690898327029107.htm

Preka, G., & Rangoussi, M. (2019). Augmented Reality and QR Codes for Teaching Music to Preschoolers and Kindergarteners: Educational Intervention and Evaluation. In CSEDU (1) (pp. 113-123).

Ptacek, R., Weissenberger, S., Braaten, E., Klicperova-Baker, M., Goetz, M., Raboch, J., Vnukova, M., & Stefano, G. B. (2019). Clinical implications of the perception of time in attention deficit hyperactivity disorder (ADHD): A review. *Medical Science Monitor*, 25, 3918–3924. DOI: 10.12659/MSM.914225 PMID: 31129679

Puncreobutr, V. (2016). Education 4.0: New challenge of learning. *St. Theresa. Journal of the Humanities and Social Sciences*, 2(2), 92–97.

Qadir, J. (2023, May). Engineering education in the era of ChatGPT: Promise and pitfalls of generative AI for education. In *2023 IEEE Global Engineering Education Conference (EDUCON)* (pp. 1-9). IEEE. DOI: 10.1109/EDUCON54358.2023.10125121

Quinn, P. O., Ratey, N. A., & Maitland, T. L. (2000). Working with an ADD coach. In Quinn, P. O., Ratey, N. A., & Maitland, T. L. (Eds.), *ADD and the college student: A guide for high school and college students with attention deficit disorder* (pp. 99–109). Magination Press.

Quinn, R.. (2022). Evaluating Bias in OpenAI's GPT-3. *Proceedings of the 2022 Conference on Empirical Methods in Natural Language Processing (EMNLP)*, 6578-6589.

Quinn, S., Sanchez, M., & Patel, R. (2022). Mitigating biases and inaccuracies in AI language models for information retrieval. *Journal of Information Science*, 48(4), 509–526.

Radianti, J., Majchrzak, T. A., Fromm, J., & Wohlgenannt, I. (2020). A systematic review of immersive virtual reality applications for higher education: Design elements, lessons learned, and research agenda. *Computers & Education*, 147, 103778. DOI: 10.1016/j.compedu.2019.103778

Raffoul, S., & Jaber, L. (2023). Text-to-Speech software and reading comprehension: The impact for students with learning disabilities. *Canadian Journal of Learning and Technology*, 49(2), 1–18. DOI: 10.21432/cjlt28296

Rahl, H. A., Lindsay, E. K., Pacilio, L. E., Brown, K. W., & Creswell, J. D. (2017). Brief mindfulness meditation training reduces mind wandering: The critical role of acceptance. *Emotion (Washington, D.C.)*, 17(2), 224–230. DOI: 10.1037/emo0000250 PMID: 27819445

Rahman, M. S., Sabbir, M. M., Zhang, J., Moral, I. H., & Hossain, G. M. (2022). Examining students' intention to use ChatGPT: Does trust matter? . *Australasian Journal of Educational Technology*, •••, 51–71.

Rai, A., (2020). Explainable AI: from black box to glass box. *Journal of the Academy of Marketing Science Volume 48*, p.p 137–141. DOI: 10.1007/s11747-019-00710-5

Raith, A., Kamp, C., Stoiber, C., Jakl, A., & Wagner, M. (2022). Augmented Reality in Radiology for Education and Training—A Design Study. *Health Care*, 10(4), 672. DOI: 10.3390/healthcare10040672 PMID: 35455849

Rajain, M., Bhatia, R., Tripathi, M., Kumar, N., & Mathur, R. (2023). Low-Frequency Repetitive Transcranial Magnetic Stimulation for Chronic Tension-Type Headache: A Randomized Controlled Study. *Cureus*, 15(2). Advance online publication. DOI: 10.7759/cureus.34922 PMID: 36938162

Rana, B., Singh, Y., & Singh, P. K. (2021). A systematic survey on internet of things: Energy efficiency and interoperability perspective. *Transactions on Emerging Telecommunications Technologies*, 32(8), e4166. Advance online publication. DOI: 10.1002/ett.4166

Randoll, D., & Peters, J. (2015). Empirical research on Waldorf education. *Educar em Revista*, 56(56), 33–47. DOI: 10.1590/0104-4060.41416

Rasidah binti Sardi. (2022). Vr Kidos – Virtual Reality Kids Pedagogy Learning System. *European Proceedings of Multidisciplinary Sciences*. DOI: 10.15405/epms.2022.10.62

Ratheeswari, K. (2018). Information communication technology in education . *Journal of Applied and Advanced research*, 3(1), 45-47.

Ratti, E. & Graves, M., (2021). Explainable machine learning practices: opening another black box for reliable medical AI. *AI and Ethics (2022) 2:801–814.* https:// doi.org/DOI: 10.1007/s43681-022-00141-z

Ray, P. P. (2023). ChatGPT: A comprehensive review on background, applications, key challenges, bias, ethics, limitations and future scope. *Internet of Things and Cyber- Physical Systems Volume 3*, 121-154.

Richards, S. (2023). Student Engagement Using HoloLens Mixed-Reality Technology in Human Anatomy Laboratories for Osteopathic Medical Students: An Instructional Model. *Medical Science Educator*, 33(1), 223–231. DOI: 10.1007/ s40670-023-01728-9 PMID: 36691419

Rodden, J. (2021 January 14). What is executive dysfunction? Signs and symptoms of EFD. ADDitude. https://www.additudemag.com/what-is-executive-function -disorder/

Roselló, B., Berenguer, C., Baixauli, I., Mira, Á., Martinez-Raga, J., & Miranda, A. (2020). Empirical examination of executive functioning, ADHD associated behaviors, and functional impairments in adults with persistent ADHD, remittent ADHD, and without ADHD. *BMC Psychiatry*, 20(1), 134. DOI: 10.1186/s12888-020-02542-y PMID: 32204708

Rosendahl, P., Müller, M., & Wagner, I. (2024). A 360° video as visual training support for independent movement acquisition—Benefit evaluation with the TAM. *German Journal of Exercise and Sport Research*, 54(3), 383–392. Advance online publication. DOI: 10.1007/s12662-023-00930-6

Roumeliotis, K. I., & Tselikas, N. D. (2023). ChatGPT and Open-AI Models: A Preliminary Review. *Future Internet*, 15(6), 192. DOI: 10.3390/fi15060192

Rovithis, E., Moustakas, N., Floros, A., & Vogklis, K. (2019). Audio legends: Investigating sonic interaction in an augmented reality audio game. *Multimodal Technologies and Interaction*, 3(4), 73. DOI: 10.3390/mti3040073

Rudin, C. (2019). Stop Explaining Black Box Machine Learning Models for High Stakes Decisions and Use Interpretable Models Instead. National Library of Medicine. https://pubmed.ncbi.nlm.nih.gov/35603010/

Rudolph, J., Tan, S., & Tan, S. (2023). ChatGPT: Bullshit spewer or the end of traditional assessmenin higher education. *Journal of Applied Learning and Teaching, 6(1), 342-363. https://doi.org/*, Journal of Applied Learning and Teaching, 6(1), 342-363. https://doi.org/.DOI: 10.37074/jalt.2023.6.1.9

Rushkoff, D. (2013). *Its not just grammar; its clear thinking.* The New York Times. Retrievedfrom:https://www.nytimes.com/roomfordebate/2012/08/13/is-our-children-learning-enough-grammar-to-get-hired/its-not-just-grammar-its-clear-thinking

Russell, S. (2023). AI likely to spell end of traditional school classroom, leading expert says. Guardian. https://www.theguardian.com/technology/2023/jul/07/ai-likely-to-spell-end-of-traditional-school-classroom-leading-expert-says

Russell, S., & Norvig, P. (2021). *Artificial Intelligence: A Modern Approach* (4th ed.). Pearson.

Ryan, R. M., & Deci, E. L. (2000). Intrinsic and extrinsic motivations: Classic definitions and new directions. *Contemporary Educational Psychology*, 25(1), 54–67. DOI: 10.1006/ceps.1999.1020 PMID: 10620381

Sadek, O., Baldwin, F., Gray, R., Khayyat, N., & Fotis, T. (2023). Impact of Virtual and Augmented Reality on Quality of Medical Education During the COVID-19 Pandemic: A Systematic Review. In *Journal of Graduate Medical Education* (Vol. 15, Issue 3). DOI: 10.4300/JGME-D-22-00594.1

Sadler, R., & Thrasher, T. (2023). XR: Crossing Reality to Enhance Language Learning. *CALICO Journal*, 40(1), i–xi. DOI: 10.1558/cj.25517

Safety Summit, A. I. (2023). Capabilities and risks from frontier AI. Department of Science, Innovation & Technology. https://assets.publishing.service.gov.uk/media/65395abae6c968000daa9b25/frontier-ai-capabilities-risks-report.pdf

Sahranavard, S., Miri, M. R., & Salehiniya, H. (2018). The relationship between self-regulation and educational performance in students. *Journal of Education and Health Promotion*, 7(1), 154. DOI: 10.4103/jehp.jehp_93_18 PMID: 30693291

Salem, S. (2022). Overcoming educational challenges in Palestine with AI: A focus on language instruction. *Middle Eastern Education Review*, 45(1), 89–104.

Salem, S., Smith, A., & Ahmed, R. (2023). Integrating artificial intelligence in education: Transforming traditional teaching methods. *Journal of Educational Technology and Innovation*, 14(2), 112–128.

Salem, S., Smith, A., & Johnson, M. (2024). Designing AI tutors for drama-based English language teaching. *Advances in Educational AI*, 15(3), 123–138.

Salem, S., Smith, J., & Brown, A. (2022). The role of drama in enhancing language learning: Evidence from recent studies. *Language Teaching Research*, 26(3), 343–359.

Salem, S., & Sumandal, A. H. (2023). Teachers' Self-Efficacy with Artificial Intelligence (AI) Based Educational Tools. *Ignatian International Journal for Multidisciplinary Research*, 1(1), 1–10.

Salem, S., & Uygun, D. (2024). Teachers' perspectives on artificial intelligence in education. *Advances in Mobile Learning Educational Research*, 4(1), 931–939. DOI: 10.25082/AMLER.2024.01.005

Salem, S., & Vygotsky, L. (2023). *Mind in Society: The Development of Higher Psychological Processes*. Harvard University Press.

Salem, S., Wang, L., Zhang, X., & Chen, Y. (2023). AI in language learning: A review of recent advancements and their impact on language proficiency. *Journal of Educational Technology & Society*, 26(2), 55–70.

Salem, S., & Wei, L. (2023). Artificial intelligence in language instruction: Impact on English learning achievement, L2 motivation, and self-regulated learning. *Frontiers in Psychology*, 14, 1261955. DOI: 10.3389/fpsyg.2023.1261955 PMID: 38023040

Salem, S., Williams, P., & Smith, J. (2022). AI and educational robotics in language learning: Innovations and challenges. *Educational Robotics Review*, 7(1), 89–105.

Salem, S., Williams, S., & Zhao, Q. (2024). Ethical Considerations and Challenges in Implementing AI in Education. *Journal of Educational Ethics*, 9(3), 89–105.

Salem, S., Yousef, L., & Ibrahim, A. (2023). Cultural integration in EFL teaching: Enhancing drama activities with technology in Palestine. *Palestinian Journal of Language and Culture*, 25(3), 102–119.

Salem, S., Zheng, X., & Wang, L. (2023). The role of AI-driven virtual reality in language acquisition through drama. *International Journal of Language and Learning Technologies*, 16(2), 98–114.

Sallam, M. (2023). ChatGPT Utility in Health Care Education, Research, and Practice: Systematic Review on the Promising Perspectives and Valid Concerns. *Health Care*, 11(6), 887. DOI: 10.3390/healthcare11060887 PMID: 36981544

Salum, R. N., Khadijah, K., & Nasution, N. L. (2024). Optimization of early childhood social emotional development through exercise methods and learning videos. [JIES]. *Journal of Islamic Education Students*, 4(1), 163. DOI: 10.31958/jies.v4i1.12281

Salvagno, M., Taccone, F. S., & Gerli, A. G. (2023). *Salvagno, M., Taccone, F. S., & Gerli, A. G. (2023). Can artificial intelligence help for scientific writing?* Santosa, A. B., Basuki, Y., & Puspita, A. M. (2019). The effectiveness of local wisdom-based teaching materials in enhancing creative writing skills of elementary school students. *Journal of English Language Teaching and Linguistics*, 4(3), 349–359.

Sancar, R., Atal, D., & Deryakulu, D. (2021). A new framework for teachers' professional development. *Teaching and Teacher Education*, 101, 103305. DOI: 10.1016/j.tate.2021.103305

Sanusi, I. T., & Oyelere, S. S. (2020). *Pedagogies of machine learning in K-12 context. In 2020 IEEE frontiers in education conference (FIE).* IEEE.

Šašinka, Č., Stachoň, Z., Sedlák, M., Chmelík, J., Herman, L., Kubíček, P., Šašinková, A., Doležal, M., Tejkl, H., Urbánek, T., Svatoňová, H., Ugwitz, P., & Juřík, V. (2018). Collaborative Immersive Virtual Environments for Education in Geography. *ISPRS International Journal of Geo-Information*, 8(1), 3. DOI: 10.3390/ijgi8010003

Saunders, S., Lindsay, R., & McGrae, B. (2023). A spotlight on Generative Artificial Intelligence (GAI) Technology. *University of Liverpool.* https://www.liverpool.ac.uk/media/livacuk/centre-for-innovation-in-education/staff-guides/generative-artificial-intelligence-technology/generative-artificial-intelligence-technology.pdf

Scalable and distributed machine learning and deep learning patterns. (2023). In Advances in computational intelligence and robotics book series. https://doi.org/DOI: 10.4018/978-1-6684-9804-0

Schaffer, A., Miranda, B., & Koyejo, S. (2023). Are Emergent Abilities of Large Language Models a Mirage? NeuralIPs. https://proceedings.neurips.cc/paper_files/paper/2023/file/adc98a266f45005c403b8311ca7e8bd7-Paper-Conference.pdf

Schiff, D. (2022). Education for AI, not AI for education: The role of education and ethics in national AI policy strategies. *International Journal of Artificial Intelligence in Education*, 32(3), 527–563. DOI: 10.1007/s40593-021-00270-2

Schlagowski, R., Nazarenko, D., Can, Y., Gupta, K., Mertes, S., Billinghurst, M., & André, E. (2023, April). Wish you were here: Mental and physiological effects of remote music collaboration in mixed reality. In *Proceedings of the 2023 CHI conference on human factors in computing systems* (pp. 1-16). DOI: 10.1145/3544548.3581162

Seagate, (2020). Rethink data. Seagate. https://www.seagate.com/content/dam/seagate/migrated-assets/www-content/our-story/rethink-data/files/Rethink_Data_Report_2020.pdf

Segal, D. (2023). Sustained attention plays a critical role in adults with and without ADHD. *Learning and Individual Differences*, 105. Advance online publication. DOI: 10.1016/j.lindif.2023.102300

Sellami, Z., Zimbra, D., & Amami, N. (2023). Ensuring information accuracy and reliability in AI language models for LIS research. *Journal of Information Science*, 49(2), 265–281.

Selvan, C., Vidhya, K., Senthil Kumar, R., & Veningston, K. (2024). *Comparative Overview of Augmented Reality*. Virtual Reality, and Mixed Reality., DOI: 10.4018/979-8-3693-1123-3.ch001

Selwyn, N. (2020). *Should Robots Replace Teachers? AI and the Future of Education*. Polity Press.

Selwyn, N. (2021). *AI in education: The challenges of classroom integration*. Oxford University Press.

Semeraro, F., Imbriaco, G., Del Giudice, D., Antognoli, M., Celin, D., Cuttitta, M., Lo Guasto, V., Giulini, G., Gnudi, T., Monesi, A., Nava, E., Tucci, R., Carenzio, A., Lo Jacono, S., Gordini, G., Gamberini, L., Coniglio, C., Baldazzi, M., Landini, C., & Tortolani, D. (2024). Empowering the next Generation: An innovative "Kids Save Lives" blended learning programme for schoolchildren training. *Resuscitation*, 194, 110088. DOI: 10.1016/j.resuscitation.2023.110088 PMID: 38101506

Senate Inquiry Australia. (2023). Inquiry into the use of generative AI in the education system. *University of Melbourne*. https://about.unimelb.edu.au/__data/assets/pdf_file/0032/396446/UoM-Submission-Inquiry-into-Generative-AI-in-Education-FINAL.pdf

Seo, K., Tang, J., Roll, I., Fels, S., & Yoon, D. (2021). The impact of artificial intelligence on learner–instructor interaction in online learning . *International Journal of Educational Technology in Higher Education*, 18(1), 1–23. DOI: 10.1186/s41239-021-00292-9 PMID: 34778540

Shahriar, S., & Hayawi, K. (2023). Let's have a chat! A Conversation with ChatGPT: Technology, Applications, and Limitations. *arXiv preprint arXiv:2302.13817*.

Shahrol, S. J. M., Sulaiman, S., Samingan, M. R., & Mohamed, H. (2020). A systematic literature review on teaching and learning english using mobile technology. *International Journal of Information and Education Technology (IJIET)*, 10(9), 709–714. DOI: 10.18178/ijiet.2020.10.9.1447

Sheard, J. K., Adriaens, T., Bowler, D. E., Büermann, A., Callaghan, C. T., Camprasse, E. C. M., Chowdhury, S., Engel, T., Finch, E. A., Von Gönner, J., Hsing, P., Mikula, P., Oh, R. Y. R., Peters, B., Phartyal, S. S., Pocock, M. J. O., Wäldchen, J., & Bonn, A. (2024). Emerging technologies in citizen science and potential for insect monitoring. *Philosophical Transactions of the Royal Society B Biological Sciences, 379*(1904). DOI: 10.1098/rstb.2023.0106

Shemshack, A., & Spector, J. M. (2020). A systematic literature review of personalized learning terms. *Smart Learning Environments*, 7(1), 33. Advance online publication. DOI: 10.1186/s40561-020-00140-9

Shen, W., & Xia, L. (2020). *Virtual reality technology*. Xiwang Electronic Publishing House.

Shen, Y., Heacock, L., Elias, J., Hentel, K. D., Reig, B., Shih, G., & Moy, L. (2023). ChatGPT and other large language models are double-edged swords. *Radiology*, 307(2), e230163. DOI: 10.1148/radiol.230163 PMID: 36700838

Shvardak, M. V. (2023). IMMERSIVE TECHNOLOGIES IN PRIMARY EDUCATION. *Scientific Herald of Sivershchyna. Series: Education.Social and Behavioural Sciences*, 2(11), 227–239. DOI: 10.32755/sjeducation.2023.02.227

Sichai, S. (2023). Is artificial intelligence advancing too quickly? What AI leaders at Google say. *60 Minutes*.https://www.cbsnews.com/news/google-artificial-intelligence-future-60-minutes-transcript-2023-04-16/

Silverstein, M. J., Faraone, S. V., Leon, T. L., Biederman, J., Spencer, T. J., & Adler, L. A. (2020). The relationship between executive function deficits and DSM-5-defined ADHD symptoms. *Journal of Attention Disorders*, 24(1), 41–51. DOI: 10.1177/1087054718804347 PMID: 30296883

Simon, F., Altay, S., & Mercier, H. (2023). Misinformation reloaded? Fears about the impact of generative AI on misinformation are overblown. *HKS*.https://misinforeview.hks.harvard.edu/article/misinformation-reloaded-fears-about-the-impact-of-generative-ai-on-misinformation-are-overblown/

Simon, M. K., & Goes, J. (2013). Scope, limitations and delimitations. Dissertation and Scholarly Research: Recipes for Success.

Slack, D., Hilbert, M., & Henderson, K. (2023). Privacy and data protection concerns in the use of AI language models. *The Journal of Privacy and Confidentiality*, 13(1), 134–153.

Šlapeta, J. (2023). Are ChatGPT and other pretrained language models good parasitologists? *Trends in Parasitology*, 39(5), 314–316. DOI: 10.1016/j.pt.2023.02.006 PMID: 36872153

Slattery, E. J., O'Callaghan, E., Ryan, P., Fortune, D. G., & McAvinue, L. P. (2022). Popular interventions to enhance sustained attention in children and adolescents: A critical systematic review. *Neuroscience and Biobehavioral Reviews*, 137, 104633. DOI: 10.1016/j.neubiorev.2022.104633 PMID: 35337900

Sobania, D., Briesch, M., Hanna, C., & Petke, J. (2023, May). An analysis of the automatic bug fixing performance of chatgpt. In *2023 IEEE/ACM International Workshop on Automated Program Repair (APR)* (pp. 23-30). IEEE. DOI: 10.1109/APR59189.2023.00012

Soler-Gutiérrez, A. M., Pérez-González, J. C., & Mayas, J. (2023). Evidence of emotion dysregulation as a core symptom of adult ADHD: A systematic review. *PLoS One*, 18(1), e0280131. DOI: 10.1371/journal.pone.0280131 PMID: 36608036

Speechify. [Computer Software]. (2024). Available at: https://speechify.com/blog/how-to-overcome-adhd-through-immersion-reading/

Sreedharan, S., Kulkarni, A., & Kambhampati, S. (2024). Explainable Human-AI Interaction: A Planning Perspective. *SLAIML*.https://link.springer.com/book/10.1007/978-3-031-03767-2

Stokel-Walker, C., & Van Noorden, R. (2023). What ChatGPT and generative AI mean for science. *Nature*, 614(7947), 214–216. DOI: 10.1038/d41586-023-00340-6 PMID: 36747115

Stracke, C. M., & Trisolini, G. (2021). A systematic literature review on the quality of moocs. *Sustainability (Basel)*, 13(11), 1–26. DOI: 10.3390/su13115817

Strain-Moritz, & Tessa, E. (2016). Perceptions of Technology Use and Its Effects on Student Writing. *Culminating Projects in Teacher Development*, 8.https://repository.stcloudstate.edu/ed_etds/8

Strobl, C., Menke-Bazhutkina, I., Abel, N., & Michel, M. (2024). Adopting ChatGPT as a writing buddy in the advanced L2 writing class . *Technology in Language Teaching & Learning*, 6(1), 1–1. DOI: 10.29140/tltl.v6n1.1168

Suarez-Angelino, L. (2024). Motion review 2024: Pros & cons, cost and who it's right for. Choosing Therapy. https://www.choosingtherapy.com/motion-app-review/

Su, J., & Yang, W. (2023). Unlocking the power of ChatGPT: A framework for applying generative AI in education. *ECNU Review of Education*, 6(3), 355–366. DOI: 10.1177/20965311231168423

Summit. [Computer Software] (2024). Your ADHD AI coach. https://www.summit.im/features/ai-adhd-coach#:~:text=Summit%20helps%20those%20with%20ADHD,to%20chronic%20illness%20and%20ADHD

Sundt, E. (2018). *The use of creative writing in foreign language teaching in Norwegian VGS.* .

Susnjak, T., & McIntosh, T. R. (2024). ChatGPT: The end of online exam integrity? *Education Sciences*, 14(6), 656. DOI: 10.3390/educsci14060656

Sutherland, I. E. (1965). The ultimate display. In *Proceedings of the International Federation of Information Processing Congress* (pp. 506-508). New York.

Sutskever, I. (2023). OpenAI Is Working to Fix ChatGPT's Hallucinations. *IEEE Spectrum*.https://spectrum.ieee.org/ai-hallucination

Swartz, S. L., Prevatt, F., & Proctor, B. E. (2005). A coaching intervention for college students with attention deficit/hyperactivity disorder. *Psychology in the Schools*, 42(6), 647–656. DOI: 10.1002/pits.20101

Sweller, J. (1994). Cognitive load theory, learning difficulty and instructional design. *Learning and Instruction*, 4(4), 295–312. DOI: 10.1016/0959-4752(94)90003-5

Sweller, J. (2008). Human Cognitive Architecture [J]. *Instructional Science*, 32(1), 9.

Taggart, S., Roulston, S., Brown, M., Donlon, E., Cowan, P., Farrell, R., & Campbell, A. (2023). Virtual and augmented reality and pre-service teachers: Makers from muggles? *Australasian Journal of Educational Technology*, 39(3), 1–16. DOI: 10.14742/ajet.8610

Taghian, A., Abo-Zahhad, M., Sayed, M. S., & Abd El-Malek, A. H. (2023). Virtual and augmented reality in biomedical engineering. *Biomedical Engineering Online*, 22(76), 1–25. DOI: 10.1186/s12938-023-01138-3 PMID: 37525193

Taguchi, N. (2021). Application of immersive data collection methods: Insights from interviews. *CALICO Journal*, 38(2), 181–201. DOI: 10.1558/cj.41136

Tai, T.-Y. (2022). The effects of mobile virtual reality on EFL learners' listening comprehension. *Language Learning & Technology*, 26(1), 1–23. https://hdl.handle.net/10125/73491

Tan, K. L., & Lim, C. K. (2018, September). Development of traditional musical instruments using augmented reality (AR) through mobile learning. In AIP Conference Proceedings (Vol. 2016, No. 1). AIP Publishing.

Tang, F. (2024). Understanding the role of digital immersive technology in educating the students of english language: Does it promote critical thinking and self-directed learning for achieving sustainability in education with the help of teamwork? *BMC Psychology*, 12(144), 1–14. DOI: 10.1186/s40359-024-01636-6 PMID: 38481260

Tang, Y. M., Chau, K. Y., Kwok, A. P. K., Zhu, T., & Ma, X. (2022). A systematic review of immersive technology applications for medical practice and education - Trends, application areas, recipients, teaching contents, evaluation methods, and performance. *Educational Research Review*, 35, 100429. DOI: 10.1016/j.edurev.2021.100429

Tan, Y., Xu, W., Li, S., & Chen, K. (2022). Augmented and Virtual Reality (AR/VR) for Education and Training in the AEC Industry: A Systematic Review of Research and Applications. *Buildings*, 12(10), 1529. DOI: 10.3390/buildings12101529

Tate, T., Doroudi, S., Ritchie, D., Xu, Y., & Warschauer, M. (2023). Educational research and AI-generated writing: Confronting the coming tsunami. *EdArXiv. January, 10*. DOI: 10.35542/osf.io/4mec3

Techcrunch, (2023). Poland opens privacy probe of ChatGPT following GDPR complaint. TC. https://techcrunch.com/2023/09/21/poland-chatgpt-gdpr-complaint-probe/

Techcrunch_A, (2024). ChatGPT is violating Europe's privacy laws, Italian DPA tells OpenAI. *TC.*https://techcrunch.com/2024/01/29/chatgpt-italy-gdpr-notification/

Terwiesch, C. (2023). Would Chat GPT3 get a Wharton MBA? A prediction based on its performance in the operations management course. *Mack Institute for Innovation Management at the Wharton School, University of Pennsylvania, 45.*

The ADHD Centre. (2023 September 27). How AI is changing the game for ADHD management. The ADHD Centre. https://www.adhdcentre.co.uk/how-ai-is-changing-the-game-for-adhd-management/

The ADHD Centre. (2023b July 9). Overcoming ADHD procrastination with these productivity-boosting strategies. [Website] Available at: https://www.adhdcentre.co.uk/overcome-adhd-procrastination-with-these-productivity-boosting-strategies/

The Scrum Guide. (2012). *Software in 30 Days*, 133–152. https://doi.org/DOI: 10.1002/9781119203278.app2

Thompson, K., Corrin, L., & Lodge, J. M. (2023). AI in tertiary education: Progress on research and practice . *Australasian Journal of Educational Technology*, 39(5), 1–7. DOI: 10.14742/ajet.9251

TickTick. [Computer Software] (2024) Available at: https://ticktick.com/?language=en_US

Timi-koleolu, S., & Atanda, O. (2023). Artificial intelligence in Nigeria: legal and regulatory guidance. Pavestones. https://pavestoneslegal.com/artificial-intelligence-in-nigeria-legal-and-regulatory-guidance/

Tirumalesh, K. V. (1991). Writing-English versus writing-in-English: New notes on an old theme . *Economic and Political Weekly*, •••, 2670–2672.

Tiwari, C., Bhat, M., Tariq Khan, S., Subramaniam, R., & Khan, M. (2023). What drives students toward ChatGPT? An investigation of the factors influencing adoption and usage of ChatGPT. *Interactive Technology and Smart Education*. Advance online publication. DOI: 10.1108/ITSE-04-2023-0061

Tlili, A., Shehata, B., Adarkwah, M. A., Bozkurt, A., Hickey, D. T., Huang, R., & Agyemang, B. (2023). What if the devil is my guardian angel: ChatGPT as a case study of using chatbots in education . *Smart Learning Environments*, 10(1), 15. DOI: 10.1186/s40561-023-00237-x

Tobias, S., Fletcher, J. D., & Wind, A. P. (2013). Game-Based Learning. *Handbook of Research on Educational Communications and Technology, 1*, 485–503. DOI: 10.1007/978-1-4614-3185-5_38

Together for Children – Sunderland. (2024) [PDF document] Retrieved from: https://www.togetherforchildren.org.uk/media/27515/One-minute-guide-Self-Regulation-and-Executive-Function/pdf/SELF_REGULATION_AND_EXECUTIVE_FUNCTION.pdf?m=637970361370630000

Tom Dieck, M. C., & Han, D. (2022). The role of immersive technology in Customer Experience Management. *Journal of Marketing Theory and Practice*, 30(1), 108–119. DOI: 10.1080/10696679.2021.1891939

Toplak, M. E., & Tannock, R. (2005). Time perception: Modality and duration effects in attention-deficit/hyperactivity disorder (ADHD). *Journal of Abnormal Child Psychology*, 33(5), 639–654. DOI: 10.1007/s10802-005-6743-6 PMID: 16195956

Tossell, C. C., Tenhundfeld, N. L., Momen, A., Cooley, K., & de Visser, E. J. (2024). *Student Perceptions of ChatGPT Use in a College Essay Assignment: Implications for Learning, Grading, and Trust in Artificial Intelligence*. IEEE Transactions on Learning.

Tripp, G. (2024 June 11). Carrots vs. sticks: The science of rewards and punishment for children with ADHD. ADDitude._https://www.additudemag.com/positive-reinforcement-reward-and-punishment-adhd/

Tritthart, M. (2021). Animated Urban Surfaces: Spatial Augmented Reality in Public Discourse. *International Journal of Film and Media Arts*, 6(2), 75–94. DOI: 10.24140/ijfma.v6.n2.05

Tsirulnikov, D., Suart, C., Abdullah, R., Vulcu, F., & Mullarkey, C. E. (2023). Game on: Immersive virtual laboratory simulation improves student learning outcomes & motivation. *FEBS Open Bio*, 13(3), 396–407. DOI: 10.1002/2211-5463.13567 PMID: 36723273

Turnlock, P., Rosen, L., & Kaminski, P. L. (1998). Differences in academic coping strategies of college students who self-report high and low symptoms of Attention Deficit Hyperactivity Disorder. *Journal of College Student Development*, 39(5), 484–493.

Tyner, A. (2023). *Think Again: Is Education Funding in America Still Unequal? Thomas B*. Fordham Institute.

Tzafestas, S. C., Palaiologou, N., & Alifragis, M. (2006). Virtual and remote robotics laboratories: Comparative experimental evaluation. *IEEE Transactions on Education*, 49(3), 360–369. DOI: 10.1109/TE.2006.879255

Uddin, S. M., Albert, A., Ovid, A., & Alsharef, A. (2023). Leveraging ChatGPT to Aid Construction Hazard Recognition and Support Safety Education and Training. *Sustainability, 15, 7121. https://doi.org/*, 1-22.DOI: 10.3390/su15097121

UNESCO. (2023). *ChatGPT and artificial intelligence in higher education.* Document code:ED/HE/IESALC/IP/2023/12. (https://creativecommons.org/licenses/by-sa/3.0/igo/)

Ustun, B., Adler, L. A., Rudin, C., Faraone, S. V., Spencer, T. J., Berglund, P., Gruber, M. J., & Kessler, R. C. (2017). The World Health Organization adult Attention-Deficit/Hyperactivity Disorder self-report screening scale for DSM-5. *JAMA Psychiatry*, 74(5), 520–527. DOI: 10.1001/jamapsychiatry.2017.0298 PMID: 28384801

V7labs, (2021). Supervised and Unsupervised Learning [Differences & Examples]. *V7labs.*https://www.v7labs.com/blog/supervised-vs-unsupervised-learning

Vadalà, R., Giugni, E., Pichiecchio, A., Balottin, U., & Bastianello, S. (2011). Attention deficit hyperactivity disorder (ADHD): From a childhood neuropsychiatric disorder to an adult condition. *Functional Neurology*, 26(3), 117–119. PMID: 22152431

Vallée, A., Blacher, J., Cariou, A., & Sorbets, E. (2020). Blended Learning Compared to Traditional Learning in Medical Education: Systematic Review and Meta-Analysis. *Journal of Medical Internet Research*, 22(8), e16504. DOI: 10.2196/16504 PMID: 32773378

Van den Berg, M., & Kuiper, O. (2020). Xai in the financial sector. Hogeschool Utrecht. https://www.researchgate.net/publication/344079379_XAI_in_the_Financial_Sector_A_Conceptual_Framework_for_Explainable_AI_XAI

Van Dis, E. A., Bollen, J., Zuidema, W., van Rooij, R., & Bockting, C. L. (2023). ChatGPT: Five priorities for research. *Nature*, 614(7947), 224–226. DOI: 10.1038/d41586-023-00288-7 PMID: 36737653

van Eck, N. J., & Waltman, L. (2010). Software Survey: VOSviewer, a Computer Program for Bibliometric Mapping. *Scientometrics*, 84(2), 523–538. DOI: 10.1007/s11192-009-0146-3 PMID: 20585380

Van Horn, G., Mac Aodha, O., Song, Y., Cui, Y., Sun, C., Shepard, A., Adam, H., Perona, P., & Belongie, S. (2018). *The INaturalist Species Classification and Detection Dataset*. Openaccess.thecvf.com. https://openaccess.thecvf.com/content_cvpr_2018/html/Van_Horn_The_INaturalist_Species_CVPR_2018_paper.html

Vaswani, A., Shazeer, N., Parmar, N., Uszkoreit, J., Jones, L., Gomez, A. N., Kaiser, L., & Polosukhin, I. (2017). Attention is All You Need. In *Proceedings of the Neural Information Processing Systems (NIPS) conference*

Veiga, F., & Andrade, A. (2021). Critical success factors in accepting technology in the classroom. *International Journal of Emerging Technologies in Learning*, 16(18), 4–21. DOI: 10.3991/ijet.v16i18.23159

Venkatesh, V., Thong, J. Y. L., & Xu, X. (2023). User-centred design in AI language models: A framework for LIS research. *Journal of the Association for Information Science and Technology*, 74(6), 1323–1340.

Voight, A. (2023). Schools right now face a choice - fight the wave of ChatGPT, or surf it. Sydney Morning Herald. https://www.smh.com.au/national/schools-right-now-face-a-choice-fight-the-wave-of-chatgpt-or-surf-it-20230118-p5cdle.html

Vreuls, J., Koeslag-Kreunen, M., van der Klink, M., Nieuwenhuis, L., & Boshuizen, H. (2022). Responsive curriculum development for professional education: Different teams, different tales. *Curriculum Journal*, 33(4), 636–659. DOI: 10.1002/curj.155

Wagner, P., & Alharthi, D. (2023). Leveraging VR/AR/MR/XR Technologies to Improve Cybersecurity Education, Training, and Operations. *Journal of Cybersecurity Education Research and Practice*, 2024(1), 1–26. DOI: 10.32727/8.2023.32

Wang, X. (2022). Computer-aided college English teaching system based on virtual reality and artificial intelligence. *Wireless Communications and Mobile Computing*, 2022, 2022. DOI: 10.1155/2022/9613587

Wardat, Y., Tashtoush, M. N., AlAli, R., & Jarrah, A. M. (2023). ChatGPT: A revolutionary tool for teaching and learning mathematics . *Eurasia Journal of Mathematics, Science and Technology Education*, 19(7), em2286. DOI: 10.29333/ejmste/13272

Warner, J. (2024). ChatGPT Cannot Teach Writing. Insider Higher ED. https://www.insidehighered.com/opinion/blogs/just-visiting/2024/01/22/arizona-state-announces-plan-give-education#

WEF. (2023). How will generative AI impact healthcare? *WEF*. https://www.weforum.org/agenda/2023/05/how-will-generative-ai-impact-healthcare/

Wei, J., Tay, Y., Bommasani, R., Raffel, C., Zoph, B., Borgeaud, S., Yogatama, D., Bosma, M., Zhou, D., Metzler, D., Chi, E. H., Hashimoto, T., Vinyals, O., Liang, P., & Dean, J. (2022). Emergent Abilities of Large Language Models. Transactions on Machine Learning Research. https://arxiv.org/pdf/2206.07682.pdf

Weigard, A., & Huang-Pollock, C. (2017). The role of speed in ADHD-related working memory deficits: A time-based resource-sharing and diffusion model account. *Clinical Psychological Science*, 5(2), 195–211. DOI: 10.1177/2167702616668320 PMID: 28533945

Weinel, J. (2020). Cyberdreams: Visualizing music in extended reality. Technology, Design and the Arts-Opportunities and Challenges, 209-227.

Weiser, W. (2000). Gender Differences in Internet Use Patterns and Internet Application Preferences: A Two-Sample Comparison. Cyberpsychology & behavior (3,2). https://www.researchgate.net/publication/250899760_Gender_Differences_in_Internet_Use_Patterns_and_Internet_Application_Preferences_A_Two-Sample_Comparison

Whalen, J., & Mouza, C. (2023). ChatGPT: Challenges, opportunities, and implications for teacher education. *Contemporary Issues in Technology & Teacher Education*, 23(1), 1–23.

Whalley, B., France, D., Park, J., Mauchline, A., & Welsh, K. (2021). Towards flexible personalized learning and the future educational system in the fourth industrial revolution in the wake of Covid-19. *Higher Education Pedagogies*, 6(1), 79–99. DOI: 10.1080/23752696.2021.1883458

Williams, C., & Beam, S. (2019). Technology and writing: Review of research. *Computers & Education*, 128, 227–242. DOI: 10.1016/j.compedu.2018.09.024

Woodruff, L. (2020, October 19). Filing cabinets don't work for ADHD minds: Help for paper pilers. ADDitude. https://www.additudemag.com/how-to-organize-paperwork-adhd-clutter/

World Health Organization. (2024 January 2). Assistive technology. World Health Organization. https://www.who.int/news-room/fact-sheets/detail/assistive-technology

Wright, S. (2014). *ADHD coaching matters: The definitive guide*. ACO Books.

Wu, J. G., Miller, L., Huang, Q., & Wang, M. (2023). Immersive virtual reality learning: An exploratory study of Chinese nursing college students. *RELC Journal*, 54(3), 697–713. DOI: 10.1177/00336882211044860

WWF. (2019). *Where do tigers live? And other tiger facts*. World Wildlife Fund. https://www.worldwildlife.org/stories/where-do-tigers-live-and-other-tiger-facts

Xiao, J., & Xie, G. M. (2019). Application and development prospects of VR technology-based educational games in English teaching. *Educational Modernization*, 6(11), 195–197.

Xue, V. W., Lei, P., & Cho, W. C. (2023). The potential impact of ChatGPT in clinical and translational medicine. *Clinical and Translational Medicine*, 13(3), e1216. DOI: 10.1002/ctm2.1216 PMID: 36856370

Yampolskiy, R., (2019). Unexplainability and Incomprehensibility of AI. International Journal of Machine Consciousness (2009 Vol. 1 Issue 1 - 2014 Vol. 6 Issue 2 https://doi.org/DOI: 10.1142/S2705078520500150

Yao, X. (2003). Research on the development of self-concept among college students. *Chinese Journal of Mental Health*, 17(1), 42–44.

Yin, H., Sun, Z., Li, D., & Shi, J. (2005). Time perception characteristic in children with attention-deficit/hyperactivity disorder. *East Asian Archives of Psychiatry*, 29(1), 22–27.

You, Y., Chen, Y., You, Y., Zhang, Q., & Cao, Q. (2023). Evolutionary Game Analysis of Artificial Intelligence Such as the Generative Pre-Trained Transformer in Future Education. *Sustainability, 15, 9355. https://doi.org/*, 1-12.DOI: 10.3390/su15129355

Young, J. C., & Shishido, M. (2023). nvestigating OpenAI's ChatGPT Potentials in Generating Chatbot's Dialogue for English as a Foreign Language Learning. *(IJACSA). International Journal of Advanced Computer Science and Applications*, 14(6), 65–72. DOI: 10.14569/IJACSA.2023.0140607

Young, S., & Thome, J. (2011). ADHD and offenders. *The World Journal of Biological Psychiatry*, 12(1), 124–128. DOI: 10.3109/15622975.2011.600319 PMID: 21906010

Yudowsky, E. (2023). Will AI Destroy Us? - AI Virtual Roundtable. Coleman Hughes. https://www.youtube.com/watch?v=xkQSiS8hpZA

Yung, R., & Khoo-Lattimore, C. (2019). New realities: A systematic literature review on virtual reality and augmented reality in tourism research. *Current Issues in Tourism*, 22(17), 2056–2081. DOI: 10.1080/13683500.2017.1417359

Zaman, A., Abir, M. R., & Mursalin, S. (2024). Extended reality in education and training: Enhancing trustworthiness. *International Journal of Science and Research Archive*, 11(1), 1705–1720. Advance online publication. DOI: 10.30574/ijsra.2024.11.1.0206

Zattoni, F., Carletti, F., Randazzo, G., Tuminello, A., Betto, G., Novara, G., & Dal Moro, F. (2023). Potential Applications of New Headsets for Virtual and Augmented Reality in Urology. *European Urology Focus*, xxxx, 10–14. DOI: 10.1016/j.euf.2023.12.003 PMID: 38160172

Zawacki-Richter, O., Marín, V. I., Bond, M., & Gouverneur, F. (2020). Systematic review of research on artificial intelligence applications in higher education. *International Journal of Educational Technology in Higher Education*, 17(1), 1–27.

Zedda, L. (2023). The importance of a transformative biodiversity education for perceiving, appreciating and supporting lichen diversity in German urban environments. *Lichenologist (London, England)*, 55(5), 161–168. DOI: 10.1017/S0024282923000312

Zhai, X. (2022). ChatGPT user experience: Implications for education. *Available atSSRN* 4312418.

Zhai, X. (2023). ChatGPT for next generation science learning. *XRDS: Crossroads. The ACM Magazine for Students*, 29(3), 42–46.

Zhang, M. H., Liu, Z. Y., & Li, Y. (2024). Can immersive virtual reality empower science education to effectively improve student learning outcomes?—A meta-analysis based on 44 empirical papers from 2011 to 2022. Modern Educational Technology, 34(6).

Zhang, M. J. (2001). Virtual reality systems (Vol. 9, No. 5). Science Press.

Zhang, J. (2022). Practical research on the assistance of music art teaching based on virtual reality technology. *Wireless Communications and Mobile Computing*, 2022(1), 8479040. DOI: 10.1155/2022/8479040

Zhang, N. (2017). A Review of the Application of Virtual Reality Technology in Specialized English Teaching. *Computer Science*, 44(6A), 43–47.

Zhang, X., Zheng, P., & Li, X. (2023). Interdisciplinary collaboration for enhancing AI language models in LIS research. *Journal of Interdisciplinary Studies*, 18(2), 187–205.

Zhao, Q. (2009). A survey on virtual reality. *Science in China Series F: Information Sciences*, 52(3), 348–400.

Zhu, S., Wang, Z., Zhuang, Y., Jiang, Y., Guo, M., Zhang, X., & Gao, Z. (2024). Exploring the impact of ChatGPT on art creation and collaboration: Benefit**s, challenges and ethical implications . *Telematics and Informatics Reports*, 100138.**

Zimmer, C. (2011). 100 Trillion Connections: New Efforts Probe and Map the Brain's Detailed Architecture. *Scientific American.* https://www.scientificamerican.com/article/100-trillion-connections/#:~:text=Join%20together%20100%20billion%20neurons%E2%80%94with%20100%20trillion%20connections%E2%80%94and,a%20human%20brain%2C%20capable%20of%20much%2C%20much%20more

Zünd, F., Ryffel, M., Magnenat, S., Marra, A., Nitti, M., Kapadia, M., . . . Sumner, R. W. (2015). Augmented creativity: bridging the real and virtual worlds to enhance creative play. In SIGGRAPH Asia 2015 Mobile Graphics and Interactive Applications (pp. 1-7). DOI: 10.1145/2818427.2818460

Zylowska, L., Ackerman, D. L., Yang, M. H., Futrell, J. L., Horton, N. L., Hale, T. S., & Smalley, S. L. (2008). Mindfulness meditation training in adults and adolescents with ADHD: A feasibility study. *Journal of Attention Disorders*, 11(6), 737–746. DOI: 10.1177/1087054707308502 PMID: 18025249

Шаповалов, Є., Білик, Ж., Атамась, А., & Учитель, О. (2019). The Potential of Using Google Expeditions and Google Lens Tools under STEM-education in Ukraine. *Педагогіка вищої та середньої школи, 51*, 90–101. https://doi.org/DOI: 10.31812/pedag.v51i0.3659

About the Contributors

Mageswaran Sanmugam acquired his B.Sc. (Hons) in Electrical Engineering from Universiti Teknologi Tun Hussein Onn in 2005 and Master's degree in Educational Management in 2012 from Universiti Utara Malaysia and Ph. D in Educational Technology in 2017 from Universiti Teknologi Malaysia. He worked as an Educator in both Primary and Secondary schools from 2008-2018, before being appointed as a Senior Lecturer in the Centre for Instructional Technology and Multimedia, Universiti Sains Malaysia, Pulau Pinang, Malaysia (from 2018- till 2021). Currently he is the Deputy Director of Research, Innovation and Networking in the Centre for Instructional Technology and Multimedia, Universiti Sains Malaysia, Pulau Pinang, Malaysia (from 2022- till now). His research interests encompass educational technology, ludology, game-based learning, gamification and game theory. He also serves as an exco-member in the Malaysian Educational Technology Association (META), and as a Senior Member under the Institute of Electrical and Electronics Engineers (IEEE)

Bosede Edwards is a seasoned academic and industry professional with over 25 years of work experience. She holds a Ph.D. in Educational Technology from Universiti Teknologi Malaysia (UTM), where she also earned a Master's degree in the same field. Her educational background is multidisciplinary, with qualifications including a bachelors in Chemistry and Masters in Organic Chemistry, which enriches her perspectives in research and teaching. Dr. Edwards has a proven track record in research leadership, particularly in instructional technology. She has authored and co-authored over 25 journal and conference articles and has been recognized with the IEEE Frontiers in Education 'Faculty Fellow' Award. Her research interests are diverse, ranging from Human-Machine In…

Nurul Maziah Mohd Barkhaya acquired her B.IT. (Hons) in Software Engineering from Universiti Malaysia Terengganu in 2011 and Master's Degree in Education (TVET) in 2013 from Universiti Tun Hussein Onn Malaysia and Ph.D in

Educational Technology in 2023 from Universiti Teknologi Malaysia. She worked as an educator in International Secondary schools from 2019 – 2023, before being appointed as a Senior Lecturer in the Centre for Instructional Technology and Multimedia, Universiti Sains Malaysia, Pulau Pinang, Malaysia started on 2024. Her research interests include the educational technology, creative multimedia, and e-Learning.

Zuheir N. Khlaif, Ph.D. in instructional design and technology, is a distinguished scholar and researcher at An-Najah National University in Palestine. As a dedicated educator, he imparts knowledge in educational technology courses and contributes significantly to the Artificial Intelligence and Virtual Reality Research Center at the University. With a keen focus on the intersection of technology and education, his research delves into areas such as technology integration, technostress, and the application of AI in educational assessment. Khlaif has delivered more than 50 lectures and workshops about e-learning and AI in higher and public education in international and local conferences and symposium. Khlaif's expertise extends to the publication realm, with more than 40 articles in peer-reviewed journals. His contributions not only advance academic discourse but also have practical implications for the effective use of technology in education in various context in the global south.

Anusha Achuthan is a Senior Lecturer at School of Computer Sciences of Universiti Sains Malaysia. She received her PhD in 2016, specializing in computer vision and medical image analysis from Universiti Sains Malaysia. Her main research publications are in computational neuroscience, knowledge-guided image analysis, biomarkers analysis for neurological disorders and aging, machine learning, and computer vision. She is a member of IEEE, ISMRM, and IACSIT. Besides research work, Dr Anusha is also very passionate about online teaching and learning resources with the integration of Generative AI.

Ismail Sheikh Ahmad, a Language Education Professor at the Kulliyyah of Education (KOED) of the International Islamic University, Malaysia (IIUM). He earned his Bachelor of Arts from Fresno State University, California (1985), Master of Science from University of Southern California (1986), Diploma in TESL, RELC, Singapore (1989) and PhD from the University of Nottingham, England (1997) and has published articles on second language reading and literacy, presented papers and invited keynote speaker in local and international language education conferences. His specialized areas include qualitative research methodology, English teaching methodology, reading literacy, critical and creative thinking skills and curriculum design. Administratively, the author has served the university in a number of posts such as the former Dean for KOED, serving the Rector's Office of IIUM as Director for Office of Corporate Communication (OCCM), Office of Corporate Strategy (OCS), Director for Centre for Professional Development (CPD), and Deputy Dean for Centre for Postgraduate Studies (CPS). In 2008, the author received a Promising Researcher Award (IIUM Quality Day Award) under the category of Social Sciences, Humanities and Business/Economic and Administration. A dynamic and versatile researcher with extensive research experience in educational and scientific settings for more than 20 years. His research activities include qualitative and quantitative perspectives in various research topics and scales involving government and private agencies' grants, worth more than RM5 million, such as MOSTI, MARA, SPA, KPM, UNICEF, ATM, FRGS (MoHE) and DBKL. Currently, he is involved in a research group in developing a new university in Kabul, Afghanistan:

Yusuf Ayodeji Ajani is an Assistant Lecturer at Al-Hikmah University, Ilorin, and a postgraduate student at the University of Ilorin, Nigeria. He has an impressive publication record, with several articles published in renowned local and international journals in the field of Education and librarianship. In 2023, he received the esteemed Emerald Literati Award in recognition of his contributions to scholarly research. For inquiries and potential collaborations, please contact Ayodeji at ajaniyusuf@alhikmah.edu.ng

Tahani R. Bsharat, Ph. D. Researcher pursuing a Ph. D. at Faculty of Major Language Studies, Universiti Sains Islam, Malaysia. Bsharat's research interest are learning pedagogy, education curriculum, development teaching experience pedagogy, collaborative learning, educational evaluation, and e-learning teaching materials.

Terri Jackson is an Assistant Professor at the University of Prince Edward Island. She holds a Doctorate in Educational Leadership from Western University (Canada) and a Master's Degree in Inclusive Education. Prior to entering academia, Terri was a public school teacher for nearly two decades. Her research interests include inclusive education, teacher education, and teacher leadership.

Rahul Joshi is an assistant professor of Journalism and mass communication at the School of Mass Media & Humanities, MRIIRS (Deemed to be University), Faridabad, Haryana. He has completed his Ph.D. (Media studies) from MRIIRS, Faridabad, and it was awarded in December 2022. He has done her Masters in Journalism & Mass Communication from GGSIP University. He was a Gold Medalist in MA (MC) at GGSIP University. He has Twelve years of academic and industry experience. His specializations are in Radio Journalism & Television Production. He has published over 15 research papers in different reputed International & National

Journals. Before stepping into academics, he was an Anchor cum Assistant producer in various National Hindi and English News Channels like ANI, India News, News X, APN News, News 18 Network and Zee News. He has also been associated with the World's biggest Radio Broadcasting House, All India Radio, as a Radio Presenter and Production executive for 5 consecutive years. His areas of research interest are Radio Production, TV Production, and Digital Media. His teaching expertise also lies in Radio and TV Production.

Amit Kakkar is an Associate Professor at Mittal School of Business, Lovely Professional University. He has a total experience of 21 years, with industry experience of more than 11 years and teaching experience of more than 10 years. He has completed his PhD from the Lovely Professional University in the consumption enhancement of VoD services. He has presented papers at more than fourteen international conferences in 10 years. He has also published more than ten papers in journals of repute. His areas of interest are strategic management, brand management, and consumer behaviour. He got the best paper awards in NICOM in 2018 and IIM Jammu in 2023.

Ruchi Kakkar is currently working as an Assistant Professor in Mittal School of Business, Lovely Professional University. She has completed her PG (MFC and MBA) from Punjabi University, Patiala and is pursuing her Ph.D. from Lovely Professional University in the domain of Management

Suman Kumari is an Academician & Researcher with over 16 years of academic experience. She holds a Post-Graduate degree in Journalism & Mass Communication and completed her PhD on the topic 'Changing Role of Female Lovers in Bollywood Hindi Movies: A Study of Social Change' from Manav Rachna International University in 2015. Her expertise spans Film Studies, Event Management, Health Communication, and Communication research. Dr. Kumari has co-organized numerous National/International academic activities, including Seminars/Conferences focusing particularly on gender issues. Additionally, she has made significant contributions to research by publishing papers in National/International Journals. She has supervised doctoral research scholars in areas such as Health Communication, Fake News, and Film Studies.

Catherine Lam Pui Kiew is a B.Sc. final-year student from the School of Computer Sciences at Universiti Sains Malaysia. She is currently working on the 'Scanimals' project which involves computer vision and image processing. Her research interests include the areas of machine learning, image analysis, deep learning, mobile application development for educational purposes, and integrating AI in Education. She is also working on managing terrestrial animal data from credible sources to be able to form a better and enriched dataset on terrestrial animals and also how to improve the accuracy of the ML model.

Galib Muhammad Shahriar Himel received a B.Sc. degree in computer science and engineering from the Ahsanullah University of Science and Technology (AUST), in 2016, an M.Sc. degree in computer science and engineering from United International University (UIU), in 2018, another B.Sc. degree in computing from the University of Greenwich (UoG), U.K., in 2021, his 2nd M.Sc. degree in computer science in intelligent systems from American International University-Bangladesh (AIUB), in 2022, and his 3rd M.Sc. degree in applied physics and electronics from Jahangirnagar University (JU), in 2023. He was a Researcher with the Bangladesh University of Business and Technology (BUBT) and Independent University, Bangladesh (IUB). He is currently involved in several researches related to biomedical image processing using machine learning and a Ph.D. Researcher with Universiti Sains Malaysia (USM). His research interests include artificial intelligence, machine learning, bioinformatics, bio-medical image analysis, and computer vision.

Krishna Pandey has over 24 years of academic and journalism experience, of which the last twelve years have been spent in academics. She has 12 years of on-field journalistic experience working in mainstream print media. She is UGC Net Qualified, and her PhD is on Crisis Communication Strategies of Cadbury's & Nestle in India: A Critical Study under the guidance of Prof CP Singh, University School of Mass Communication, Guru Gobind Singh Indraprastha University. She has an MA in Mass Communication from Guru Jambheshwar University of Science & Technology and a Gold Medalist in Political Science from Benaras Hindu University. Dr Pandey is also an alumnus of the Indian Institute of Mass Communication (IIMC), New Delhi, and a graduate of the prestigious Sophia Girls College Ajmer.

Manpreet Kaur Riyat is a dedicated and innovative PhD scholar at Lovely Professional University (LPU). With a strong academic background in marketing, including an MBA, she has developed a keen interest in the cutting-edge fields of gamification, the metaverse, consumer engagement, and education technology. Her research aims to bridge the gap between traditional marketing approaches and the rapidly evolving digital landscape. Manpreet's gamification work focuses on using game-like elements to motivate and engage users in a variety of contexts, from marketing campaigns to educational settings. Her exploration of the metaverse delves into how virtual and augmented reality can transform consumer experiences and create new opportunities for brands.

Tella Adeyinka is a Professor and Head of the Department of Library and Information Science at the University of Ilorin, Nigeria. He holds a Bachelor's Degree in Guidance and Counselling/Political Science from the University of Ibadan, where he graduated with Second Class Honour, Upper Division. He earned two Master's Degrees from the same institution, one in Counselling Psychology and another in Library and Information Studies. His academic journey continued with a Ph.D. in Test, Measurement, and Evaluation, initially at the University of Ibadan and later supported by a Commonwealth Scholarship at the University of Botswana, Gaborone. Tella has authored numerous articles in international journals and contributed chapters to books, focusing on ICT4Development, e-learning, information literacy, and information ethics. Tella serves as Editor for several prominent journals and has edited multiple reference texts in Information Science. Recognized for his scholarly contributions, he has received awards such as the Dr. T.M. Salisu Award for Most Published Librarian in Nigeria and the Emerald Literati Award for Outstanding Paper. He holds a C2 researcher rating from the South African National Research Foundation and has been appointed a Distinguished Professor at the University of South Africa. Beyond academia, Tella is actively engaged in mentoring Ph.D. candidates and masters students across several universities and welcomes opportunities for research collaboration both locally and internationally. He is committed to his family and his faith, believing that "WITH HIM, ALL THINGS ARE POSSIBLE".

Wan Ahmad Jaafar Wan Yahaya has been Director at the Centre for Instructional Technology and Multimedia (PTPM), Universiti Sains Malaysia (USM), Penang, Malaysia since 2016. He is the Chairman of the Technical Committee on Audio, Video, Information Technology, and Communication Equipment at the Department of Standard Malaysia, Ministry of International Trade and Industry Malaysia. He holds a PhD in Multimedia Education from the University of Leeds and a Master of Education (Computer in Education) from the University of Manchester, United Kingdom. He has published more than 170 articles/papers in indexed journals like Scopus, Web of Science (ISI), SCSI, SSCI, conference proceedings, and book chapters, primarily in the field of multimedia education. He is the recipient of the Thailand Association International Honor Award for Outstanding Contributions to the field of Educational Communications and Technology for the year 2023.

Seema Yadav has academic qualifications of M. Sc. (Chemistry), M Ed (RIE), UGC-NET, and PhD in Education. Presently she is working as an Assistant Professor, in the Department of Education at The Bhopal School of Education, Bhopal. She has almost 18 years of teaching experience. Her area of specialization is educational psychology, environmental education, and curriculum development. She has published articles in UGC-CARE Journals and Scopus index books.

Khaled A.Yaseen is an EFL teacher who has been working for the Ministry of Education in Palestine since 2027. Recently he has achieved his MA thesis from An-Najah National University in Nablus.

Index

A

C

M

O

P

R

S

T

U

W

www.ingramcontent.com/pod-product-compliance
Lightning Source LLC
LaVergne TN
LVHW080346011026
849161LV00009B/2205